Scripps Institution of Oceanography: Probing the Oceans 1936 to 1976

SCRIPPS INSTITUTION of OCEANOGRAPHY:

PROBING THE OCEANS
1936 to 1976

Elizabeth Noble Shor

Tofua Press
1978

Copyright © 1978 by Elizabeth Noble Shor
All rights reserved.
Library of Congress Catalog Card Number 78-52598
ISBN: Cloth, 0-914488-17-1; Paper, 0-914488-18-X

Printed in the United States of America.
Published by Tofua Press, 10457-F Roselle Street,
San Diego, California 92121.

Other books by Elizabeth N. Shor: *Fossils and Flies*
                                    The Fossil Feud
                                    Dinner in the Morning

Editor: Elizabeth Rand
Designer: Diane Polster
Cover Design: Jo Griffith

The following publishers have graciously granted permission to quote material from their publications: American Association for the Advancement of Science (for excerpts from *Science*); Doubleday and Company; W. H. Freeman and Company (for excerpts from *Scientific American*); Little, Brown and Company; McGraw-Hill Book Company; Naval Research Reviews; the Oceanic Society (for excerpts from *Oceans*); and John Wiley and Sons. McIntosh and Otis, Inc., provided permission to quote an excerpt from John Steinbeck.

On the cover: Sonderdruck / Auszug

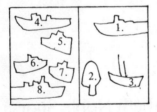

1. *Argo*
2. *Flip*
3. *E. W. Scripps*
4. *Spencer F. Baird*
5. *Scripps*
6. *Paolina-T*
7. *Alpha Helix*
8. *Melville*

Frontispiece: The *Spencer F. Baird* leaving port on Transpac Expedition, 1953.

*In memory of Helen Raitt,
the first chronicler of Scripps*

The man of science . . . goes down to the sea, and the questions there about big things and little pour in upon his mind in a resistless flood. So he hurries away to find somebody to help him with money and laboratories and boats and microscopes and thermometers and nets and balances and measuring sticks and books and chemicals and all sorts of things in order that he may do his best to cope with his flood of questions.

William E. Ritter*

*"The Marine Biological Station of San Diego" (San Diego: Frye & Smith, 1910), 21.

Contents

Preface viii

Part I. The Early Years

1. Harald Sverdrup Sets the Sails 3
2. Crossing the Threshold 19

Part II. The Research Units

3. The Greatest Fishing Expedition in History:
   The Marine Life Research Program 43
4. Making and Chasing Echoes:
   The Marine Physical Laboratory 79
5. Seeing the Light: The Visibility Laboratory 111
6. A Potpourri: The Institute of Marine Resources 117
7. Watching Waves in Land and Sea:
   The Institute of Geophysics and Planetary Physics 149
8. A Different Point of View:
   The Physiological Research Laboratory 167
9. "Let's Visit Scripps":
   The Thomas Wayland Vaughan Aquarium-Museum 185

Part III. The Disciplines

10. Life in the Sea: Studies in Marine Biology 201
11. The Ocean in Motion:
    Studies in Physical Oceanography 241
12. Sand, Silt, and Sea-floor Spreading:
    Studies in Marine Geology 269
13. Within the Waters and Muds:
    Studies in Marine Chemistry 321

Part IV. Out to Sea and Back Again

14. Marine Facilities and the Fleet 345
15. Oceanography is Fun:
    A Glimpse of the Expeditions 373
16. Back on the Beach 439

Appendices 479
Index 491

Preface

I was fascinated by the first tide pool I saw, when I was sixteen, and vowed to become a marine biologist. But I was distracted by a seismologist, who, seven years later, turned from the oil industry and earthquakes to oceanography at the Scripps Institution of Oceanography.

I continued to like tide pools, but at times I hated the ocean — because it took him away for weeks at a time to delightful-sounding ports and islands, while I stayed home. Luckier than many other left-behind wives, I *was* able to meet him several times for brief vacations when he left the ship in Chile, Hawaii, Mexico, and Australia.

I didn't really like the ocean until I sailed on it myself. The first trip, Quartet Expedition, only eight days in January 1965, persuaded me that I wasn't chronically seasick, and introduced me to Scripps' most accomplished ship, the *Horizon*. The second trip, leg 10 of Antipode Expedition in 1971, across the Indian Ocean, taught me how to stand watch in a noisy, confusing, shipboard lab — and that there is no color more stunning than aquamarine on a sunny calm day at sea. The third trip, legs 1 and 2 of Eurydice Expedition in 1974, made me feel really part of a working team, and took me to my first atoll, remote Fanning Island, and to the spontaneous, lively *fia fia* in western Samoa for Aggie Grey's seventieth birthday. Now, any time, I am ready to sail again: to watch flyingfishes, man-o'-war birds, PDRs, computer printouts, and sunsets.

Even when I didn't like the ocean, I always liked Scripps Institution. When George and I arrived in 1953, the institution still had a feeling of one big happy family (and a lot of gossiping). Some of that has been lost as Scripps has grown and UCSD has arrived. But there is still rapport, camaraderie, and always pride in belonging to the Scripps Institution of Oceanography.

So when, in December 1972, Bill Nierenberg asked me to bring up to date the history of the institution, I did not hesitate long. I did talk with Helen Raitt, who filled the unofficial role of historian of Scripps. She encouraged me to go ahead. I had almost finished my manuscript when, to the great sorrow of her many friends, Helen died on 26 March 1976. I am very grateful that she had read, and approved, a major portion of my manuscript.

Bill Nierenberg, Bob Fisher, John Isaacs, and Jeff Frautschy read the completed manuscript and made helpful suggestions. Other Scrippsians have read short or long sections, and I want to thank them here for their time and their comments: Lynn Abbott, Vic Anderson, Bob Arthur, Dru Binney, Ted Bullock, Harmon Craig, Quimby Duntley (who provided all the material for chapter 5), Denis Fox, Ed Goldberg, Baird Hastings, Fran Haxo, Sam Hinton, Doug Inman, Martin Johnson, John McGowan, Bill Menard, Walter Munk, Russ Raitt, Joe Reid, Ellen and Roger Revelle, Dick Rosenblatt, Pete Scholander, Dick Schwartzlose, Elizabeth and Fran Shepard, Jim Snodgrass, Noel and Sally Spiess, Jim Stewart, Vic Vacquier, Don Wilkie, and Claude ZoBell.

Many of the above, and the following, have kindly granted me permission to quote from their letters and their publications: Tanya Atwater, Ed Barr, Teddy Bullard, K. O. Emery, Giff Ewing, Jim Faughn, Nelson Fuller, Walt Garey, George Harvey, Joel Hedgpeth, Alan Jones, Gene LaFond, Wheeler North, Mel Peterson, and Harris Stewart. Thank you all.

I am also grateful to others who have made special contributions: Ralph Roblee, for his help on the dates of Scripps ships; Bill Goff and Paul Leverenz, for their help in the Scripps archives; Larry Ford for several of his photos and for copying all the others; Betty Stover, for help in compiling the list of students; Nelson Fuller, Jackie Janke,

Gert Schlegel, and Nancy Mancino for information and access to the files of the Public Affairs Office; Martin Johnson, for his 1937 Christmas card and permission to use it; Jim Faughn, for the loan of his account of Naga Expedition and his personal log of the construction of the *Alpha Helix* and her first expedition; Baron Thomas and Harris Stewart for the loan of personal logs of expeditions.

I have also appreciated the help of many other people who allowed me access to their files and of all those who have told me "tales out of school."

Scripps Institution is always on the move, shifting, rearranging, progressing, Events in this book are carried only to the end of 1976, not into the year of publication.

Elizabeth N. Shor
1 December 1977

Part I
The Early Years

In the beginning...
of the Scripps Institution there were biologists William E. Ritter and Charles A. Kofoid, and there were newspaper owner E. W. Scripps and his philanthropic half-sister Ellen Browning Scripps, and there were medical doctor Fred Baker, who collected sea shells, and his liberated wife Charlotte, and there were the other San Diegans who became members of the Marine Biological Association. There was also a University of California then, to which Ritter belonged. That university in those primordial years was at Berkeley, and its campus was green in spring, golden in summer and fall, and had no automobiles.

The early history of the Scripps Institution has been well detailed in *Scripps Institution of Oceanography: First Fifty Years,* by Helen Raitt and Beatrice Moulton,* which covers the founding years, the directorship of William E. Ritter from 1903 to 1923, and also the directorship of T. Wayland Vaughan from 1924 to 1936, and more.

The present account begins in 1936, with some overlap of the account by Raitt and Moulton. It is intended to cover the era of expansion in oceanography from boats along the coast to ships at sea, as has been carried out, with verve, by the Scripps Institution of Oceanography.

*Los Angeles: Ward Ritchie Press, 1967.

1. Harald Sverdrup Sets the Sails

Oceanography is not so much a science as a state of mind. So is the Scripps Institution of Oceanography.

Scripps is the oldest institution of oceanography in the United States (organized in 1903 and dedicated to research in oceanography in 1925), the first to award a doctorate in oceanography (1930), and the largest oceanographic institution in the United States (approximately 1,100 staff and students in 1976). It is officially defined as a research and graduate school of the University of California, San Diego.

As a state of mind, Scripps Institution is a group of scientists and students whose interests lie in the ocean itself. They began as biologists, geologists, chemists, physicists, or even engineers, but they call themselves oceanographers. Their talk of science is sprinkled with Navy terms and fishermen's jargon. Geography of the edges of the sea is as natural to them as was that of the Mediterranean to the Phoenician sailors. Like any sailors, their gaze goes suddenly faraway when they speak of certain ports.

They do not really love the ocean.

"To oceanographers," said William A. Nierenberg, "the sea is an enormous and restless antagonist. The work is nowhere near as glamorous as it's supposed to be — it's tough, rough, very difficult."[1]

H. William Menard, who has been going on expeditions for a quarter of a century, has called the ocean "little more than a nuisance to a marine geologist. It provides a convenient medium for transporting equipment, although it is regrettably unstable. Otherwise it seems to be an unnecessary filter which obscures every bit of information that one manages to collect."[2]

That restless mass of opaque water hides the mountains and troughs, conceals the fishes and the sea serpents, wobbles the sound waves, and churns the stomach. But its power and its mystery hold some people in a spell, and among those are oceanographers. They devise instruments to penetrate the obscurity of seven miles of swirling water. Slowly they wrest secrets from an alien world that holds them back.

"I have often wondered," mused Roger Revelle, "why it is so pleasant to be on a small, oily, and uncomfortable ship, far from the nearest land. . . . I am convinced that it is because on shipboard both the past and the future disappear — only the present is left."[3]

The lowering and return of a rock dredge can take many hours. Time stands still. One can muse over the effortless roll of the porpoise, wonder how the flyingfishes fly, watch the rim of the setting sun turn emerald green. There is time for geography, for history, for listening to the ideas of a colleague in another discipline. Time and the vastness of the sea create oceanographers — interdisciplinary, international, and interesting raconteurs.

It was not always so at Scripps. The institution actually began in the tide pools, and many years passed before it reached the middle of the sea. In 1892 William E. Ritter, then the new chairman of the zoology department at the

"The sea is an enormous and restless antagonist," as illustrated by Willard Bascom on Capricorn Expedition, 1953.

Scripps Institution of Oceanography: Probing the Oceans

University of California at Berkeley, began his search for a seaside field station for summer studies. His vision was to make "a biological survey of that part of the Pacific Ocean adjacent to the coast of California." By 1907 he had found his location: an isolated piece of barren pueblo land just north of the village of La Jolla, on the dusty outskirts of the city of San Diego. In California's southernmost city Ritter had found enthusiasm and financial support for his vision, in Fred Baker, E. W. Scripps, Ellen Browning Scripps, and the other members of the Marine Biological Association, which was founded on 26 September 1903 to sponsor the Scripps Institution for Biological Research.

Like many others since then, Ritter came under the spell of La Jolla and the Scripps Institution. He moved from Berkeley to La Jolla to become the institution's first director, serving from 1903 until 1923. In 1912 he helped negotiate the transfer of the small, self-sufficient marine station to the University of California. The scope of this philosopher-biologist was broad, and he realized that other sciences were necessary for his biological survey. So he brought a physicist and a chemist to the young institution. Indeed, he laid the foundations deep in the sands of La Jolla for a many-faceted study of everything about the ocean. Ritter's breadth of interests in the ocean created the diversity that characterizes the Scripps Institution.

As his retirement approached, Ritter urged the institution into encompassing all of oceanography, and he favored the appointment of geologist T. Wayland Vaughan as his successor. Vaughan, then with the United States Geological Survey, was already well known in scientific circles for his work on foraminifera, on corals, and on coral reef formation.

In his directorship from 1924 to 1936, Vaughan drew the institution into national and international ventures. For ten years he chaired the International Committee on the

Harald Sverdrup Sets the Sails

Oceanography of the Pacific. He served as an active member of two National Research Council committees and especially of the Committee on Oceanography of the National Academy of Sciences. One of his goals through that committee was to expand oceanic studies by United States scientists in both oceans. Vaughan expected the Scripps Institution to take a major role in Pacific oceanography. As a beginning he arranged that observations and measurements from Navy and Coast Guard ships, from steamship companies, and from lighthouses be sent to the institution.

Since the sale of the *Alexander Agassiz** in 1917, the institution had not owned a research ship. This Vaughan remedied in 1925 by purchasing a 64-foot purse seiner, the *Thaddeus,* which was renamed the *Scripps.* She was a small vessel to face the entire Pacific Ocean. Vaughan's intent, however, was to pursue oceanographic studies by sending Scripps staff members on ships of other agencies. The most extensive program of this kind was the participation of chemist Erik G. Moberg in the cruise of the *Carnegie* for two months in 1929, from San Francisco to Honolulu.

That same year Vaughan began arrangements for a scientific program on the *Carnegie* over a wide area of the Pacific, to be carried out by the Scripps Institution after the last voyage of that ship for its own organization, the Carnegie Institution of Washington, in 1931. Moberg, in consultation with other staff members, prepared a detailed program of work. In November of 1929, however, the *Carnegie* was destroyed in Apia, Samoa, by an explosion which also killed her captain. Vaughan wrote:

> Those members of the staff of the Scripps Institution who had been associated with the staff of the

---

*To avoid confusion with later ships, see the list of Scripps ships in chapter 14.

Scripps Institution of Oceanography: Probing the Oceans

> 'Carnegie' were paralyzed when the tragic news first reached us. We both admired and were personally fond of Captain [James P.] Ault and in addition to the grief we felt over his untimely death, there was the regret over the interruption of one of the finest programs of oceanographic research in modern times and the delay of further work on the oceanography of the Pacific.[4]

Vaughan, urged by Moberg, briefly considered carrying out the proposed researches on the *Scripps*. But, as he said in a plea to Robert P. Scripps: "All of us who are familiar with the 'Scripps', notwithstanding its being excellently adapted to short cruises along and off the coast, are of the opinion that it is rather small for making real sea voyages."[5]

Robert P. Scripps, the son of E. W. Scripps, did respond with interest to Vaughan's plea, and in 1930 advised University President Robert G. Sproul that he could provide $150,000 for constructing a ship to be used for research by Scripps Institution and other institutions. Before this offer could be carried out, however, Robert Scripps found that his resources had been "affected by the financial depression," and university officials, in consultation with Vaughan, concluded that "the income of the Scripps Institution [was] not adequate for the operation of a sea-going research vessel."[6]

Director Vaughan's hopes for an oceanographic program in the Pacific were dashed. Even the 64-foot *Scripps* was not used extensively, according to biologist Winfred E. Allen, who noted in 1931: "In the last six years the Scripps Institution boat has been in operation on about 130 days, in the last four years about 60 days, in the last two years not more than ten days."[7]

Vaughan was commissioned in 1932 by the Committee

Harald Sverdrup Sets the Sails

on Oceanography of the National Academy of Sciences to tour oceanographic institutions around the world. Upon his return in 1933, and facing retirement in three years, he devoted considerable time to selecting a worthy successor.

On 1 September 1936, Harald U. Sverdrup became the third director of Scripps Institution. Born in Sogndal, Norway, on 15 November 1888, Sverdrup had already had a distinguished career. His native country was among few in the world then that called oceanography a science. At the University of Oslo, Sverdrup had enrolled in the encompassing course called physical oceanography and astronomy, but he was turned toward meteorology and oceanography by Professor Vilhelm Bjerknes, whom he followed to Leipzig (during World War I) where in 1917 he completed his Ph.D. dissertation: *Der Nordatlantische Passat,* the North Atlantic Trade Wind. Under explorer Roald Amundsen, from 1917 to 1925, Sverdrup was in charge of the oceanographic and meteorological work on the Norwegian research ship *Maud,* which was intentionally frozen into the ice in order to be carried across the polar regions for scientific observations. He also served as navigator and occasionally as cook on the *Maud,* lived for seven months with the native Siberian Chukchi tribe, and wrote a great many reports while icebound. In later years Sverdrup said that he was most proud that "after seven years on the *Maud* he parted friends with his shipmates."[8]

He forecast his future, too, while icebound:

> During the last winter in the Arctic, in 1924-25 [he wrote many years later], we used to discuss what we wanted to do after returning to civilization. One of our party wanted to go to Peru, cross the Andes and, instead of drifting with the ice, to drift down the Amazon River on a raft. He did. I used to say that I should like an opportunity to do oceanographic work

9

in the Pacific Ocean. It took me much longer to reach that goal.[9]

Before getting to the Scripps Institution, Sverdrup in 1931 accompanied Hubert Wilkins on the first attempt at exploration beneath the polar ice in the submarine *Nautilus*. Of that trip, Sverdrup said that "he didn't mind living aboard a submarine, but not that submarine. He had expected to find a perfect machine for the business, but instead was greeted with an ex-naval vessel, the '0-12,' which had laid a number of years idle in Philadelphia. The expedition operated to the north of Spitzbergen, and soon called it quits."[10]

Sverdrup became a professor at the Geophysical Institute in Bergen, and in 1931 he became a research professor at the Christian Michelsen Institute there. He became acquainted with the United States by spending two six-month periods at the Department of Terrestrial Magnetism of the Carnegie Institution of Washington, where he worked on data collected by the *Maud* and also data from the seventh cruise of the research ship *Carnegie*.

Vaughan spoke aright when he called Sverdrup "one of the foremost living authorities on dynamic oceanography" — for Sverdrup helped to make it dynamic. He was, however, a self-effacing person, as indicated in a letter to Vaughan in which he agreed to be the director of Scripps for three years:

> I must confess that I have hesitated in assuming the responsibility which the directorship of a great institution like Scripps must involve. My hesitation is, I believe, rooted in my deep respect for scientific research. . . . It does not worry me that I am not an expert within many of the branches of oceanography which are represented at the Scripps Institution, since

Harald Sverdrup Sets the Sails

it is impossible to find anyone who would have such qualifications, but I do not want to disappoint those who think me able to carry your plans further and, according to these, to help making Scripps Institution a centre also of dynamic oceanography.[11]

Vaughan had paved the way, for in 1925 he had completed the transformation, in projects and people as well as in name, from the Scripps Institution for Biological Research to the Scripps Institution of Oceanography. So, in the fall of 1936, at his new location Sverdrup found the full spectrum of oceanic studies:

Director Vaughan was completing a report on the fossil foraminifera of Trinidad before his retirement to Washington, D.C.

Professor Francis B. Sumner, a philosophical biologist who had given up his 17-year study of the genetics of deer-mice *(Peromyscus)* in order to fit into oceanography, was studying how fish change color and what protection their colors afford them.

Professor George F. McEwen, a physicist concerned with ocean currents and temperatures, was predicting a "moderate excess" of precipitation for the season, from his analysis of the many years of temperature measurements recorded daily, usually by Stanley W. Chambers, who also took tide records, plankton samples, and water samples for density and chemical analysis.

Assistant Professor Winfred E. Allen, expert on diatoms, was suggesting that river runoff influenced the productivity of marine plankton, and was statistically analyzing collections of phytoplankton from the daily pier collections at Scripps and at Point Hueneme.

Assistant Professor Erik G. Moberg, chemist and also skipper of the *Scripps,* had just directed one cruise on the Fish and Game Commission ship *Bluefin* and was continuing

Scripps Institution of Oceanography: Probing the Oceans

chemical analyses of samples collected from the *Scripps,* from stations five and ten miles out to sea.

Assistant Professor Denis L. Fox, physiologist, who had worked as an oil company chemist before earning his Ph.D. at Stanford, was gathering mussels from the pier for studies of their feeding habits and the carotenoids that turned them orange, as well as working with Sumner on the pigments of fishes.

Assistant Professor Claude E. ZoBell, microbiologist, who had come to Scripps from the Hooper Foundation in San Francisco after receiving his Ph.D. at Berkeley, was working on a statistical approach to the distribution and characteristics of bacteria that live in the sea, and analyzing their effect upon sea-floor muds.

Instructor Martin W. Johnson, zoologist, who had earned his Ph.D. at the University of Washington and had come to Scripps from the International Fisheries Commission on the east coast, was making biweekly collections of zooplankton along the pier, having recently returned from a cruise on the Coast and Geodetic Survey ship *Guide,* and was continuing to work out the life histories and distribution of pelagic copepods.

Associate in Oceanography Percy S. Barnhart, who had recently completed a book on the marine fishes of southern California, was managing the popular public aquarium and bemoaning the damage to his colorful display of mounted fishes that had been loaned to the California-Pacific Exposition in San Francisco.

Instructor Richard H. Fleming, chemist and a Scripps alumnus, had led two trips on the Fish and Game Commission boat *Bluefin* to set out 4,000 drift bottles for current studies, and was trying to determine seasonal changes in the coastal waters.

Easter Ellen Cupp, who had received her Ph.D. in 1934 for work at Scripps, was completing a taxonomic monograph

Harald Sverdrup Sets the Sails

on plankton diatoms of southern California.

Visiting investigator Francis P. Shepard, geologist from the University of Illinois, was rowing out from the pier every few months to take lead-line soundings to measure changes in the Scripps submarine canyon; a commerical model of the canyon had been constructed shortly before from his earlier soundings, and one was on exhibit in the museum.

Visitor La Place Bostwick, formerly a jewelry designer and pearl buyer in St. Louis, was cultivating pearls in abalones.

Visitor Robert T. Young was collecting parasites from fishes and birds, and tracing their life histories.

Fred Baker, M.D., civic leader and a most active founder of the Marine Biological Association of San Diego, held an honorary curatorship of mollusks and owned one of the largest private shell collections in the world, from which he frequently donated shells to the institution's museum.

Five graduate students were registered for studies at Scripps; one of them, Roger Revelle, had received his Ph.D. in May 1936, with a dissertation on bottom samples collected by the *Carnegie* before her explosion. Six other students were on campus, chiefly for summer work but not registered. Five technical and clerical assistants and six maintenance workers completed the institution's roster, and in addition a number of workers assigned by the Work Projects Administration helped in everything from laboratory analysis to bookbinding, painting, and paving.

Such was the modest, yet diverse, nucleus that Sverdrup found for delving into the mysteries and workings of the ocean.

The assorted researches were carried out in the institution's three main buildings: George H. Scripps Laboratory, William E. Ritter Hall, and the Library-Museum, which was connected to Scripps Laboratory by a second-story bridge. The library, under the care of Ruth Ragan, housed 15,000 volumes, which were being recatalogued according to the Library of Congress system by one of the WPA workers, a trained librarian. The public aquarium, whose register was signed by 5,306 visitors in 1935, was housed in a wooden structure just north of Scripps Laboratory. The Scripps pier extended 1,000 feet seaward to reach clean sea water and to provide a platform to accommodate instruments for plankton and chemical sampling, for current, tide, and weather measurements. The *Scripps* was berthed at the San Diego Yacht Club in San Diego Bay.

The director and his family were housed in a two-story residence on campus. Many of the staff, the students, and some visitors lived in the 24 one-story frame cottages on the campus — "terrible cottages," thought Mrs. (Gudrun) Sverdrup, and she later admitted that she cried when she first saw them. "The tight colony," recalled Denis Fox long afterward, "like most such communities, received the pulse of any personal news items with the rapidity of an electric current crossing a wire network, and gossip could be extensive at times. It was not easy to keep clear of the busy, sometimes inventive, 'grapevine' in those early days."[12]

An annual income of about $800,000, of which approximately half came from the state of California, kept the research work going. The greatest needs of the institution, according to Director Vaughan in mid-1936, were a new aquarium building and repairs to the pier.

On 13 November 1936, just two months after Sverdrup's arrival, the institution's small ship, the *Scripps*, burned at its berth following an explosion in the galley. Martin W. Johnson, unhappily landlocked, then penned a Christmas wish for the institution:

Harald Sverdrup Sets the Sails

IF SANTA WERE A SAILOR

If Santa were a sailor, he'd bring us all a ship.
He'd fill it full of lab supplies and take us on a trip.
He'd bring a lot of citrate flasks, and silver nitrate too,
And heaps of sparkling glassware of curious shapes,
    and new.

He'd give us Nansen bottles, to bring up ocean brine,
And messengers and what-not, and miles and miles of
    line.
There'd be a long slide-rule on board to figure out the
    flow,
For Santa sure would like to know just where oceans
    go.

He'd bring us nets, and dredges too, to send down in
    the deep,
To fetch back fish and other things all in a squirming
    heap.
He'd say, "Now boys, I'd like to know, do mermaids
    really be?
Do awful scaly serpents torment them in the sea?
It's up to you, my gallant lads, to check these weird
    tales,
So hoist your flag, your anchors weigh, and set your
    shining sails."[13]

    Sverdrup, who had already been thinking of acquiring a vessel large enough for truly oceanic work, promptly requested an appointment with Robert P. Scripps, who this time agreed to provide $50,000 for a new ship. Sverdrup moved quickly; he found the 100-foot graceful sailing schooner *Serena* available for sale, and on 5 April 1937, Robert P. Scripps bought her for the institution. In gratitude for many favors, the staff of the institution renamed her the *E. W. Scripps.*

Martin Johnson demonstrated his enthusiasm with a jaunty sketch of the ship on his 1937 Christmas card for the institution's bulletin board.*

From that moment, the direction for the institution was: out. That is, out to sea. From 1937, with a total of 19 vessels over the years through 1976, the Scripps Institution scientists and ships have logged 3,738,526 nautical miles, from Mazatlán to Mahé, from Hakodate to Punta Arenas — and dozens of other ports — in the oceanographers' pursuit of happiness.

*See illustration. A longtime custom at Scripps was a Christmas card bulletin board for staff members to use to exchange season's greetings throughout the campus. The savings on postage were contributed by many participants to the Red Cross. The bulletin board has been revived in recent years.

"If Santa were a sailor, he'd bring us all a ship. . . ." Christmas card drawn by Martin W. Johnson, 1937, after the purchase of the *E. W. Scripps*.

NOTES

1. Quoted in: Allan C. Fisher, Jr., "San Diego: California's Plymouth Rock," *National Geographic,* Vol. 136 (July 1969), 145.

2. *Marine Geology of the Pacific* (New York: McGraw-Hill, 1964), Preface. Used with permission of McGraw-Hill Book Company.

3. Quoted in: Helen Raitt, *Exploring the Deep Pacific* (New York: W. W. Norton, 1956), xiv.

4. Letter to Harry L. Smithton, 7 December 1929.

5. Letter of 25 January 1930.

6. Letter to ship broker Daniel H. Cox, 15 January 1932.

7. Letter to State Director of Finance Rolland C. Vandegrift, 7 October 1931.

8. Roger Revelle and Walter Munk, "Harald Ulrik Sverdrup — An Appreciation," *Journal of Marine Research,* Vol. VII, No. 3 (1948), 127.

9. "Response by the Medallist," *Science,* Vol. 90 (14 July 1939), 26.

10. Frederick C. Whitney, "Sverdrup of Scripps," *La Jollan,* Vol. 1, No. 8 (4 December 1946), 18.

11. Letter of 11 April 1936.

12. "Again the Scene," Manuscript in SIO Archives, 1975, 124.

13. In SIO Archives, *History in the News, 1936-40,* 17.

2. Crossing the Threshold

Harald Sverdrup's philosophy was that throughout the broad Pacific Ocean a great many things awaited study. As soon as the *E. W. Scripps* was converted into a research vessel, he set the institution to studying those things. His own interests were particularly in currents and the relationship of the ocean to weather. But he also favored an all-out attack on problems, one that used the talents of every specialist. A recurrent theme in Sverdrup's talks and reports was his aim to draw together the varied disciplines toward studying The Ocean.

Mapping the currents along the coast not only contributed to understanding weather but also could be combined with finding where the California sardines spawned, in a project undertaken with the U.S. Bureau of Fisheries in 1938. For six months of each year until mid-1941, the *E. W. Scripps* plied a grid of 40 stations within the California Current system. Plankton samples were taken, temperatures were recorded, and oxygen and phosphate analyses of seawater were made regularly at each station. From these data Sverdrup was satisfied that "a definite relationship was established between the ocean currents, the character of the water masses, and the regions of maximum spawning of the sardines."[1]

One-half of the operating expenses of the *E. W. Scripps* in her first year came from the Geological Society of America as a grant to research associate Francis P. Shepard, so the geologists sailed off to take bottom cores and to set current meters on the floor of the ocean, after which they proudly announced the "largely unexpected" discovery of a current as fast as one-half mile per hour along the bottom off San Pedro, and constantly shifting currents in many other places.

In 1939 came the institution's first major expedition, when the *E. W. Scripps* in February and March carried venturesome geologists and naturalists to explore the mysteries of the Sea of Cortez, the gulf between Baja California and mainland Mexico. The explorers found the floor of the Gulf of California to be unexpectedly rough topography, and the waters to be extremely rich in marine life, reddened in patches with dense plankton blooms. They also found the Seri Indians of Tiburon to be no longer (if ever) cannibals.

From October to December of 1940 geologists led another expedition on the *E. W. Scripps* to the Gulf of California. The "fathometer echo-sounding machine" was run almost continuously, bottom samples and long cores — sometimes of green stinking mud — were taken, and the geology of islands and mainland was explored. Clear blue water greeted the explorers in the fall, but by December patches of red water again showed the richness of the gulf nutrients.

Those trips were exercises in ingenuity and cost-effectiveness, according to Shepard's students Robert S. Dietz and K. O. Emery:

> Prior to the cruises we built dredges, grab samplers, sediment traps, and corers. . . . We purchased junk

Most of the Scripps campus and all of the fleet — the *E. W. Scripps* — just before World War II.

lead at 3¢ per pound, used scrap 2½-inch pipe, and built two corers for about $50 each. . . . The scientific party [seven members] was expected to be sailors to run the ship and technicians to operate oceanographic winches, assemble and use the water and bottom samplers, and do various shipboard analyses for water chemistry. Among our duties while steering the ship was to tabulate by hand the water depth every two minutes. We did this with great enthusiasm since we had installed aboard the latest Submarine Signal Co. fathometer, which indicated the depth on a revolving red-flashing neon light. Graphic recorders had not yet been invented, so this instrument represented to us a remarkable advance over the sounding lead.[2]

Armed with new facts, collections, and theories from the many hours of work at sea, Sverdrup still could but conclude diffidently in 1940: "We are only crossing the threshold to the ocean world."[3]

In the lull before the fury of World War II, Scripps was still small and isolated, but busy in widely varied researches. Some say that the establishment was then actually run by Sverdrup's secretary Tillie Genter, who had been with the institution since 1919. Just outside the director's office in Scripps Laboratory was the only telephone, to which a staff member, when necessary, was summoned in person (and in obvious disapproval) or via the hand-cranked telephone system in Ritter Hall and the Library. In 1941 the Aquarium-Museum registered 8,453 signed visitors, and the library held 18,550 bound volumes. Each March the staff held an afternoon open house for the town in honor of the university's charter day. Gudrun Sverdrup made quantities of cookies every Christmas for a holiday party for the children in the campus cottages.

Sea-going staff members gathered samples from the

Crossing the Threshold

E. W. Scripps or hitchhiked on ships of the Coast Guard, the Navy, the California Fish and Game Commission, or the U.S. Fish and Wildlife Service. They supervised the handful of graduate students, who were registered at Berkeley for their work at Scripps. Most of the students were in residence only intermittently; when Walter H. Munk began his doctoral work, "the entire student body of this subtropical University outpost consisted of one graduate student. This was quite pleasant; I would be asked to all faculty parties and introduced: 'and here is our student.'"[4] In the late 1930s the administration of Scripps students was transferred from Berkeley to UCLA, and some of the oceanographers drove weekly to and from the "Westwood campus" to teach courses there also.

Staff members also spoke to civic groups frequently, and patiently answered reporters' questions. The cutting of kelp, they repeatedly said, does *not* bring more seaweed onto the beaches. Rip currents are *not* caused by heavy onshore winds, and a swimmer can escape rips by swimming parallel to shore. The warm Japan Current does *not* bring unusually warm weather to California, for it does not reach this coast by many miles. Local weather and fishing prospects, they said, were becoming better understood through detailed studies of the ocean. News items of that time recount Sverdrup proclaiming that the sea is not level, Winfred E. Allen repeating that discoloration in the ocean is caused by myriads of microscopic plants, Roger Revelle theorizing that the Gulf of California might be forming a future oil field. A 17-foot core, the longest then taken, came in for a bit of bragging, with due credit to Dietz and Emery, who had improved the design of the corer.

In addition to his own research, everyone was helping in projects of others. Denis Fox was analyzing pigments in the mud cores that Francis Shepard and Roger Revelle had brought up from the floor of the Gulf of California, and

Claude ZoBell was culturing bacteria from some of those cores. Revelle waded out to his six-foot, four-inch height to act as a wave staff in the buffeting surf for a study of waves and rip currents by not-as-tall Shepard, Eugene LaFond, and K. O. Emery.

Many at Scripps contributed to the monumental volume, "four pounds and all muscle": *The Oceans: Their Physics, Chemistry, and General Biology*,[5] by Harald U. Sverdrup, Martin W. Johnson, and Richard H. Fleming, described soon after its publication in 1942 as "epoch-making" and "without doubt the most comprehensive and authoritative treatise or text which has thus far appeared in this field." Because of its "comprehensive treatment of valuable information" during international war, the government limited the sale of the book to the United States and Canada for the first year.

Sverdrup hoped for even broader vistas for the staff, as he noted in the annual report in mid-1941:

> Emphasis has again been placed on attendance at scientific meetings by members of the staff because the isolated location of the Scripps Institution makes it desirable for the staff to retain contacts with other workers in their fields and to establish new contacts.

Instead of scientific meetings, the new contacts came from a quite different direction in that year of 1941: the Navy's need-to-know. For 35 years that union has been a fruitful one of mutual cooperation.

The Navy's greatest need in mid-1941 was to meet what was called the menace of the German submarine. The National Defense Research Committee set up a project on the west coast through the university at the Navy Radio and Sound Laboratory on Point Loma, soon named the University of California Division of War Research (UCDWR). For that project the Navy chartered the *E. W. Scripps* for research cruises. Harald Sverdrup, Martin Johnson, Richard Fleming, Eugene LaFond, Walter Munk, Francis Shepard, and Roger Revelle were among the early Scripps participants.

Physicists led the frontal attack on antisubmarine warfare at first, as Revelle recalled later:

> . . . [Ernest] Lawrence and his friends, reasoning with some justification that oceanographers were bumbling amateurs, quickly decided that underwater sound was a poor way to catch submarines and that optical methods should be used instead. They constructed an extremely powerful underwater searchlight and sewed together a huge black canvas cylinder which could be towed underwater to imitate a submarine. Unfortunately, it turned out that when the searchlight was directed on this object, it could be detected out to a range of about 100 feet. Shortly thereafter many of the physicists disappeared from UCDWR. The rest of us did not learn until after the war that they had gone off together to design and build an atom bomb.[6]

The oceanographers stayed, and the staff of UCDWR grew to 600 persons. Some of those came fortuitously: Jeffery D. Frautschy, for instance. Professor Franklin B. Hanley from the University of Minnesota had joined UCDWR early in the war, and he telephoned his former geology student to say that a job was available for him in San Diego. In those days of wartime secrecy, he could not say what the job was, but he assured Frautschy that it was in his line. "By the way, do you get seasick?" asked Hanley. "I never have," said Frautschy, without adding that he had never been within a thousand miles of the ocean. On the morning after his arrival in San Diego he went to sea for

UCDWR on the U.S.S. *Jasper** — and did not get seasick. He has stayed with oceanography ever since.

The researchers at UCDWR experimented with transmitted waves — radio, infrared, ultraviolet — and within the first year they concluded that the impervious ocean would yield only to sound waves. They quickly learned that the "silent" sea is instead a cacophony. Whales moan, porpoises squeal, some fish grunt, others whistle or sing, and some just grind their teeth. In warmer waters of the Pacific, a bedlam of crackling noise that overwhelmed all other sounds was soon identified by Martin Johnson as snapping shrimp in great colonies; then the little crustacean's habits and distribution could be predicted, so that, later, patches of them served as noisy havens for American submarines in Japanese waters.

Scientists transmitting sound pulses into the sea found echoes from the often-rugged topography of the ocean floor, reverberations from everywhere, and blockages and channels for sound waves in the far-from-uniform water of the ocean, which swirls and eddies, dilates and contracts under the influence of currents, tides, salinity, temperature, and pressure. They investigated the thermocline (first described in freshwater lakes in 1897): the discontinuity between warm surface water and the cold depths, a transition that is found at different depths in different regions, and that deflects sound waves passing through it. The depth to the thermocline could be measured, so that it too provided a haven for submarines.

Near Guadalupe Island off Baja California, in 1942, UCDWR researchers Russell W. Raitt, R. J. Christensen, and Carl F. Eyring found the phantom bottom of the sea, that registered on the scope of the echo-sounder as if it were the floor of the ocean itself. But, where the bottom was known

*See *Stranger* in the ship list, chapter 14.

to be 12,000 feet deep, the phantom registered at 1,500 feet. The peculiarity was first thought to be created within the instruments but was soon confirmed as real. For a time it was designated the ECR layer, for the initials of the puzzled discoverers,* but physicists began calling it the deep scattering layer, because it scattered the sound waves. Martin Johnson was again called upon, to see if the "phantom" was alive. He predicted that, if the layer were made up of living organisms, it should rise in the evening and then disperse and form again the next morning. On a round-the-clock trip at sea, on 26 and 27 June 1945, he confirmed his prediction. More detailed studies of the deep scattering layer were postponed until after the war.

To solve the complications of the sea's noises and echoes, as they were unscrambled by workers at UCDWR and other similar laboratories — at Harvard; Massachusetts Institute of Technology; New London, Connecticut; Woods Hole Oceanographic Institution; and Orlando, Florida — the Navy began installing hydrophones on ships to pinpoint underwater sounds as far away as ten miles, gain controls to screen out some echoes, and bathythermograph recorders to find the thermocline and to make continuous records of the always-varying ocean temperatures and pressures. Sonar charts were prepared for the Pacific and Indian oceans, as were topographic charts, and supplements to the Navy's sailing directions for submarines. On all of these, UCDWR worked closely with the Fleet Sonar School in San Diego, and after 1943 with the Pacific Submarine Force. In 1945 the laboratory was placed under the Navy's Bureau of Ships.

By 1944 the menace of the German submarine was

---

*An alternate story says that the initials stood for the Echo-ranging group of UCDWR, usually called the ECR group, of which those three were members.

Harald Sverdrup, adjusting a current meter at the end of the Scripps pier, probably in the late 1930s.

ended. "When I look back on the whole thing," said Admiral Jonas H. Ingram then, "I'd say that what we won on was the ability of the American boys to learn faster than the Germans how to become expert in using the stuff scientists put out."[7] UCDWR turned to improving the capabilities of American submarines, by continuing the development of sonar devices, echo-sounders, and bottom-scanners.

Separately from UCDWR, those who remained at Scripps were helping wherever and however they could with national defense. Sverdrup's own role was affected by his sorrow over his homeland. He had originally taken a three-year leave of absence from the Christian Michelsen Institute to serve as director of Scripps until 1939, and then had accepted a two-year extension. In May 1940, while his country was being overrun by foreign troops, Sverdrup asked University President Robert G. Sproul for permission to stay indefinitely at Scripps, on the basis that his own country would be unable to participate in any research for some time, and that he would only be a "liability" there, as his "ability and training" were not such that he could "render active help."[8] After the war Sverdrup said: "I thanked God that I was here in the United States where I might better fight the evil with knowledge, rather than being just another old man with a musket."[9]

But soon after beginning with the work at UCDWR, Sverdrup was withdrawn from it on the regulation forbidding "a person with relatives in an occupied country to work with classified material."[10] Although disappointed, he continued as director of Scripps, advised on other war-related projects, and taught courses for Air Force personnel. His clearance for classified work was later reinstated. During the war the Sverdrup family became United States citizens.

Sverdrup also helped with the wave-forecasting project. In 1938 U. S. Grant IV of UCLA and Francis Shepard had begun a study of coastal breakers with a specially built wave

Scripps Institution of Oceanography: Probing the Oceans

machine, in the hope of establishing a warning system for destructive waves on California beaches. In 1940, when such waves struck Los Angeles beach communities, Sverdrup guessed that they might have come from a severe storm in the North Pacific Ocean. The state of the art took such sudden forward leaps that by 1943 wave forecasting eased the landings of Allied troops on European beaches.

The wartime project began with Scripps student Walter H. Munk, who had transferred from the California Institute of Technology in 1939 to study under Sverdrup. Early in his wartime Army service, Munk proposed a means of forecasting sea and swell on beaches, based on a formula that incorporated wind velocity, fetch, and duration. He had become concerned when he learned that practice landings at the Army training camp in North Carolina were discontinued whenever breakers were more than five feet high, while he knew that average breakers along European coasts were regularly six feet and more. Sverdrup helped persuade military authorities that the calculations of his young student were valid, and that forecasting would be feasible.

> A great deal of the work [Sverdrup said later] was devoted to the study of the transformation of waves as they travel from deep to shallow water and to the relation between swell and surf. The studies covered such subjects as the velocity of progress in shallow water, the effect of shoaling bottom in the bending of waves (refraction), height of breaker in relation to height of swell, and relation between height of breaker and depth of breaking.[11]

These were not simple problems. Munk and his colleagues were fortunate in having advice from one of UCDWR's experts, Carl Eckart — a physicist characterized by preciseness, shyness, and quiet, deliberate speech. Munk said that when presented with a problem Eckart's custom was to respond with few questions or comments, but on the next day he would return with a "beautifully elegant set of handwritten notes . . . written down as they occurred, without corrections, so well worded, so carefully annotated, so beautifully placed on paper that six months or a year later . . . it was impossible to improve on the simplicity and presentation that he made."[12]

Not only the expertise of UCDWR, but even the beauty of La Jolla contributed to the war effort, for photographs of wave patterns along the La Jolla coast, distinctive for its bilobed submarine canyon, became the models for analyses of waves on beaches across the world.

In the fall of 1943, in cooperation with Woods Hole Oceanographic Institution and the engineering department at Berkeley (where John D. Isaacs and Willard Bascom participated), Scripps engaged in a program toward forecasting sea and swell from weather maps, forecasting breakers and surf, using aerial photographs to examine surf, and determining water depths from aerial photographs. Throughout the war more than 200 military officers (including some British ones) were trained at Scripps in these techniques. Beaches on Pacific islands, isolated peaks in a broad deep ocean, posed a separate problem, but the technique of working backward from aerial photographs made predictions of landing conditions feasible there also. At Palau a trained observer found several unknown passages through the reef by noting the pattern of breaking waves from his spotting plane.

When describing the wave-forecasting project after the war, Sverdrup said:

> The weather situation the day of the Normandy landings was unsettled, but the forecasts were reliable and no mishaps occurred because of incorrect estimates

of weather and surf. . . . Two former Scripps Institution students, Capt. John Crowell and Capt. Charles Bates, were in charge of surf-forecasting. When the landings on Sicily took place the Fleet ran into a severe storm . . . and it appeared possible that the Fleet would have to turn back. However, the aerologist forecast that the wind would die down and that only moderate surf would be encountered next morning at H-Hour. . . . This turned out correct and was one of the most brilliant predictions made during the war.[13]

In other wartime contributions, Denis Fox and Claude ZoBell served as consultants to industries on fouling organisms and microorganisms on seaplanes. Separately, Fox found a means of precluding fouling animals from settling on aluminum panels by applying an organic detergent dissolved in petrolatum. Robert H. Tschudy and Marston Sargent, and later postdoctoral student C. K. Tseng, gathered and cultured the red alga, *Gelidium cartilagineum,* in an effort to reduce this country's dependence on agar from Japan, vital to hospitals as a medium for culturing infectious organisms. J. Frederick Wohnus studied the beds of giant kelp, which was being considered more intensively as a possible agar substitute and as a stabilizer and binder in food and chemical products. ZoBell studied also the oxidation of natural and synthetic rubber by microorganisms, advised on public health problems, and gave first aid classes — as did Fox, who also worked with Thomas W. Whitaker of the nearby Department of Agriculture field station on developing vitamin-rich vegetables for the armed forces. McEwen and Sverdrup provided requested oceanographic information to the Army Air Force, and various staff members gave courses in oceanographic subjects to military personnel.

ZoBell noted after the war that "all of the personnel

and facilities in physical oceanography and marine meteorology at the SIO have been devoted to special war projects during the last five years, resulting in the accumulation, application and dissemination of a great wealth of oceanographic data of importance to the nation in times of peace and war."[14]

The Scripps library lent many of its books to defense projects and provided some to the British Admiralty that it had been unable to find elsewhere. The library also compiled a bibliography of its holdings on physical oceanography for the Weather Bureau, and it provided to staff researchers, for a Navy project, its file of temperatures, salinities, and current data in the Pacific Ocean gathered before the war by Japanese oceanographers. The institution improved its fire-fighting equipment as a defense measure, and it endured the deterioration of grounds and buildings from lack of manpower — to the degree that a reporter who visited the campus just after the war spoke of a "weather-beaten, musty smelling group of ramshackle buildings"[15] (while praising the research work). The pier, not yet repaired in spite of Vaughan's plea in 1936, and further damaged by a heavy storm in December 1942, was closed to the public. But the daily pier records were continued, and Martin Johnson pursued round-the-clock studies of underwater animal noises from the pier, where he kept a cot for night catnaps. The pier, in the absence of the *E. W. Scripps,* became the main source of replacements for the aquarium exhibits, both by line fishing and by means of a large trap kept set at the deep-water end.

The war, of course, was a time of heartache, endurance, and inconvenience for everyone. At its end, many at Scripps who had been drawn into special tasks returned with relief to their preferred work.

But a change was becoming evident. In a remarkable understatement in March 1945, a Scripps committee concerned with building needs at the institution commented: "There is a likelihood that after the war the Navy will continue its interest in the Pacific." The Navy did so, and it remembered the west coast institution that had provided it with a great deal of scientific aid.

The Scripps Institution was, after all, the only place in the United States at which a higher degree in oceanography was offered in 1946. Several of the officers who had taken the wartime courses in wave forecasting were eager to enter this new profession, and the Navy had a growing need for trained oceanographers. The Navy also had money for ocean research and, equally important, it had ships.

"The war showed how oceanography could be applied," said Sverdrup in September 1946. "This fall we shall have 25 men studying for their doctorate [by November there were 37]. For the first time there are enough opportunities in oceanography so that we feel justified in urging young men to study it."[16]

The golden age of global oceanography was beginning.

Sverdrup himself did not stay to see the outcome of what he had helped to stimulate, for in March 1948 he returned to Norway to become director of the Norsk Polar Institutt. Just before departing he commented, "The last 11 years with the University of California have been among the happiest of my life."[17]

Sverdrup had been good for the institution, too. He had expanded the horizons and opened new doors to studying the ocean. His philosophy of directing the laboratory was to make himself accessible by leaving his door open on Thursday afternoons. He also established the sociable and informative weekly staff luncheon.* Sverdrup is remembered as a charming person, a small and wiry man who

---

*Wednesday staff luncheon continued into the 1960s.

Crossing the Threshold

livened parties by telling of his life among the Chukchis — and by standing on his head, or doing an abrupt backward flip. He enjoyed fly fishing in the Sierras, and camping every New Year holiday with Guy and Margaret ("Peggy") Fleming (Guy was an active conservationist who helped establish Anza-Borrego State Park and Torrey Pines State Reserve, and who had given valuable advice to Vaughan on campus planting in the 1920s). Sverdrup built by hand the stone fireplace still standing behind the former director's house at Scripps.

Colleagues Revelle and Munk felt that Sverdrup's outstanding characteristic was "internal harmony between the diverse aspects of a complex personality, between the explorer and the scholar, the naturalist and the theorist, the teacher and the administrator." They credited him with "humor, kindliness, insight and self-discipline," and remarked that "behind that self-discipline one occasionally glimpses a Norse emotionality, intensity and recklessness."[18]

"Reasonable" was Sverdrup's favorite English word, said Munk, who also defined his mentor as an unsophisticated kind of scientist who had "an inordinate amount of common sense" and was best at synthesizing material. His 1940 paper on the transport equation, which Sverdrup himself did not consider important at the time, is considered a classic in the explanation of the equatorial current system. Chapter 15 of *The Oceans*, on "The Water Masses and Currents of the Oceans," was his outstanding summary, concluded Revelle and Munk; in that chapter, "Sverdrup's 'feeling' for data, his common sense interpretation of physical laws, his experience and remarkable memory [had been] brought to bear upon the problem of dealing with all oceans from a common point of view."[19]

The sailing was not all smooth for Sverdrup at Scripps. He was sorely hurt when his security clearance was withdrawn early in the war for a year and a half, causing an awkwardness in his relations with the military classified projects. An administrative lapse was also unfortunate: during his tenure as director his salary was never increased. This was suddenly noticed by university officials when Sverdrup announced his impending departure, and an offer was promptly tendered for him to continue as director, with the salary doubled. But Sverdrup had not meant to bargain. He and his family were anxious to return to their homeland, which was "struggling to get back on its feet after five years of oppression." There he served as director of the polar institute until his death in 1957.

Sverdrup left Scripps Institution in remarkably sound condition for studying The Ocean. In 1946, at least partly because of Roger Revelle, then on Navy duty with the Bureau of Ships, the Navy had created the Office of Research and Inventions, soon renamed the Office of Naval Research (ONR), to pursue promising leads from the wartime researches. On 1 July 1946, the first ONR contract with the Scripps Institution began; it was for $120,000 for the first year and was fully oceanic:

> Conduct surveys and research, analyse and compile data and technical information, prepare material for charts, manuals, and reports, and foster the training of military and civilian personnel in the following fields of oceanography: permanent currents; interaction of the sea and atmosphere (including wind waves, swell and surf); the distribution of physical properties; the distribution of chemical properties; the distribution of organisms; the characteristics of the sea bottom and beaches; tides, tidal currents and destructive sea waves; the physics and distribution of sea and terrigenous ice. Such a program shall include both geographical investigations (surveys), experiments in the laboratory and

at sea, pertinent theoretical studies and necessary travel.[20]

The following year the state of California appropriated $300,000 to Scripps for each of three years to help solve the state's greatest oceanographic problem: where had the California sardines gone?

From available Navy ships, without charge (and with considerable correspondence between the institution and Revelle, still on Navy duty in Washington), Scripps was able to select a 143-foot seagoing tug and a 134-foot minesweeper, "on the basis that the Navy would be under no financial obligations to recondition or maintain the vessels or to pay any part of the operating expenses thereof."[21] Through a contest on campus, the ships were renamed *Horizon* and *Crest.* *

The original intent had been to acquire and modify three Navy ships, but the cost of conversion proved too high, so Scripps instead bought, with funds provided by the Navy, the instantly usable 80-foot purse-seiner *Paolina-T*. She was the first in use to supplement the war-weary *E. W. Scripps,* which was returned by UCDWR in 1947. The *Crest,* although sound enough otherwise, proved to have faulty engine shafts, so she went into a shipyard in Long Beach for extensive repairs. That took funds away from converting the *Horizon* to oceanographic service, so Scripps ship personnel did that work themselves whenever they could find bits of time and money. James L. Faughn, who became technical marine superintendent in 1948, found himself dealing more often with ailing than with sailing ships for some time.

Four seagoing ships and access to the newly developed wartime electronic equipment opened new doors to studying

---

*See list of ships in chapter 14.

Scripps Institution of Oceanography: Probing the Oceans

the ocean, especially when the institution's income in 1948 was five times that of 1936. These opportunities and a staff of 111 people* of richly varied background combined to create the diversity that characterizes Scripps today. The history since 1945 becomes almost beyond recounting. The only way to attempt to do it justice is to break up the story into the units that today make up the institution.

*Excluding the ships' crews.

NOTES

1. SIO Annual Report, 1941.

2. "Early Days of Marine Geology," *Oceanus,* Vol. 19, No. 4 (Summer 1976), 20.

3. *San Diego Union,* 30 March 1940.

4. "The Nth Campus Problem," *Bear Facts* (May 1966), 2.

5. New York, Prentice-Hall, 1942.

6. "The Age of Innocence and War in Oceanography," *Oceans Magazine,* Vol. I, No. 3 (March 1969), 8.

7. *San Diego Union,* 30 September 1946.

8. Letter of 1 May 1940.

9. Quoted in: Frederick C. Whitney, "Sverdrup of Scripps," *La Jollan,* Vol. I, No. 8 (4 December 1946), 18-19.

10. Letter to University of California President Sproul, June 1942.

11. SIO Biennial Report, 1944-46, 12.

12. Memorial service for Carl Eckart, 3 November 1973.

13. *Los Angeles Times,* 15 January 1946.

14. SIO Biennial Report, 1944-46, 10.

15. Frederick C. Whitney, "Sverdrup of Scripps," *La Jollan,* Vol. I, No. 8 (4 December 1946), 18.

16. *San Diego Union,* 8 September 1946.

17. *La Jolla Light,* 26 February 1948.

18. "Harald Ulrik Sverdrup — An Appreciation," *Journal of Marine Research,* Vol. VII, No. 3 (1948), 127.

19. *Ibid.,* 130.

20. ONR Progress Report No. 1, SIO Reference 46-7, 1946.

21. Minutes of Marine Research Committee, 13 March 1947.

Part II
The Research Units

3. History's Greatest Fishing Expedition: The Marine Life Research Program

Sardinops sagax* — the Pacific sardine, sometimes called the pilchard — is a tasty member of the herring family. It is a glittering dark green to blue above and iridescent silvery below. Only occasionally reaching twelve inches in length, it is a smallish fish to have caused so much furor. But it does seem as if the Pacific sardine since 1945 has spawned more initials than eggs: i.e., MRC, MLR, CalCOFI, DCPG, WWD, IKMT. The search for it poured a great deal of money into studies of California fisheries.

The figures are staggering: in the 1936-37 fishing season, 726,000 tons of sardines were hauled into California harbors in wee-daylight hours — one-quarter of the total tonnage of fish caught in the United States that year; in the 1946-47 season, even with more fishermen searching, only 234,000 tons could be found.

Why? Of course, everyone had a pet theory: changing currents were carrying off the eggs; water temperatures were changing; adult sardines were migrating to greener

---

*Formerly *Sardinops caerulea*, and thus designated throughout most of the sardine study.

pastures; the increase in gray whales was somehow reducing the numbers of sardines; sea lions were eating them; there was less salt in the sea; maybe it was all because of the atom bomb tests.

At stake was the livelihood of three thousand fishermen and one hundred canneries in Monterey Bay, Los Angeles, Newport, and San Diego. So in January of 1947 representatives of the fishing industry called a meeting of experts to look into the problem. To that meeting the Sardine Products Institute sent William C. Moorehead, Irwin Isaacs, David Oliver, and Julian G. Burnette; Montgomery Phister, who acted as chairman, represented the tuna fisheries; Oscar E. Sette and John C. Marr attended from the U.S. Fish and Wildlife Service; Frances N. Clark and Richard Croker represented the California Bureau of Marine Fisheries; Robert C. Miller and Wilbert M. Chapman represented the California Academy of Sciences; and Harald U. Sverdrup attended from Scripps. Sverdrup was essentially nominated by Chapman, who wrote Phister in December 1946: "In regard to a representative from Scripps . . . Dr. H. U. Sverdrup, Director, is in my opinion the most competent oceanographer now working in the world. . . . He is the key man in Pacific Oceanography."[1]

Within a year the California legislature had established the Marine Research Committee (MRC) of nine members, five from the fishing industry, one public representative, one from the California Fish and Game Commission, and two from the California Division of Fish and Game. The legislature also guaranteed $300,000 for the first year of a three-year study to the University of California, to be used by the Scripps Institution (this was increased to $400,000 the next year). The fishermen, through the California Sardine Products Institute, augmented this by persuading the legislature to impose a tax of 50 cents per ton on sardines for four years, to be collected by the

Marine Life Research Program

California Department of Fish and Game and administered by the Marine Research Committee. There was indeed money available to learn the secrets of the sardine.

The organizational meetings were not smooth, as each participating group worked within its own restrictions and needs to establish its role in a complex undertaking. Chapman admitted candidly to Don T. Saxby of the California Packing Company after the first MRC meeting: "The whole damned thing is likely to fly apart before it gets started" — and then worked furiously to see that it didn't.

Sverdrup was a vital participant in setting up the study although he had already announced his intention of returning to Norway. He commented at the time that he might not have left Scripps had he known that such an all-encompassing oceanographic project was about to begin.

The following year, 1948, Scripps was represented by its new director, Carl Eckart, at the second meeting of the Marine Research Committee, held in La Jolla, and he was accompanied by colleagues Carl L. Hubbs, Martin W. Johnson, Roger Revelle, John D. Isaacs, and graduate student J. Laurence McHugh. Revelle had just returned to Scripps from his Navy service and was designated as associate director and given charge of the Scripps portion of the sardine project.

Thanks to the groundwork laid by studies that had begun in the 1920s, including those in which Scripps had participated in cooperation with the California Division of Fish and Game (1937) and with the U.S. Fish and Wildlife Service (1938 to 1941) throughout the California Current, the organizations that came to the aid of the sardine fishermen "were able to get to work with little waste motion. They knew what they were after and the best way to get it." They already knew that "the sardine is a restless and far-traveling creature. When California, Canada, Washington, and Oregon were conducting tagging operations, large sardines tagged off Southern California in February were retaken off British Columbia in the following July. Fish tagged in Sebastian Viscaino Bay [Baja California] were found as far north as the Columbia River. . . . The fact is, the sardine respects neither state lines nor national boundaries."[2]

But only the larger and older fish drifted far north; their numbers dwindled first, and were not replaced in a succession of poor spawning years. Canneries in Canada, then in Washington and Oregon closed, soon followed by those in San Francisco. Supported by fish from fairly good spawning in 1946 to 1948, Cannery Row in Monterey hung on. But it too was doomed.

The sardine project that began in 1948 was very much a cooperative one. The U.S. Fish and Wildlife Service, through its South Pacific Fishery Investigations, undertook studies on the spawning, survival, and recruitment of sardines, using its ship *Black Douglas*. The California Division of Fish and Game, which had been studying the slippery sardine for thirty years, set out to determine its availability to fishermen by studies of the animal's abundance, distribution, migration, and behavior, using its ship *N. B. Scofield* and later also the *Yellowfin*. The California Academy of Sciences began laboratory studies on the behavior and physiology of sardines, and Stanford University's Hopkins Marine Laboratory joined the project in 1951 to study the oceanography of the Monterey Bay area in detail.

The Scripps Institution — on which we shall concentrate here, with apology to the others — from the beginning was given responsibility for gathering general oceanographic data in the sardine habitat, as well as information on the organic productivity of the ocean. Sverdrup gave the term Marine Life Research program (MLR) to the Scripps part of the project.

Marine Life Research Program

There was not space within the walls of the four research and office buildings on the Scripps campus to contain the envisaged project. But the close ties between Scripps and the Navy research groups on Point Loma made it possible to arrange for facilities there. So the Scripps MLR group was first located in Navy barracks buildings at Point Loma in 1948. The personnel were gradually moved onto the Scripps campus as space became available.

A laboratory of the U.S. Fish and Wildlife Service was first located at Stanford University, but in February 1954 Revelle extended an invitation to John C. Marr to move that research group to Scripps, "to facilitate the integration of the California Cooperative Oceanic Fisheries program." That same summer part of the group moved into the director's house (T-16) at Scripps, and others into facilities at Point Loma, where a branch laboratory of the Fish and Wildlife Service was already located. In 1964 the spectacular Fishery Oceanography Center* was completed on land provided by the university at the north end of the Scripps campus, and the Bureau of Commercial Fisheries of the U.S. Fish and Wildlife Service moved into it. (The building was renamed Southwest Fishery Center in 1970, and the Bureau was renamed the National Marine Fisheries Service, an agency of NOAA, the National Oceanic and Atmospheric Administration.**)

The statewide research project under the Marine Research Committee was referred to as the California Cooperative Sardine Study until 1953, when it became the California Cooperative Oceanic Fisheries Investigations (at

---

*Sometimes called the "Fish Hilton" because, Sally Spiess recalls, while it was under construction passing tourists often asked when the new hotel would be open.

**When this organization is mentioned throughout this book, its name at the time cited will be used. The building is generally called the Fisheries Building by Scripps people.

Retrieving the bottom fish-trap, 1950; John D. Isaacs sitting on rail, Lewis W. Kidd second from left.

Marine Life Research Program

first abbreviated CCOFI, now more commonly CalCOFI and pronounced calcoffee). The terms CalCOFI and MLR are used almost interchangeably (but not correctly) around Scripps, and it is worth emphasizing that the Marine Life Research program is the institution's name for its own broad study, which is under the auspices of the state-supported CalCOFI, which includes all of the participating laboratories.

Thanks to the state's contribution and the Navy's interest in general oceanographic studies, the sardine study could be tackled with a fleet. In addition to the ships of the other participating agencies, Scripps Institution's *Horizon* and *Crest* were assigned specifically to MLR, and indeed their conversion to research vessels was mostly accomplished with MRC funds and university money for that purpose. In addition, the *Paolina-T* and sometimes the *E. W. Scripps* were used for occasional special investigations or when one of the regulars was laid up.

The plan of attack devised in 1948 (slightly modified in 1950) was a bold one. It was to survey 670,000 square miles of ocean, from the mouth of the Columbia River to halfway down Baja California and extending outward 400 miles. Through this region flowed the California Current, the mass of water from the great clockwise circulation of the North Pacific that moves southeastwardly at a speed of less than half a knot parallel to the American west coast, warming in the sun, until at about 25° north latitude it swings westward to join the North Equatorial Current. The California Current is complicated by countercurrents and eddies, intermittent and permanent, and by regions of upwelling.

To study the region systematically, a grid was laid out by drawing a line roughly parallel to the coast from which right angle lines were drawn at 120-mile intervals. Along these lines stations were spotted every 40 miles. Each station was to be occupied once a month by one of the participating ships. At each station were taken a plankton tow, a hydrographic cast, a bathythermograph record, and a phytoplankton cast. Dip-netting for fish was done at night, notes on marine birds and mammals were recorded, and weather observations were sent to the U.S. Weather Bureau four times a day.

All this was more easily planned on land than carried out at sea, especially during the winter cruises, and the time schedule often had to adapt to nature's whims.

On one early CalCOFI cruise, the U.S. Fish and Wildlife ship, *Black Douglas* — a former sailing yacht — was called to aid a burning lumber boat in heavy seas. Directed by Coast Guard helicopters, the yacht reached the scene of distress at night, to find that the ship had broken up and its crew were all drifting in two lifeboats. One of the lifeboat occupants was heard to exclaim: "Christ — a goddamned yachat!" In spite of 25-knot winds, the plunging *Black Douglas* was held steady alongside the first lifeboat, and as waves raised the boat about level with the gunwales, the survivors were yanked onto the deck of the *Black Douglas* one at a time. The second boat was awash and beginning to founder when finally located, but its wet, cold occupants were quickly pulled aboard the *Black Douglas,* which cut short its survey to get the rescued back to port.

Results from the monthly cruises began coming in very promptly. "The oceanographic approach to the sardine problem," said the progress report of 1950, "is the feature which makes the present work unique; it has never been tried on such a scale before anywhere in the world. What — very briefly — the scientists hope to do is to correlate changes in water conditions with sardine spawning, availability, and abundance."[3]

Marine Life Research Program

The talk by the participants was soon of year-classes, upwelling regions, phytoplankton, and weather. Two major centers of spawning were outlined, one near Cedros Island off Baja California, and a larger one off the boundary between California and Baja California. Both were areas of upwelling, separated by an almost barren region. Two other known spawning areas, in the Gulf of California and off southern Baja California, were considered to be outside the sources of the California catch. Water temperatures between 12.5° and 16° C. proved to delimit sardine spawning. There was talk of last year's weather becoming the key to this year's forecast of spawning because of its effect on currents and upwelling. Certain years were found to produce the majority of caught sardines for several years; the year-classes could be readily identified by growth counts on the scales (now the lines of growth on the otoliths, or "ear bones," are also used). Experiments ashore showed that antibiotics added to sea water greatly increased the hatch of sardine eggs, suggesting that disease might be affecting the numbers. Laboratory tests showed that sardines could be "herded" by electrical currents. The lateral-line system of fishes came under scrutiny. Attempts were made to determine individual races of sardines so that their origin and extent of mingling could be unraveled.

As graduate student J. L. ("Laurie") McHugh had commented when the project was being established in 1948: "The resources of both the physical oceanography and the recruitment research shore sections will be taxed to the limit in handling the material collected at sea." Within the first year of the sardine project, the staff of Scripps was increased 42 percent. It was at that time that the Data Collection and Processing Group (DCPG) made its appearance. Their task was to analyze hydrographic data from approximately 1,500 stations each year, quickly enough for the results to be used in planning the next year's program. At first, "this was found impossible to do, largely because there were not enough skilled oceanographers available for the work of processing the data. A new system had to be worked out in which relatively unskilled workers under the direction of trained men could process the data."[4] A boost was given to the program by its gaining access to the Institute of Numerical Analysis at UCLA, "where ingenious 'thinking machines' do in 20 hours what it took skilled men some weeks to do before."

Thanks to the breadth of the sardine project, MLR money supported some projects and personnel in almost every aspect of ocean studies at Scripps, but chiefly, of course, in biology, physical oceanography, and instrumentation, beyond the considerable costs of running the ships.

WWD made its appearance too — not an abbreviation in this case, but the call letters of the lifeline to the ships, the radio station. Having learned the usefulness of steady radio communication through the work for UCDWR by the *E. W. Scripps*, the institution wanted its own voice between home base and the fleet of ships running the station grid. Persistent correspondence led to permission in 1948 for the institution to use the frequencies of the U.S. Fish and Wildlife Service for MLR work and Navy frequencies for other traffic. The Fish and Wildlife Service finally persuaded the Interdepartmental Radio Advisory Committee to assign them a station for the sardine project, to be operated by Scripps personnel, at first in a temporary building south of Scripps Laboratory (where it interfered with campus telephone calls whenever it was transmitting). Frank Berberich opened "Willie Willie Dog" in August 1949, and the voice at first was feeble, limited by edict to 125 watts on the 500-watt Navy equipment obtained, while it reported on storms and new fishes and sometimes troubles at sea, but it did speed the work and was a comfort to those ashore. In 1952 WWD was relocated on the hilltop eastward of the main

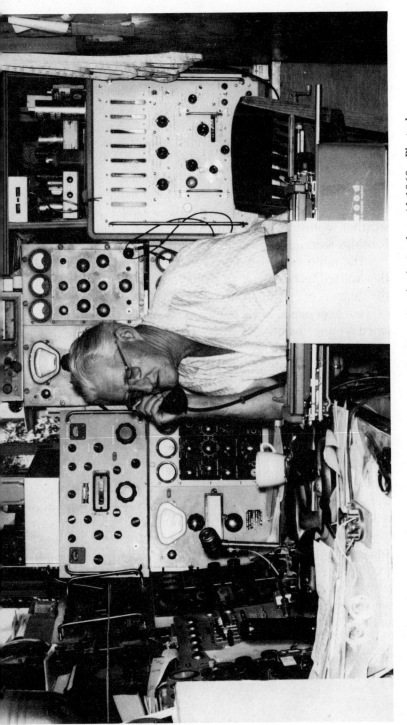

The lifeline to the ships: "Nick" Carter in the campus radio shack, in the mid-1960s. Photo by J. D. Frautschy.

Scripps Institution of Oceanography: Probing the Oceans

part of the campus, among the ground squirrels and rattlesnakes. In 1969 much more powerful transmitters were installed twelve miles inland, on former Camp Elliott land, but the receivers and personnel have remained on the Scripps hilltop. The station serves the entire Scripps fleet, anywhere in the world, and the National Marine Fisheries Service ship *David Starr Jordan;* it also handles traffic for the Deep Sea Drilling Project's *Glomar Challenger,* provides weather reports to the fishing fleet, and transmits messages for ships working in cooperation with Scripps.

The radio shack has been renovated in the 1970s, under the leadership of Donal C. Crouch. For many years it was the domain of Arthur B. ("Nick") Carter — "a professional of vast experience with the enthusiasm of the most dedicated young 'ham' operator." In those days the quarters were cluttered and noisy, with fragments of code and garbled voices drowning each other out, bells and telephones ringing, and teletype clattering. From the apparent chaos Carter would emerge with a bluff, hearty greeting for a visitor that said you were the one person he most wanted to see. Along with the greeting came an offer of thick coffee in a thick mug. After many years, Carter still marveled at the thousands of miles of ocean that his key could span in an instant and draw a response, so that somehow the ocean seemed smaller in that friendly room. In Navy tradition, the typed messages come out of the radio station in cryptic sentences in block letters, and with little punctuation. The spelling is sometimes startling, but excusable when one takes into account the fading of distant transmission and the technical terms used — polysyllabic scientific names of biological specimens, chemical terms, geological descriptions, names and numbers of parts for gears and engines and winches.

The first years of the sardine project were, of necessity, devoted to defining the problem and collecting data. The cruises went out month after month on the station grid, fair weather or foul. They trained a generation of biologists and physical oceanographers, and provided them with a lifetime of memories . . . of beer-filled nights on Cedros Island . . . of bars in Mazatlán at the south end of the grid . . . of storms and surge from Point Conception northward . . . and of camaraderie.

There were special cruises as well. In 1955, MLR, along with other groups at Scripps, participated in "the most ambitious oceanographic survey ever attempted" to that time: Norpac (for North Pacific) Expedition. A cooperative synoptic survey of the Pacific Ocean north of 20° latitude was first proposed by Joseph L. Reid, Jr., in 1953, and it grew to include 19 research vessels of 14 institutions from the United States, Japan, and Canada. From July to September, 1,002 hydrographic stations were occupied, and more than 2,000 plankton samples were gathered, for an almost simultaneous picture of the vast region from Acapulco to Alaska and across to Japan. August had been selected as the time least likely to be disturbed by weather, and indeed the "expeditions were remarkably free from bad weather and breakdowns," reported coordinator Reid,[5] who also recalls that the Scripps ships were among his headaches. *Horizon* sailed for Norpac on 3 August 1955, *Spencer F. Baird* on 8 August, and the newly acquired *Stranger* finally left on 9 August after delays for repairs. From home base Reid monitored messages requesting permission to return because of minor problems, which he firmly denied, so the ships sailed on. Two of the participating Japanese ships rode out typhoons "but pressed on to complete their planned work." It appeared to one awed reporter that "the ships are measuring anything that might be pertinent to a better understanding of the mysterious sea,"[6] but the observations and collections were actually of temperature, salinity, dissolved-oxygen content, inorganic phosphate-phosphorus, zooplankton, and phytoplankton. Norpac was a very successful introduction for Scripps into international cooperative oceanographic studies. It set a standard not always matched in later ventures; records of all the data, with preliminary processing completed, were exchanged by all the participating agencies less than five months after the end of the field work.

In 1957, the Marine Research Committee paused to see what it had done in a decade. A special committee summarized the accomplishments, which included: defining four spawning areas of sardines; developing methods of estimating numbers of eggs, larvae, juvenile, and adult sardines; obtaining information on north-south migrations; studying behavior patterns; determining that common nutrients are not limiting to phytoplankton; defining zooplankton areas; accumulating temperature, salinity, oxygen, and phosphate data; and gathering information on the abundance and location of the eggs, larvae, and adults of anchovy, jack mackerel, Pacific mackerel, saury, and hake.

The special committee, which consisted of John D. Isaacs of Scripps, John Radovich of California Fish and Game, and chairman John C. Marr of the U.S. Fish and Wildlife Service, concluded that further studies of the sardine populations and the oceanography of the spawning region should be continued. CalCOFI, they felt, had "made real contributions toward a better understanding of the fisheries but [lacked] effective coordination."[7] In a reorganization the members of the "three-Johns committee" became the governing technical committee and Garth I. Murphy was appointed CalCOFI coordinator in 1958 (until 1965). John Isaacs also became in 1958 the director of the MLR program at Scripps. Revelle had included the direction of MLR among his duties as director of Scripps, but various people (for example, Paul L. Horrer) had handled the routine running of the project for several years.

Marine Life Research Program

John Dove Isaacs — the big man with the beatific smile and courtly air, distinguished with and by white hair and beard — was the guiding hand of MLR until late 1974, when Joseph L. Reid, Jr., became director of MLR and Isaacs continued with the Institute of Marine Resources (chapter 6). If one were to recommend a training course for a future oceanographer, it could well copy Isaacs's. Being a bit of a maverick helps, as does a love of literature and a keen interest in questioning nature. Isaacs has these. He once described himself — accurately — as "a naive, enthusiastic sort of person."[8] Born in Spokane, Washington, in 1913, he began in oceanography as a commercial fisherman, until his boat tangled with a jetty in the Columbia River and he had to swim ashore. With Willard Bascom during World War II, Isaacs tested crashing breakers and measured beaches in DUKWs and other landing craft until the Coast Guard concluded that it was too dangerous. He graduated in engineering at Berkeley in 1944, worked on wave refraction with Sverdrup and Walter H. Munk, and joined Revelle in Operation Crossroads in 1946 (see chapter 15). Isaacs began at Scripps as an associate oceanographer, at Revelle's invitation in 1948, when the MLR program was just beginning. By 1961 he had become a professor of oceanography. Isaacs is phenomenal in proposing ingenious solutions to old problems. His motto may be "Why not?" And indeed, he is often right.

Considering Isaacs's participation in the special committee of 1957, it is surprising that, in summarizing the results of the first decade, more emphasis was not placed on the major contribution of CalCOFI in developing new instruments for use at sea. The classics of marine biological collecting were invented, tested, and modified during the early years of the California sardine program, many of them in the Scripps unit called Special Developments, headed by Isaacs and James M. Snodgrass.

57

Scripps Institution of Oceanography: Probing the Oceans

Equipment at sea is the oceanographers' greatest headache. Of things put over the side, some refuse to go down and others refuse to come up. Towed items flutter to the surface, overhead balloons dive into the waves. Cables kink, winches jam — and suddenly the air turns blue. But, there has to be a way. . . .

Sam Hinton pointed out one of the problems faced by inventors in the early days of CalCOFI: "A good part of the area under consideration is stormy and rough, and the devisors of new equipment aim for a product that can be operated by a seasick technician who must hold on to a stanchion with both feet and one hand while doing his work!"[9] Help was already on the way; James Snodgrass commented in 1949 that "the ability of a man to do competent work at sea is tremendously improved with the use of . . . the new motion-sickness drug 'Dramamine.' "[10] Sir Edward Bullard once called seasick pills "probably the greatest scientific contribution of the twentieth century to oceanography."[11]

On the CalCOFI cruises, plankton samplers were not catching the fleetest of the animals, the ones that darted from the disturbance of the towed cable away from the lagging net. So by 1950 the high-speed plankton collector was devised, with a spherical cable clamp that allowed the net to project forward of the point of attachment. A depth-flow meter, which could draw a continuous record of the depth and flow of water through the net, was designed to use with the collector. Also, for underway sampling, a bronze depressor was built to hold the towed cable and collecting equipment steadily and uniformly beneath the surface. But strained cables can snap, so a hydraulic dynamometer was added to measure cable stress.

The floating fish-larvae trap was another early invention; this was an inverted pyramid of wire mesh, pierced by

Bringing aboard the Isaacs-Kidd Midwater Trawl, Transpac Expedition, 1953.

small holes, and with an attached light to attract animals at night. It proved much more successful than the customary dipnet. A plankton-sample splitter was built, to obtain aliquot divisions in the samples sorted ashore. An automatic servo-operated photometer was designed for colorimetric analyses of chemicals in sea water. In fact, every piece of equipment taken to sea and every laborious shipboard procedure was scrutinized by ingenious inventors in the early days of MLR, and constant improvements were made.

California Fish and Game researchers took advantage of equipment developed for wartime — a sonar set and recording echo-sounder — to try locating schools of fish. Although the equipment was considered bulky and expensive and required a trained operator, early tests proved that it could indeed locate schools up to 800 yards away in depths beyond 50 fathoms. The goal was to expand sardine fishing hours, as the custom was to seek the schools only on moonless nights when flashes of phosphorescence betrayed them.

In 1951 another of oceanography's classical tools was being tested, and proving to be a valuable means of exploring ocean depths: the Isaacs-Kidd midwater trawl (IKMT). Although surface fishes and plankton had been regularly collected previously with surface nets, as had some bottom-dwelling fishes and invertebrates with bottom trawls, the vast middle area had only occasionally yielded specimens to science, more by accident than design, and never in the Pacific Ocean. The deep scattering layer that troubled the physicists on their echo-sounders during the war would continue to baffle biologists until a means could be designed to collect within it.

John Isaacs and Lewis W. Kidd took on the design problem, with the experienced advice of Carl L. Hubbs. The trawl consisted of "a net of special design attached to a wide, V-shaped, rigid, diving vane. The vane keeps the mouth of the net open and dives, thus maintaining the net at a predetermined depth for extended periods at comparatively high ship speeds."[12]

Early tests delighted Hubbs. From the *Horizon* on Northern Holiday Expedition in August 1951, Isaacs queried by radio:

> Took a Macropinna-like fish at 1100 meters in 2400 fathoms X Small fins and entirely flat belly with forward projecting foot suggesting attachment a-la remora X Shall I keep him? X He is not very large X Hauls medium to excellent.

Hubbs replied: "Bring fish back or don't come back yourself." A few days later he continued:

> Macropinna-like fish apparently Opisthoproctus / New for Pacific / One of rarest and strangest types / Two species in Atlantic / Other captures obviously unusual and exciting / Wish I could see them come in / Good luck.

Apart from the novelty of the new species of midwater fishes, all the other common fishes of the sardine region were also under study in the program. For some of the fishes were known to be predators of sardine eggs and young, and others were obviously competitors. There was hope, too, of encouraging other fisheries, particularly the northern anchovy, *Engraulis mordax,* found in similar numbers and sites as the sardine. The fishermen had little choice, as sardine spawning continued at a low level, and in 1952 the combined catch of "substitute sardines" — anchovy, jack mackerel, and Pacific mackerel — exceeded that of sardines. The tax for research of 50 cents a ton was raised to one dollar and was imposed on those fisheries that year (it has since been extended to anchovetta, herring, and squid).

More than fishes were under scrutiny. To study the countercurrents along the coast, Paul L. Horrer began a drift-bottle project in 1954. Commercial salad dressing bottles ballasted with sand were released on each CalCOFI station every month in the early years, usually twelve bottles on each station. A red postcard enclosed in the bottle explained in English and in Spanish the purpose of the study and requested the finder to send to Scripps the place, date, time of finding, and finder's name and address. Only a small percentage was returned, most of them released not more than 20 miles from shore. But from these returns Richard A. Schwartzlose was able to define the Davidson Countercurrent, 50 miles wide, from central California north to British Columbia, that began in the fall months, built up to speeds of 0.5 to 0.9 knots for several hundred miles, and disappeared in the spring. Elsewhere, he found short countercurrents and intermittent eddies at irregular times. Drift-bottle releases have been continued. Recoveries vary from zero to 23 percent (average 3.4), from points as distant as Alaska and Acapulco, and in 1971 one was picked up on the island of Hawaii.

The second decade of CalCOFI produced preliminary findings on the region that it called its own, and opened doors to speculation. As John Isaacs noted in a meeting of the MRC, as each question was answered, new ones arose.[13] But questions *were* being answered, in a multitude of reports, scientific papers, and student dissertations, and in that second decade a new philosophy was emerging in MLR.

Nature stepped in to help — or confound — the scientists in 1957, simply by warming the ocean. Not much, only 2° to 4° C. But — "Hawaii had its first recorded typhoon; the seabird-killing *El Niño* visited the Peruvian Coast; the

Marine Life Research Program

ice went out of Point Barrow at the earliest time in history; and on the Pacific's Western rim, the tropical rainy season lingered six weeks beyond its appointed term."[14]

Suddenly, off California, sport fishermen were catching more bonito, yellowtail, barracuda, skipjack, and dolphinfish than ever before recorded, and farther north. Salmon trollers were catching white seabass; marlin and sailfish were reported, and swordfish were caught off Monterey Bay. Green sea turtles were seen off the Farallon Islands, and many hammerhead sharks were sighted off California beaches.

And the sardines? They moved northward, spawned farther north (but not in greater quantity), and in the 1958-59 season were caught in the greatest numbers since 1951: 102,000 tons, *vs.* 22,000 tons in 1957-58.

Because of the unusual conditions throughout the Pacific Ocean that year, Scripps hosted a symposium on "The Changing Pacific Ocean in 1957 and 1958" at Rancho Santa Fe in June of 1958. Attending were physical, chemical, geophysical, and biological oceanographers, fisheries personnel, meteorologists, and an astrophysicist. They came from east coast and west coast, and one came from Japan. For three days they talked of warmed islands and beaches and shifting currents and fishes, until even expert-in-meteorology Jerome Namias expressed it for all: "It is certainly much more complex than I ever dreamed."[15]

Symposium editors Oscar E. Sette and John Isaacs summarized:

> It appears to the Editors that one of the most valuable results of the Symposium is to have pointed out clearly and unequivocally, and from a wide range of evidence, that locally observed changes in ocean conditions, marine fauna, fisheries success, weather, etc., are often the demonstrable result of processes

Retrieving a free-vehicle current meter from the *Melville* on Benthiface Expedition, 1973. Photo by Tom Walsh.

acting over vast areas. In the case of local Pacific conditions, the changes obviously often are only a part of changes involving the entire North Pacific if not the entire Pacific or the entire planet.

It appears that this realization should emancipate many provincial marine investigations and stimulate much thought and inquiry into these vast and critical events that so profoundly influence the local areas of the Pacific.

This is to say, for example, that a basic understanding and subsequent basic forecasting of the fluctuations of a coastal fishery probably can be best achieved by a *thoughtfully limited* study of the entire ocean, in addition to concentrated concern with the immediate area of the fishery.[16]

The symposium marked a turning point in MLR, carrying its scope beyond the California Current to all of the North Pacific Ocean. CalCOFI in 1960 redefined its objectives:

To acquire knowledge and understanding of the factors governing the abundance, distribution, and variation of the pelagic marine fishes. The oceanographic and biological factors affecting the sardine and its ecological associates in the California Current System will be given research emphasis. It is the ultimate aim of the investigations to obtain an understanding sufficient to predict, thus permitting efficient utilization of the species, and perhaps manipulation of the population.[17]

In line with its broadened goals, the Marine Research Committee adopted the system of awarding contracts to special projects submitted to it by the participating agencies, with the emphasis to be on analysis and synthesis rather than on continuing field studies. The cruise plan was also changed in 1961, from monthly to quarterly coverage of the station grid, and farther out to sea. This continued until 1967, when a program of eight cruises every three years began. It seemed as if enough basic data had been gathered to establish seasonal and annual variations, and ship operating costs were continuing to rise. But special MLR cruises, separate from the station grid, have been occasionally scheduled.

The old ships were showing signs of wear. In fact, MLR was an important contributor to wearing out the *Crest,* the *Paolina-T,* the *Horizon,* the *Stranger,* and the *Orca.* So, also in 1961, a new ship entered the program, when Scripps acquired a 180-foot light-freight vessel that had previously been operated by the Army Transportation Corps. Following an old tradition, the ship's new name was selected through a contest at Scripps: *Alexander Agassiz,** in memory of one of Scripps Institution's first ships as well as in honor of the noted oceanographer who had visited the La Jolla marine station in 1905 and had contributed to its library and its morale. Through funds provided by the Regents of the University of California, the *Agassiz* was outfitted especially for the CalCOFI program, with laboratories and enough deck space for handling the bulky midwater trawl and other collecting gear, and with glass-covered viewing ports below the waterline.

Having acquired many years of oceanographic basic data — in "warm years, cold years, monotonous years, years with strong countercurrents, years with invasion of tropical waters, etc." — MLR could at last turn to "thoughtful studies of the samples and data already obtained . . . and

---

*See list of Scripps ships in chapter 14.

to special investigations to answer the more discriminating questions that can now be asked."[18]

The sardine, which had created CalCOFI and its progeny, was, after all, "a rather minor part of the biomass . . . like the gold of California, a conspicuous, valuable, easily-harvested element in the midst of less-conspicuous riches of far greater potentialities but requiring painstaking development."[19]

The phenomenally high catches of sardines during the 1920s and 1930s were historically the result of a major propaganda campaign during the first World War to eat more fish (Ritter had helped on that), at a time when sardines happened to be plentiful, followed by an artificially created market for that same "inexhaustible" resource for fish meal and oil. "What is done with a resource," pointed out Frances N. Clark and John C. Marr, "is essentially governed by the needs or desires of Man and, in a sense, has little or no connection with the resource *per se.*"[20] Sardines were caught and canned in tomato sauce originally because it was patriotic — and later because thirty thousand people made their living that way.

But CalCOFI's studies showed that other fishes can gain ascendancy over the sardine, especially the northern anchovy, which appears to have a spawning advantage at lower ocean temperatures, as in the remarkably uniform decade from 1948 to 1957. The larvae of the anchovy, which also spawns farther north than the sardine, also are more likely to be maintained within the California Current and not carried off to unfavorable climes. In 1951 anchovy larvae outnumbered sardine larvae in net hauls on CalCOFI cruises by three to one, but in eight years this had leaped to 45 to one.

The sardine, concluded CalCOFI researchers, "can prosper when there is much variation in the environment, but . . . under steady conditions, or at least under steady cool conditions, the sardine, perhaps abetted by the pressure of man, gives way to its competitors."[21] Revelle had anticipated this in 1949, when he predicted that "if the sardine disappears, some other fish will occupy the former's place at the sea's food table."[22]

What may have been the final chapter of the sardine story — but not of the MLR story — came out in the latter 1960s. In an enclosed anaerobic basin off Santa Barbara, cores were drilled through the undisturbed accumulation of sediments that represented a record covering one thousand years. From those cores, Andrew Soutar identified and counted the numbers of fish scales from Pacific sardine, northern anchovy, and Pacific hake, which together constituted 80 percent of all fish scales. The hake, *Merluccius productus,* already known to be a major component of the biomass, proved to be consistently the most common fish, followed by the anchovy; both were much more uniform in numbers throughout the centuries than was the sardine, which fluctuated through highs and lows. But only once previously, about 800 years ago, did the core record indicate an abundance of sardines like that of the 1920s and 1930s, when California fishermen found their pot of gold.

To adapt to the changing populations, the fishermen changed their ways and, for example, in 1961 landed one million pounds of hake — to feed to mink, to produce fur coats!

Throughout the years, all the varied aspects of the original CalCOFI program continued and expanded. DCPG constantly improved its methods of processing hydrographic data and became the model for all other oceanographic laboratories and for the National Oceanographic Data Center, a credit to the meticulousness of Hans Klein and his training program for technicians. Historical studies were begun, to find documents and records that pertained to the climate of the past. Studies on the sediment cores were enlarged to look into the biological productivity and distribution of plankton.

More sophisticated instruments were developed, especially for extreme depths, including the whole family of free-vehicle equipment. The key to this series was simply a link made of magnesium, which, as John Isaacs had noted on 7 April 1947 after conducting laboratory tests, held promise as a release mechanism, as magnesium is corroded by electrolysis in seawater.

> Lew Kidd and I [wrote Isaacs in 1967] first worked on autonomous free-fall instruments at Scripps in 1949. Our first model bore a weak but effective light, a piece of yellow phosphorus in a "tea egg" on the mast (this spontaneously ignited after emergence and produced clouds of smoke), a five gallon can of gasoline for buoyancy, a magnesium link timer, and an aluminum mast. Operating around the Channel Islands of Southern California, we had very poor success in retrieval because of the many small boats that also showed masts, lights, and smoke.[23]

One of the early MLR free-vehicle units was a fish trap designed by Isaacs and George B. Schick and put into use in 1959. This device, the benthic trap, is weighted with ballast, and dropped to the sea floor; a gasoline-filled float raises the unit to the surface when the magnesium link on the weight has been corroded through. The trap has proved to be a very effective means of gathering denizens of the deep. For example, Carl L. Hubbs has obtained enough hagfish with free-vehicle traps to identify several new species from the Pacific coast. Hubbs also urged the addition of setlines to the device, by which a number of deep-water fishes have been collected.

Scripps Institution of Oceanography: Probing the Oceans

By 1966 George Schick and Meredith Sessions had developed the first free-falling camera system, which on its first testing snapped photographs at fifteen-minute intervals down to 3,000 meters. The camera ballast weight carried frozen mackerel to attract potential photogenic models, and suddenly the presumably barren depths proved well populated. MLR's deep-sea cameras have been routinely used down to 7,000 meters, beyond which special housing is necessary because of the extreme pressure. "The thousands of pictures make it clear [wrote Isaacs and Schwartzlose] that much of the deep-sea floor teems with numerous species of scavengers: vigorous invertebrates and fishes, including some gigantic sharks, that are supported by a marine food web whose extent and complexity is only beginning to be perceived."[24] In the photos have appeared some undescribed deep-sea species and some unexpected ones, such as Greenland sharks, usually found in Arctic waters but caught by the camera in cold depths off Baja California. Sablefish and tanner crabs showed up in sufficient numbers from 600 to 2,000 meters to represent potential large new fisheries.

When the bait and camera reach the sea floor, the authors continue:

> Usually the number of fish gathered around the bait increases slowly, reaching a maximum after a few hours. Often the scene develops into one of furious activity, with several species of fish competing for the bait, thrashing and tearing at it and sometimes attacking one another. Shrimps, brittle stars, amphipods and other invertebrates encroach on the melee. In almost half of the sequences from drops down to 2,000 meters the party ends abruptly after three to eight hours, when some creature, usually a large shark, moves in, frightens off the other fish and consumes the

bulk of the bait. In any case the time comes when most of the bait has been eaten. The fish depart, and slowly crabs, sea urchins, snails and other such creatures arrive to complete the task of sanitizing the sea floor.[25]

MLR engineers have also developed a free-vehicle sediment trap, for determining the rate at which surface debris accumulates on the sea floor, and deep-sea current meters. The latter have been used to trace the flow of cold Antarctic bottom water into the Pacific, to measure the flow of water through Drake Passage from the Pacific to the Atlantic, to determine that a slow current flows along the bottom of the Tonga Trench, and in various other deep-water sites. The recovery rate of these meters has been exceptionally high, especially for finicky oceanographic instrumentation. Jery B. Graham, for example, who has methodically readied and lowered at least one hundred current meters, has recovered all except one of them. Various release mechanisms have been developed for the several kinds of free-vehicle equipment, as have been a variety of devices for signaling to and from the surfaced item for retrieving it.*

Equipment development continues apace; for example, in the early 1970s:

> Daniel M. Brown has developed four new systems for sampling marine organisms. One is a conversion of the Isaacs-Kidd mid-water trawl into an opening-closing net. Another is a closing, vertically towed net

---

*A story, perhaps apocryphal, recounts that someone from Scripps dropped a free-vehicle instrument into Lake Tahoe on a research project — and after waiting long after the release time suddenly recalled that fresh water does not corrode magnesium.

built to reduce the handling and "scarring" problems of this style of sampling. The third is a trap that can capture live fish at great depths, hold them in their own cold water, and keep them under pressure. The fourth is a low-cost free-vehicle, drop-camera system developed to photograph schools of fish detected on sonar, thus providing a simplified method of identifying pelagic fish stocks.[26]

Plankton studies have long been an intensive part of CalCOFI studies. Isaacs noted in 1969 that "there is now in the archives at Scripps the greatest and most complete plankton collection of any area in the world."[27] Since 1967, Abraham Fleminger has been curator of this vast assemblage — more than 60,000 samples. As the collections have grown, taxonomic studies of individual groups have been completed: calanoid copepods by Fleminger; chaetognaths by Angeles Alvariño; Euphausiacea by Edward M. Brinton; pelagic molluscs by John A. McGowan; and Thaliacea by Leo D. Berner.

For many years the spiritual leader of the researchers on plankton was Martin W. Johnson, who became a staff member at Scripps in 1934 and continued to emeritus status in 1962. Besides his wartime work on identifying snapping shrimp cracklings and other underwater noises and on determining that the deep scattering layer was composed of living creatures (see chapter 2), Johnson became interested in identifying water masses through the composition of the plankton organisms within them. These studies were especially cited by the National Academy of Sciences in 1959, when it awarded to Johnson the Alexander Agassiz medal. His particular specialty has long been the larvae of spiny lobsters and of slipper lobsters; through the years he has identified and described the pelagic larvae of all species of lobsters known from the Pacific coast and Hawaii — no mean task — and has presented his results in a number of publications. This work has led to ecological studies of the lobsters. Johnson guided the researchers on invertebrates into analyses of the plankton communities.

John A. McGowan joined the Scripps staff immediately after receiving his Ph.D. at the institution in 1960. Born in Oshkosh, Wisconsin, McGowan had received his bachelor's and master's degrees at Oregon State College. For three years beginning in 1956 McGowan was staff marine biologist for the Trust Territory of the Pacific Islands, headquartered in Palau. Under the Marine Life Research program, he has been especially interested in gaining more precision in plankton sampling, in order to determine the detailed structure of plankton communities in time and space. Daniel M. Brown and McGowan designed the Opening and Closing Paired Zooplankton Net, nicknamed the "bongo" net. It consists of two parallel nets (that resemble bongo drums) with no bridle, towline, or cable ahead of them that might warn fleet creatures of the approaching net. The nets can also be opened and closed at any desired depth, instead of the former custom of hauling nets from the start of the tow to the surface. From such precise sampling, McGowan and colleagues were able to conclude that communities of animals are found together in a particular water mass, and that boundaries can be drawn for the animal populations just as they can for the separate water masses. By using a drogue that performed diurnal vertical migrations, McGowan and his coworkers were able to sample for the species structure of individual plankton masses and simultaneously for the physical characteristics of the surrounding water.

The region of intensive study during the 1970s, by McGowan and colleagues — including Elizabeth ("Pooh") Venrick, Lanna Cheng (Lewin), and members of the Food Chain Research Group — has been the North Pacific Central

Gyre. This constitutes almost a closed system, in which the community structure and the influences on it can be analyzed on repeated cruises, during which replicate profiles of a number of parameters have been taken. "The data have established that during the summer months the relative proportions of abundances of macrozooplankton and mesopelagic micronekton species are less diverse and more stable than in the California Current."[28] Alumna and staff member Elizabeth Venrick has also been investigating the seasonal and long-term fluctuations in the species composition, especially of the phytoplankton, in the uppermost level of the gyre.

Edward M. Brinton, also a Scripps alumnus, joined the staff in 1959 and has also participated in detailed plankton studies, with emphasis on euphausiid and sergestid crustaceans. Euphausiids — also known as "krill" — are a significant food source in the ocean, the chief food of the baleen whales and of various commercially valuable fishes. Brinton's group has devoted considerable effort to the taxonomy of these crustaceans, including the identification of the multi-formed larval and adolescent stages. Brinton has also analyzed plankton samples from two expeditions, Piquero in 1969 and Aries in 1971, that were carried out in the same area of the South Pacific at opposite seasons. Using multiple sets of collecting nets, Brinton and colleagues gathered samples from as many as ten depths simultaneously to determine the vertical distribution of plankton within and along the equatorial currents. In the Peru-Chile Current, Brinton found that the "production of life in the surface waters is so great that, upon its dying, sinking, and then decomposing, oxygen is almost completely depleted from a layer beneath the surface."[29] A localized group of species has developed in this area that is adapted to migrate up and down through the almost anaerobic layer.

Another aspect of plankton studies has been carried out

by the Biomass Laboratory of MLR. There, about twenty major groups of zooplankton are measured by volume in each sample, to determine the total organic component of the water. Wide fluctuations in the composition of the zooplankton have been found between cold years and warm ones. The major factor that determines the composition of the zooplankton proved to be the California Current itself, not the available phytoplankton or the chemical composition of the water.

A new form of summary publication appeared in 1963: the first volume of the CalCOFI Atlas series, under the direction of Hans Klein, "a man of dignity and loyalty, intensely proud of his association with Scripps,"[30] who headed DCPG from 1956 to 1968. The first atlas presented temperature and salinity data for a depth of ten meters for the California Current region, on the data from 1950 to 1959. Others in the series, which in 1976 had reached 24 volumes, presented summaries on the distribution of several zooplankton groups, on the distribution of the larvae of a number of species of fishes, on geostrophic flow, on zooplankton biomass and volume, on temperatures and salinities, on drift-bottle results, and on seasonal sea-level pressure patterns.

Thanks to the fickle sardine, the CalCOFI committee has felt justified in declaring that "the oceanography, the biology of the California Current system, and the variations in these are now the best documented and best understood of any oceanic area in the world."[31]

To understand that current in its entirety, it seemed appropriate to look even farther, to the forces that push and shape the California Current that in turn determines where the plankton will be that in turn become food for all the living resources from California waters. MLR since the mid-1960s has enlarged its scope to include studies of

Scripps Institution of Oceanography: Probing the Oceans

the interaction between the upper layer of the ocean and the lower levels of the atmosphere, far at sea. Meteorologists and physical oceanographers have joined in these studies, to determine whether temperature fluctuations at the ocean surface are coupled with weather fluctuations on land. Some of that continuing story is covered in chapter 11.

Sophisticated equipment and many years of experience and observations — as well as ingenuity — are bringing us closer to real answers to the fluctuations in the weather and in the ocean, to which the glittering sardine responds without knowing why.

NOTES

1. Letter in MLR archives.

2. CalCOFI Progress Report (1950), Introduction, 7-8.

3. *Ibid.,* 5.

4. Letter in MLR archives.

5. *Oceanic Observations of the Pacific: 1955, The NORPAC Data* (Berkeley and Tokyo: University of California Press and University of Tokyo Press, 1960), Introduction.

6. *New York Herald Tribune,* 25 August 1955.

7. Minutes of Marine Research Committee, 19 December 1957.

8. CalCOFI Reports, Vol. VII (1 January 1960), 26.

9. News release, 10 January 1950.

10. Letter to Andrew R. Boone, 14 July 1949.

11. Quoted *in* T. F. Gaskell, *Under the Deep Oceans* (London: Eyre and Spottiswoode, 1960), 73.

12. John D. Isaacs and Lewis W. Kidd, "A Midwater Trawl," SIO Reference 51-51 (November 1951), ii.

13. Minutes of MRC, 4 October 1960.

14. CalCOFI Reports, Vol. VII (1 January 1960), 21.

15. *Ibid.,* 211.

16. *Ibid.,* 214-15.

17. Minutes of MRC, 4 October 1960.

18. CalCOFI Reports, Vol. IX (1 January 1962), 9.

19. CalCOFI Reports, Vol. VIII (1 January 1961), 7-8.

20. CalCOFI Progress Report, 1 July 1953 to 31 March 1955, 13.

21. CalCOFI Reports, Vol. VIII (1 January 1961), 7.

22. *Los Angeles Times,* 22 June 1949.

23. "Remarks on Some Present and Future Buoy Developments," *Transactions,* Second International Buoy Technology Symposium, Marine Technology Society (1967), 504.

24. "Active Animals of the Deep-sea Floor," *Scientific American,* Vol. 233 (October 1975), 85.

25. *Ibid.,* 86.

26. SIO Annual Report, 1974, 29.

27. CalCOFI Reports, Vol. XIII (1 January 1969), 8.

28. CalCOFI Reports, Vol. XVII (1 October 1974), 15.

29. SIO Annual Report, 1970, 14.

30. CalCOFI Atlas 17 (June 1972), iii.

31. CalCOFI Reports, Vol. XI (1 January 1967), 5.

4. Making and Chasing Echoes: The Marine Physical Laboratory

Another project of broad spectrum began in San Diego under the auspices of the University of California even earlier than the Marine Life Research program, but it did not become a unit of Scripps Institution until several years later. This was the Marine Physical Laboratory, the peacetime successor to the University of California Division of War Research (see chapter 2).

The possibility of university scientists' pursuing postwar studies of underwater sound was proposed to President Sproul by the Chief of the Bureau of Ships on 31 January 1946, when UCDWR was about to be dismantled. Roger Revelle, who was serving in the Bureau of Ships office at that time, recalled later that the establishment of the Marine Physical Laboratory "required a pledge of long-term support from the Bureau. This was a radical departure from previous Navy practice and it required much prayerful consideration by the chief of the Bureau, Vice Admiral Edward L. Cochrane."[1]

Already some of UCDWR's staff members had completed their projects; Francis P. Shepard and Eugene LaFond, for instance, and some of their workers had returned to Scripps in 1945. Many of the others were scheduled to join the Navy Electronics Laboratory that was being established on Point Loma, "to effectuate the solution of any problem in the field of electronics, in connection with the design, procurement, testing, installation and maintenance of electronic equipment for the U.S. Navy."[2] Simultaneously, by agreement between the Navy and the university, the Marine Physical Laboratory was created from UCDWR to study "purely scientific problems in underwater physics which it is appropriate for a university to investigate without regard to possible naval applications."[3]

It opened on 1 July 1946 with a scientific staff of five people. Under Task 10 of Contract NObs-2074, on 25 November 1946, the Navy Bureau of Ships assigned as research problems of the Marine Physical Laboratory:
- (a) Theoretical and experimental investigations of the physical principles governing the generation and propagation of sound in the sea;
- (b) Studies of related phenomena as necessary to provide a broad scientific foundation for the above principles;
- (c) Investigations of the principles governing the recognition of signals, with special emphasis on underwater sound signals of all kinds.

Carl Eckart, who liked a university environment and had encouraged the establishment of this university laboratory, became the first director of MPL. In 1942 he had joined UCDWR from the physics department of the University of Chicago, where he had made noteworthy contributions in quantum mechanics, particle physics, and thermodynamics. At UCDWR Eckart turned to the complexities of sonar system concepts and also became associate director there in charge of planning and coordinating research.

The new laboratory was set up on Point Loma, alongside

Marine Physical Laboratory

NEL, from which it was given access to equipment and supplies, and with which it continued a cordial relationship. The MPL staff also worked closely with Scripps Institution. MPL provided accommodations for graduate students, and the staff of the laboratory offered courses at Scripps and at UCLA. They also set up a joint seminar for scientists from NEL, Scripps, and MPL.

The first research projects at MPL set the pattern for the nature of the laboratory's studies. Senior Research Associate Russell W. Raitt pursued studies on the reflection of supersonic waves from the sea bottom, assisted by Research Fellow William C. Kellogg, Jr. Research Associate Robert W. Young began work on image interference in the presence of refraction, a project that he soon transferred to NEL. Research Associate Leonard N. Liebermann joined the MPL staff in its first year from Woods Hole Oceanographic Institution, at Eckart's invitation, to study the effect of small-scale inhomogeneities in the sea — very small temperature changes, salinity gradients, and even gas bubbles — on the propagation of sound.

Director Eckart, whose first project was described as a study of the absorption of sound by liquids, gave considerable impetus to all the laboratory's projects. He was especially interested in determining means of finding sound signals buried in other noise, the field that came to be known as signal processing. He also investigated the problem of how sound is scattered from a rough surface. Eckart was the spiritual leader of MPL from its founding until his death in 1973. He tackled each problem with keen insight — and occasional whimsy, such as defining Kelvin's First Law of Oceanographic Instrumentation as: "There is ample driving power in the sea."

A mainstay of the laboratory from its beginning was Finn W. Outler, who began as the marine supervisor in 1946 and continued as the technical superintendent and business

The four directors of the Marine Physical Laboratory in April 1968. From left: Fred N. Spiess, Sir Charles S. Wright, Carl Eckart, and Alfred B. Focke. Photo by Elizabeth Shor.

Marine Physical Laboratory

manager until his retirement in 1968. Outler had been mustered out of the Navy in San Diego as a Chief Warrant Electrician during the war (for a disability), and he applied promptly to Eckart for a job at UCDWR. For MPL, his first task was directing the conversion of two large Navy vessels into "seagoing laboratories." The ships, *PCE(R)-857* and *PCE(R)-855* were assigned to NEL, and were also to be readily available to MPL researchers. Unruffled, soft-spoken, firm when necessary, and ingenious, Outler over the years proved to have a phenomenal ability in obtaining any piece of equipment and in sending it when necessary to the ends of the earth. He knew "channels" and ways around them.

When Sverdrup left Scripps in the spring of 1948, Carl Eckart became director of the oceanographic institution, and simultaneously MPL became part of Scripps. Eckart considered his Scripps post temporary, and yielded it readily to Revelle as acting director in March 1950.* He continued as director of MPL until his sabbatical leave in 1952. Eckart and Revelle invited Sir Charles S. Wright to serve as Eckart's replacement; they had known this physicist, who was born in Toronto in 1887, in his longtime position as Director of Scientific Research for the British Admiralty and, after the war, as scientific advisor on the British Naval Staff in Washington, D.C. Wright's earliest scientific work had been with the ill-fated Scott Antarctic expedition from 1910 to 1913. During World War I he was occupied in "wireless (Radio now) and later 'Wireless Intelligence.'" With the British Admiralty, Wright said that "most of my time involved Geophysics, detectible properties of ships & submarines & their countermeasures. Mines,

---

*Some say he forced the decision by notifying President Sproul that he would no longer be at the director's desk after a certain date. See also chapter 16.

83

torpedoes & their queer habits & countermeasures. In fact the whole works."[4] He retired to Canada in 1947, from which place he emerged to become the British scientific advisor in Washington, and had retired again when Revelle and Eckart sought him out. Wright served as acting director and then director of MPL until May 1955. He called his MPL period "the happiest time of my life (and the easiest) . . . where I was left with nothing to do because you [the MPL staff] were all so competent." His colleagues said of Wright that he repeatedly "demonstrated his ability to generate and assimilate new ideas and to give sound advice concerning their reduction to practice." After his third retirement, he served as consultant to the Canadian Pacific Naval Laboratory in Esquimault, British Columbia, for many years; in 1965 he returned briefly to his old haunts in the Antarctic. Wright died in 1975.

At MPL in 1955 his successor was Alfred B. Focke, who had joined the laboratory's staff in January 1954 from NEL, where he had worked on studies of very low frequency airborne sound and on the effects of underwater explosions. Focke directed the nuclear depth charge project, Wigwam, soon after assuming the MPL directorship. In 1958 he became technical director of the Pt. Mugu Naval Air Missile Test Center, and from there became a professor at Harvey Mudd College. Fred Noel Spiess, who had joined MPL in the fall of 1952, became director of the laboratory in 1958. A change in research emphasis in the Bureau of Ships led to shifting the major support for MPL to the Office of Naval Research at about that time, so among Spiess's first tasks was separating the facilities of the laboratory from those of NEL, then still funded by the Bureau of Ships. Spiess, like Eckart, served double duty to institution and laboratory, for he continued as director of MPL while he was acting director of Scripps from 1961 to 1963 and director of Scripps from October 1964 to June 1965.

Marine Physical Laboratory

From the handful with which it began, the Marine Physical Laboratory has expanded to about 150 people, has of course outgrown its even enlarged facilities at Point Loma, and now also has offices on the La Jolla campus, where most of the senior staff teach and continue to direct graduate students. The program of this laboratory, which was generated by military need for basic science studies of a complex liquid medium that covers three-quarters of the globe, was defined in 1970 by MPL's senior staff:

> ... For over twenty years it [the Marine Physical Laboratory] has supported the work of a group of physicists challenged by a desire to understand the oceans and the earth's crust beneath them and interested in the ways in which man can best work on and within the sea. . . . The program of the laboratory is generated within the staff, giving due consideration to the relevance of the work to basic marine science and to the national interest. In particular this includes concern for the deep ocean problems of the Navy and the basic understanding of the environment needed for it to operate intelligently.[5]

The key to MPL's work is "naval relevance," a term defined by Robert W. Morse, then Assistant Secretary of the Navy for Research and Development, when he spoke at the dedication of the Nimitz Marine Facilities in March 1966:

> In oceanography especially it will often be the scientist who first senses naval relevance in his research. We must make it clear that the term 'naval relevance' is not synonymous with military applied oceanography and its possible security restrictions. The scientist must understand that the Navy's problems

with the environment are as large as the oceans themselves and as long range as science itself.[6]

MPL has concentrated on underwater acoustics:

> The study of the manner in which sound travels through the sea and the nature of the noises which occur within it. Such studies bring one quickly into contact with the physical, chemical, and biological nature of the sea itself, as well as to a broader look into two related fields — marine geophysics, in which sound signals provide a major tool, and signal processing, in which the goal is to understand the principles which provide the basis for design of sonar systems.[7]

In its early years a major proportion of MPL's work was classified research. The complications and delays in obtaining security clearances for new employees, and the impact of Communist-hunting Senator Joseph McCarthy on these delays, were frequent topics of conversation throughout the 1950s. The situation gradually eased, and MPL staff members contributed to the easing by urging the declassification of many reports and by encouraging publication of research results in nonclassified locations whenever possible. The proportion of classified work is now relatively small. Rapport with the U.S. Navy continues to be very close.

In its pursuit of sound in the ocean — which Revelle once called a "badly designed auditorium" — MPL has produced a good deal of sound itself, and sometimes fury, usually signifying something. "The sound pulses generated by marine physicists," George Shor said, "have undoubtedly provided more information about the contents and contours of the sea and sea floor than any other single research tool." The physicists have recorded and analyzed the background noise of the sea, have measured reverberations, have studied the fluctuations in the propagation of sound waves through the water, and have tried to determine the distribution of the sound scatterers throughout the water. Some have measured the earth's magnetism and the fluctuations throughout time in the magnetic field, some have recorded the flow of heat through the sea floor and the land, and others are working closely with geologists as they map the configurations of submarine canyons and abyssal fans and determine the nature of materials beneath the sea floor.

SIGNAL PROCESSING

MPL has distinguished itself in a field that had not been christened when the laboratory began: signal processing, "the theory and practice of applying spatial and/or temporal transformations to samples of an acoustic wavefield to enhance the measurement of a desired signal in the presence of an interfering background."[8] In other words, how to select the noise that one wants to hear among many others. Signal processing has led to the development of highly sophisticated sonar equipment.

In the 1940s the field was waiting for new technology, something to substitute for bulky and expensive vacuum tubes. High-speed digital circuits appeared in the mid-1950s, and signal processing was fairly launched. MPL was among the pioneers, thanks to Eckart's leadership.

A major contributor was Victor C. Anderson, a lanky physicist who arrived at MPL as a UCLA student in 1947 to investigate the deep scattering layer under Raitt's direction. Anderson proved to be fantastically adept in electronics as well as highly capable of assimilating theoretical approaches. He combined forces with quiet, unobtrusive

Philip Rudnick, a physicist especially good in statistics and in applying mathematics to a given project.

Rudnick expanded the concept of "hard clipping," which was a continuation of the principle of representing the desired signal merely as present or absent, in a continuing series. From this, Anderson in the 1950s, while working on a postdoctoral with Frederick V. Hunt at Harvard, developed a digital technique of rapidly repeating the stored information. Anderson called his system the DELTIC correlator, for *De*lay *L*ine *Ti*me *C*ompressor. The first one was developed for a project in architectural acoustics, but studies of underwater sound could use the technique, and the Navy was promptly interested.

"Those were truly the primitive days of digital technology compared to today," said Anderson in 1972. "Quantity prices for vacuum tube shift registers were in the 10 to 20 dollars per bit category. Compare that with today's prices for LSI [large scale integrated circuit] random access memories, which are capable of operating at 10 times the speed of those early vacuum tube shift registers and which are now selling in quantity for one cent a bit. We've seen a reduction of a factor of 1000 to 1 in cost alone, let alone the factor of 10 in improved operating speed within the last 15 years."[9]

The Navy's interest was in ambient noise in the water, and they encouraged other developments in the study of underwater noises. DIMUS, the *Di*gital *Mu*ltibeam *S*teering unit, for example, was assembled at MPL. "This was an extension of the two-channel DELTIC correlator concept into a multiple-channel clipped waveform processor," said Anderson. "Signals from an array of hydrophones are converted into sequences, or one-bit samples, which are delayed in digital memory circuits and then combined in multiple 'beams,' each 'beam' representing a reinforcement of signals arriving from a specific direction." Of a series of these built by Anderson's group, one was for a surface ship sonar unit, two were for submarine hydrophone sets, and one for an experimental noise-measurement instrument. For that unit, Anderson devised an array of 32 hydrophones arranged for listening simultaneously in all directions, a cumbersome piece of equipment labeled "the great stellated icosohedron." It was the first unit to employ signals and power multiplexed into coaxial well-logging cable, since used in a number of other pieces of equipment. This device was transported usually on the fantail of the *Paolina-T* for underwater noise surveys along the California coast.

"By 1960," said Anderson, "we were dealing with multiple channel processors using solid-state electronics as discrete components. By mid-1960s, the integrated circuits were filling the gap, and now, of course, we have the MSI [medium scale integrated circuit] and LSI [large scale integrated circuit] technology which has opened new horizons in the implementation of the more sophisticated signal processing techniques."[10]

It is a far cry indeed from prewar oceanography, when, as Revelle said: "We even had a slogan that the best oceanographic instruments should contain less than one vacuum tube per unit."[11]

Anderson noted: "Reverberation in the sea presents a major obstacle to the performance of ASW [anti-submarine warfare] sonar systems on the one hand, while it stands as an open portal to oceanographic research on the other. Regardless of which viewpoint is chosen, the incentive for intensive study of this complex phenomenon remains strong."[12] He has carried out measurements of the distribution of volume-scattering coefficients as a function of depth and frequency in the hope of classifying the larger contributors, such as schools of fishes and dense masses of plankton. He concluded that bubble resonance appeared to be a "dominant scattering mechanism in the deep scattering layer."[13]

Scripps Institution of Oceanography: Probing the Oceans

MARINE GEOPHYSICS

The marine geophysical work at MPL was begun by Russell W. Raitt in 1946 with a project to determine how sound waves are reflected from within the sea floor. He first used oscillograms of bottom echoes gathered by Jeffery Frautschy while at UCDWR. Raitt quickly determined that the bottom reflects sound waves diffusely — indeed complexly. By 1947 Maurice Ewing (then at Columbia University, before Lamont Geological Observatory was established there) had shown that low frequency sound waves would penetrate the bottom and that reflections from layers beneath the sea floor could be obtained, thus providing information on the structure.

In 1948 Raitt began using TNT charges to study sub-bottom layers. At that time, and for a number of years afterward, excess explosives were readily available through the Navy, without cost, for research purposes. Raitt began with five-pound charges, but he soon enlarged his scope to fifty-pound TNT bombs. His first attempt at refraction measurements was with Sofar bombs set to sink and detonate when the ship had moved about ten miles away, but the charges proved to be detonating too deep to produce a signal in the low frequencies needed. Raitt then devised a successful refraction method by using a whaleboat as the shooting ship, by using slow-burning time fuses to set off the charges at shallow depth, and by floating the hydrophones at nearly neutral buoyancy one hundred to two hundred feet beneath the surface, streamed well behind the noisy ship. Those who wallowed in the whaleboat, however, did not enjoy it, and Raitt admitted that "experience of several cruises demonstrated that this technique was practicable only when restricted to areas near islands or other land points where lee from wind and waves provided sufficient protection for whaleboat operation."[14]

Russell W. Raitt throwing a one-half-pound TNT charge from the *Spencer F. Baird*, Capricorn Expedition, 1952.

Scripps Institution of Oceanography: Probing the Oceans

So the *Paolina-T* or the *Saluda* (a racing yacht of NEL, also used for research) or the *E. W. Scripps* was called into service as the shooting ship to accompany NEL's *PCE(R)-857*, and Raitt began accumulating refraction profiles throughout the southern California borderland. He then joined Midpac Expedition in 1950 for deep-sea refraction work — on which he recorded 1,200 miles of seismic profiles — and for seismic surveys of atolls. Two years later he also sailed on Capricorn Expedition, where he enthusiastically called for larger and larger shots until he had used up 41,409 pounds of TNT, including one single outburst of 480 pounds. As the empty powder boxes accumulated, some of the adept members of the group occupied their spare time with fashioning furniture from them.

George G. Shor, Jr., joined the MPL staff in the fall of 1953, having become interested in Raitt's reports on his deep-sea results, and because the chance of getting to exotic spots like Adak and Rapa seemed much more exciting than analyzing seismic data for an oil company probably based in Houston. He hasn't changed his mind in twenty-odd years. Raitt and Shor, wanderers both, have separately and together crisscrossed the Pacific Ocean from the Aleutians to New Zealand and the Indian Ocean from Mombasa to Djakarta. They are usually planning the next expedition before they have worked up the results of the previous one(s) — and have often departed without even a remorseful glance at the unfinished data. In this they are not alone among oceanographers.

These geophysicists have bombarded the sea floor with sound waves, in the interest of delineating the features within the crust and to the mantle of the earth. Whenever possible they are seeking a signal returning from the Mohorovičić discontinuity, the boundary between crust and mantle. The most satisfactory method for producing the necessary high-energy, low-frequency waves continues to be explosives, although other methods have been tested.

For a while the Arcer was favored; its energy source was electricity. One version — with 60 capacitors and 60 rectifiers producing 10,000 volts — was put together with "the perfection of a fine Swiss watch" by Harold Sammuli and Baron Thomas in 1968 for use on Styx Expedition. Sammuli, who is of Finnish ancestry, believed in reincarnation and approached the Arcer with less trepidation than most others. The Arcer was expanded to reach 18,000 volts, too hazardous for comfort, and it turned out to be not the best sound source, so it was abandoned for other methods. It was spectacular in use, especially at night, for it lit up a patch of sea behind the ship with a stunning electric-blue flash.

Another method, the airgun, has proved to be a very useful tool. Simple and versatile, it creates a succession of bubble pulses under water by means of an air compressor. The records derived from the continuously operating airgun draw a profile of reflections from the upper layers of the sea-floor sediments.

Sonobuoys offer another means of tracing sub-bottom layers while the ship is under way. The sonobuoy was originally devised as a submarine detector to be used from an airplane. It consists of a listening device (a hydrophone) and a radio transmitter to relay the hydrophone output to the airplane. An array of sonobuoys of various radio frequencies can be dropped from an airplane to monitor noises in the ocean. When flung off a ship, a sonobuoy provides a receiver for sound waves generated at the ship and transmitted through the sea floor as the ship moves away. The Navy has been generous with sonobuoys for such researches.

The various seismic reflection and refraction techniques, used by Scripps and by other institutions, have shown that the structure of all the ocean basins is quite similar.

Scripps Institution of Oceanography: Probing the Oceans

According to Shor: "The main basins of each ocean have a structure that consists of a variable amount of sediment (thickest in areas where sediments from the continents can reach the ocean without obstacles such as ridges and trenches), about 1 km of 'second layer' or 'basement,' about 5½ km of 'oceanic crust' with a velocity close to 6.8 km/sec, and mantle with velocity near 8.1 km/sec. Deviations from this occur in predictable places; areas that are topographically unusual, and areas close to land."[15]

Raitt's early refraction studies established the average thickness of the sediment column throughout the Pacific Ocean, results that were unexpected in the 1950s and were only understood much later when geologists realized that sediments are constantly destroyed as the sea floor shifts outward from spreading centers and into the trenches.

In 1964, Harry Hess of Princeton noted that some seismic records off California and Hawaii indicated that the velocity of sound within the mantle is faster in an east-west direction than in a north-south one. Anomalously, this does not correspond with the structural trends of the Pacific coast. Raitt and Shor set out to confirm the existence of what is technically called anisotropy (which Raitt pronounces "ani so' tropy" while Shor says "ani' sotropy"). The task requires an elaborate pattern of shooting and receiving, such that at least one-half the arc of a circle 30 miles in radius is included in the recording. (Twenty years ago, in the open ocean, a ship wasn't always sure of its position within 30 miles! An expanding Loran network and satellite navigation have made possible much more precise navigation now.)

The most intensive effort at proving anisotropy was in 1966, on Show Expedition, which had five ships, and personnel, equipment, and expertise from four institutions (*S*cripps, University of *H*awaii, *O*regon State University, and University of *W*isconsin — hence the expedition's name). Shor described it later as much like trying to conduct NATO fleet exercises in Turkish.

There is a Jekyll-and-Hyde effect in shipboard communication, which is probably caused by Neptune's resentment at the invasion of his realm. Somehow it seems that any perfectly reasonable oceanographer — a fine fellow on land to swap sea stories with, and who sounds most rational there when discussing intricate details of equipment — becomes an impossible tyrant and tyro across a few miles of ocean when he is condensed into a voice or message over the radio. All of one's own shipboard complications are so straightforward to explain; all of the other guy's are beyond comprehension, and could be so easily resolved.

Such a Neptune effect complicated the multiple Show Expedition, as it sometimes has others. Ships became lost; buoys drifted contrary to supposedly known currents; confusion reigned. But the records *did* confirm anisotropy.

The complications and the costs of multi-ship expeditions have led to designing means of recording anisotropy from a single ship, using moored sonobuoys and balloon-suspended radio transmitters. The rigmarole of simultaneously setting out moorings down to the sea floor as deep as three miles, releasing upward a 20-foot balloon, and casting outward a hydrophone array is very much like a three-ring circus, although rather more colorful in language.

Gerald B. Morris coped with the complications of the balloon-and-buoy system on Scan Expedition in 1969; Morris became chief scientist abruptly, after Russell W. Raitt had to be airlifted to Tahiti when he broke his leg while boarding a longboat at Pitcairn Island. Morris found numerous problems: "The balloons occasionally fell victim to intense squalls, sharks bit through the mooring line setting the buoy adrift, and fish actually ate large parts of the styrofoam floats."[16]

Scripps Institution of Oceanography: Probing the Oceans

Raitt's enthusiasm for seismic work at sea has never flagged. From Tasaday Expedition in 1973 he wrote to Shor:

> Tasaday leg 6 was extraordinarily successful refraction-wise, having completed 44 refraction stations. . . . In the western part of the Bay of Bengal, many sonobuoys were rendered useless by a tremendous low-frequency noise, evidently caused by a large current shear between the surface and hydrophone depth, which produced strumming noise on the hydrophones. We had considerable success with primitive hydrophone balancing in which the ship's galley-force were very cooperative by providing us with a fantastic variety of bottles for hydrophone and cable floats. The most successful combination seemed to be a ketchup bottle supporting the hydrophone and Tabasco sauce bottles supporting the leader. Unfortunately, consumption of Tabasco sauce is very small on the T. Washington in spite of my efforts to popularize it and we were forced to resort to unsatisfactory substitutes, such as mustard, pickle and garlic-salt bottles.

Other exciting aspects of the cruise were being chased successively by a Burmese gun-boat, which was seen to man its guns but didn't fire on us, and then by a severe tropical storm, which may have been called a typhoon in other parts of the world. . . .

HEAT FLOW

The first successful instrument for measuring the flow of heat through the ocean floor was laboriously devised at

Scripps by Edward Bullard, James M. Snodgrass, and Arthur E. Maxwell in 1949 (see chapter 15, Midpac Expedition). After Maxwell's departure, the clumsy temperature probe was next put into service by graduate student Richard P. Von Herzen in the latter 1950s, then under MPL sponsorship. The use of the probe has continued to be a project of MPL, in recent years under the direction of Victor Vacquier and for a time also John Sclater.

One of the disadvantages of the original Bullard probe was the necessity of taking a separate sediment core for thermal conductivity measurements immediately after the probe lowering, which not only required considerable extra time but also could occur some distance from the probe measurement because of ship drift. Scripps and its frequent competitor in creativity, Lamont Geological Observatory, naturally devised slightly different techniques for solving this problem. The Scripps approach was to attach a conductivity needle to a slider on the probe, which measured the conductivity at an average depth of 50 cm below the surface of the sediment. Lamont developed a heat probe attached to the piston corer.

The streamlined instrument devised by Charles Corry, Carl Dubois, and Victor Vacquier in 1968 has proved to be a convenient device at sea, even in weather scarcely comfortable for its handlers: "with the ship rolling up to 35° and in wind speeds up to 60 knots."[17] The heat probe appears to be more susceptible to breakage or bending on deck than on the sea floor.

The results from the first measurements of heat flow through the sea floor were among several surprises gathered on the first major Scripps expedition, Midpac, in 1950: the temperature gradient was very similar to that measured on land, whereas it had been expected to be considerably less. "The only adequate source of heat that has been suggested is radioactivity within the earth," noted Bullard,[18] and oceanic basalts are considerably less radioactive than continental rocks. The source of heat, therefore, must be deeper within the earth, presumably beneath the Mohorovičić discontinuity.

Other surprises were in store, as scattered measurements of heat flow were taken on various expeditions in Atlantic and Pacific. From Capricorn Expedition in 1952 Maxwell and Revelle reported an unusually high value on the broad topographic feature first known as the Albatross Plateau (part of the East Pacific Rise), which was not recognized until later as having been taken on a ridge. About the same time, Bullard, then at the National Physical Laboratory in England, found a high value in the Atlantic Ocean. As measurements continued, a pattern began to emerge. Heat flow on the crests of the East Pacific Rise and the Mid-Atlantic Ridge, and in the Gulf of California, was generally higher than elsewhere. Geologists were examining these regions closely and by every means at their disposal, as these anomalous regions appeared to be the foci of intensive sea-floor activity. As theories developed around the concept of sea-floor spreading, the measurements of heat flow fitted into the emerging picture: high heat flow indicated areas of more intensive crustal activity.

The development of a convenient shipboard device for measuring the flow of heat resulted in the gathering of more measurements at sea than on land. In 1967 Von Herzen (by then at Woods Hole Oceanographic Institution) and Vacquier estimated that only 11 percent of the 1,300 measurements to that time had been taken ashore. In an ingenious use of the shipboard technique, these researchers carried a temperature probe to Africa and obtained "land" heat-flow measurements aboard a Fisheries Research Unit vessel in Lake Malawi in the east African rift zone. Later Vacquier and John Sclater used a temperature probe in Lake Titicaca in South America. The heat-flow enthusiasts

Marine Physical Laboratory

also used the more laborious method of drilling boreholes in geologically distinctive land areas, and thus found high heat-flow values in the basin-and-range province of southwestern United States. They have not closed the gap, however; heat-flow measurements at sea now exceed 6,000.

MAGNETISM

Recording the magnetism harbored in sea floor rocks was a pioneering venture of Scripps and MPL. Jeffery D. Frautschy participated in the early stages of compiling a set of equipment from parts of surplus units, in consultation with the Naval Ordnance Laboratory. In 1952, Ronald G. Mason joined Capricorn Expedition, at the invitation of Revelle, following a meeting of the Institute of Geophysics in La Jolla. On that trip Mason recorded more than 4,000 miles of magnetic profiles, although he noted that such lines "have limited application since they give no indication of the lateral extent or direction of anomalous magnetic trends and therefore provide no basis for quantitative geological interpretation."[19] Also on Capricorn Expedition, Mason made detailed magnetic surveys of small areas of special interest, such as the Tonga Trench.

Sir Charles Wright particularly encouraged the continuation of a magnetic program at MPL, and Scripps ships began towing magnetometers frequently. In August 1955, Scripps researchers seized an opportunity for a detailed magnetic survey by way of the U.S. Coast and Geodetic Survey ship *Pioneer* on its hydrographic survey off southern California on a series of east-west lines about five miles apart. Arthur D. Raff and Maxwell Silverman, both alumni of Capricorn Expedition, alternated aboard ship to keep the equipment working. The magnetometer was described by Mason as a "gimbal mounted flux gate oriented in the direction of the Earth's magnetic field by two other flux gate elements which control orienting servo motors."[20] It was mounted in a streamlined "fish" and towed about 500 feet behind the ship.

The chart of magnetic intensities from the *Pioneer* survey startled geologists. Before them lay evidence of great north-south lineations and a single right-lateral offset of 155 kilometers along the Murray fracture zone off southern California. The survey was "the first attempt to make a detailed magnetic map of an extensive area of the oceans," said Bullard and Mason. "The results are of exceptional interest in that they reveal major structural trends of which there is little or no indication in the topography, and they provide evidence for unsuspected horizontal displacements along some of the faults of the north-east Pacific greater than any that have so far been observed over the continents."[21]

Mason returned to the Imperial College of Science in London in 1962, but the magnetic program continued at MPL under the guidance of Victor Vacquier. During World War II he had developed the flux-gate magnetometer, the most sensitive method then known for measuring a magnetic field — which, among other uses, could detect submarines. Later, with Sperry Gyroscope Company, Vacquier developed highly accurate gyrocompasses. When he joined the MPL staff in 1957, from the New Mexico Institute of Mining and Technology, he promptly relieved sea-weary Raff on the *Pioneer*. He then began his own modifications of a borrowed bulky proton magnetometer, which he first improved by separating the amplifying and tuning equipment from the towed "fish" and placing it aboard ship. To get away from magnetic objects such as automobiles while calibrating and testing equipment, Vacquier found a quiet spot under the eucalyptus trees near radio station WWD, and there he and his associates set up shop in two plywood geodesic domes equipped with brass fittings and covered with aluminum paint. They built half a dozen magnetometers for regular shipboard use before an engineering company began supplying them commercially.

In 1958 Vacquier made a magnetic survey along the Pioneer Ridge north of the Murray fracture zone and there found the magnetic anomalies offset 265 kilometers — in the wrong direction, i.e., left lateral or opposite to that of the Murray. The next discovery was the vast Mendocino fracture zone where the greatest offset of all was found: 1,185 kilometers, also left lateral. Revelle later called these results "the most important geophysical discovery of the past ten years."[22]

The structural geologists found that the magnetic lineations were a valuable addition to the data on the history of the sea floor. In 1963, while analyzing magnetic surveys in the Indian Ocean, F. J. Vine and D. H. Matthews at Cambridge University "suggested that [Harry] Hess' idea that a strip of new ocean floor was continually being formed on the axis of the mid-ocean ridge would provide a double tape recording of the intensity and the reversals of the earth's magnetic field. Each magnetic stripe was magnetized when that piece of ocean floor was formed in the central valley on the ridge axis."[23]

During the 1960s the Scripps magnetic program became part of the routine shipboard measurements and the data were coordinated and tabulated with other geological information by the Geological Data Center. MPL has continued development work on equipment for magnetic measurements and has carried out special programs using magnetometers. Vacquier, for example, conducted detailed surveys of seamounts for measuring the orientation of their magnetization, surveys that showed that the Pacific plate had moved northward by thirty degrees of latitude since the seamounts had formed in the Late Cretaceous period. Fred N. Spiess and John D. Mudie have incorporated a magnetometer into the Deep Tow package in order to make measurements very close to the sea floor.

PLATFORMS AND VEHICLES

MPL's staff includes several people who, as Spiess said, "want to build something and take it out and run it." Over the years they have produced an astonishing array of devices, under a collection of contrived acronyms — the kind of ocean-going gear that frequently draws puzzled queries from passing ships.

Flip is the most widely known: the *F*loating *I*nstrument *P*latform. It is defined as a manned ocean buoy, and its purpose is to gain access to the relatively calm water below the wave-churned surface. This it does by standing on end, or flipping upright, in which position it exposes 55 feet above water and extends 300 feet below the surface. Upright it becomes a stable platform that moves up and down only a small fraction — about five percent — of the height of the passing waves. Designed originally, and successfully, for fine-scale studies of sound sources in the water, it has proved "more valuable to the Navy than when it was first launched." Those who work from *Flip* appreciate her stability for their own sake.

Submariners had long known of the advantages of working beneath the wave zone. But submarines are expensive research tools, and do not hold a set depth except under way. Allyn Vine at Woods Hole Oceanographic Institution once suggested standing a submarine on end for research studies. In the late 1950s war-experienced

Flip, the floating instrument platform of the Marine Physical Laboratory, in its working position.

Scripps Institution of Oceanography: Probing the Oceans

submariner Fred Spiess at MPL began turning that notion into a new piece of equipment. It was a group effort: Frederick H. Fisher became the project officer and Philip Rudnick worked out the theory for the configuration to minimize the craft's motion. They began with scale models, one of which flipped to vertical satisfactorily — but disconcertingly returned upside down. "For a while," said Fisher, "we were building and testing a new design every week." Naval architect Lawrence M. Glosten of Seattle completed the final design work, and Spiess called back from retirement an acquaintance from submarine days, Cdr. Earl D. Bronson. In 1962 *Flip* was built in Portland, Oregon, under the shipyard supervision of Bronson, who became *Flip*'s first captain and maintained operational responsibility for it until his second retirement in 1973.

The first trials (with some trepidations) took place in Dabob Bay in Puget Sound in late July of 1962. There the flipping on end and the return to horizontal were entirely successful, and so they have continued.

When being towed in the horizontal position, the craft is stabilized by concrete and steel ballast well below the horizontal center line. The transition to vertical is accomplished by filling ballast tanks with sea water, which takes about twenty minutes. Personnel on *Flip* always stay on the outside platform during a flip, for safety's sake, each with one foot braced on the deck, the other on the bulkhead, as the craft shifts. All permanent equipment is on trunnions so that everything can manipulate a 90-degree turn. Mistakes rarely happen, but once an unnoticed can of food became lodged under the galley range during a flip, which prevented that unit from turning, so, of course, the simmering stew was flung all over the neatly made bunks.

Flip has chiefly been used for studies of sound propagation in the water; she was designed, in fact, and used by Fred Fisher for studies of the "twinkle" of a sound source

Marine Physical Laboratory

under water. *Flip* also provides an excellent quiet platform for mooring various kinds of equipment. The craft's stability has proved useful for seismic-refraction studies, for which it serves as an especially quiet recording point. *Flip* has contributed to studies of internal waves in the ocean and also studies of surface waves, one of which was measured as 20 meters high as it washed over the vertical platform. Most of *Flip*'s research time has been in the Pacific, from California to Hawaii, but in 1969 she was towed through the Panama Canal to work in the Caribbean as part of the multi-vehicle study called the Barbados Oceanographic and Meteorological Experiment (BOMEX). *Flip*'s role was for studies by Charles W. Van Atta and Carl H. Gibson on air turbulence at the sea surface. Right after BOMEX, Fred Fisher used *Flip* in a sound-fluctuation project carried out jointly with *Spar,* the Naval Ordnance Laboratory's unmanned craft of design similar to *Flip.*

ORB and RUM are a pair of MPL vehicles that often work as a team. The Remote Underwater Manipulator (RUM) was first intended to work alone, crawling about on the sea floor at depths down to 6,000 meters to gather objects and samples, to take photographs, and to install deep-sea instruments. Victor C. Anderson began assembling it in 1958, starting with a Marine Corps self-propelled rifle carrier; to this he added a boom and a steel claw that could be pivoted in any direction out to about five meters to pick up objects. The gasoline engine was replaced with a pair of heavy electric motors in an oil-filled compartment. Sonar was installed, and a powerful light and four television cameras for sea-floor surveillance from a portable shore station (actually a bus). Power for RUM and sensor signals were provided by way of a coaxial cable 8,000 meters long. Early tests in shallow water were only moderately successful, and RUM was set aside for other projects.

Scripps Institution of Oceanography: Probing the Oceans

By December 1967, ORB (Ocean Research Buoy) had been developed as a platform for suspending equipment and particularly as a service vehicle for RUM. ORB is a barge 45 feet by 65 feet with a large center well through which the ten-ton RUM is operated by means of a constant-tension winch. It has two laboratories, a galley and messhall, and sleeping quarters for twelve people. "Loading RUM is a somewhat unconventional operation," its designers wrote. "RUM is first lowered to the bottom of the bay by a crane. Then ORB is moved to a position over RUM, divers attach the strain cable, and RUM is lifted up through the well doors."[24] Unconventional or not, it does work. RUM has been used for taking cores at depths down to 1,900 meters, for measurements of sediment properties in place, for underwater photography, for recovering equipment at depths down to 1,260 meters, and for sampling deep-sea biological communities. It has the advantage of being able to stay on the sea floor at work much longer than manned submersibles. On one of its earliest sea trials, in 1970, RUM placed two small sonar reflectors on the sea floor, crawled away from them, and returned to find and retrieve them. It also found a third sea-floor object:

> ... a can of a well-known brand of stewed tomatoes. ... The can was found to be the dwelling of a small and very frightened octopus. We feel [said RUM's inventors] that this is one of the first times that a mobile biological specimen has been selectively retrieved by a remotely controlled manipulator as well as record of the first sea-going anti-pollution effort by such a unit.[25]

Anderson also developed the Benthic Laboratory, first used as a communications center for Sealab II in 1965 (see chapter 6). The laboratory housed electronic equipment which transmitted the many signals of voice, of television, and of instruments back to the shore station through a single multiplex wire. The first housing was concrete, but it leaked. "We had one of the wildest summers I've ever seen," commented Spiess, for to meet Sealab's deadline a steel housing had to be built very rapidly. The Benthic Lab was lowered to the sea floor beyond the Scripps pier in 60 meters of water, near the aquanauts' living quarters, and was pumped full of kerosene to protect the sensitive electronic devices from sea water. Repairs to the dome's contents during the six-week experiment in underwater living were solved by means of a mechanical "hand" that could be instructed to check electronic circuit cards for a faulty one and to replace it with a spare.

Another of MPL's novel contrivances is the Deep Tow (an attempt to name this FISH, for Fully Instrumented Submersible Housing, has not been entirely successful, although almost any slender object towed behind a ship is called a fish anyway). The Deep Tow is a mapping and navigation system in the form of a package of instruments that can be lowered and towed close to the sea floor to make detailed surveys. Development of such a unit began early in the 1960s, chiefly by Maurice S. McGehee and Dwight ("Tony") E. Boegeman, Jr. The first design problem was in the motion of the towed unit, so a roll meter, a pitch meter, and a flow meter were developed. Upward-looking and downward-looking sonars were added, and these were followed by a variety of oceanic instruments.

In 1967 one of the units was lost in 3,000 meters of water when the tow wire parted. Its exact location was known, so "after quick development of some special equipment," Spiess and others returned to the spot six months later and successfully retrieved their unit with the aid of a second one — no small feat in ocean navigation. A special

Scripps Institution of Oceanography: Probing the Oceans

Deep Tow was used in the summer of 1967 to locate an ancient shipwreck off the coast of Turkey for archaeologists from the University of Pennsylvania. For that search a compact unit with two side-looking sonars was sent to Turkey and installed on a local fishing boat. After the Deep Tow had located the target, observations from a two-man submarine confirmed that the wreck was an ancient Roman ship.

Surveys by Deep Tow have been made in trenches and canyons, over seamounts and fans, in the Pacific and in the Atlantic. The array of instruments can include precision navigation equipment, side-looking sonar, a low-frequency sound source for seismic studies, underwater cameras, television cameras, thermometers, and a magnetometer towed behind the "fish." All these instruments can be put to work on command and be monitored from the towing ship by way of the coaxial tow cable. For precise navigation of the ship and the Deep Tow, acoustic transponders, which answer to sound pulses from the towed unit, are placed on the sea floor. The precision capability of the instrument package was well demonstrated in 1971, when it was able to pinpoint the wreckage of five munitions ships that had been pulverized during munitions disposal. Geologic features as small as ten meters across have been mapped.

As with many of MPL's devices, the Deep Tow has more than one application: its capability of fine-scale surveying makes it equally useful to geologic mapping and to locating objects, such as shipwrecks or patches of manganese nodules, on the ocean floor.

NOTES

1. "The Age of Innocence and War in Oceanography," *Oceans Magazine,* Vol. I, No. 3 (March 1969), 12.

2. *San Diego Union,* September 1946.

3. *San Diego Union,* 10 October 1946.

4. Letter to Victor C. Anderson, 13 August 1975; written two and one-half months before Wright died.

5. Memorandum of 5 May 1970.

6. Quoted *in* Larry L. Booda, "Research – But Emphasis on Education," *Undersea Technology* (May 1966), 37.

7. SIO Annual Report, 1967, 29.

8. V. C. Anderson, "The First Twenty Years of Acoustic Signal Processing," *Journal of Acoustical Society of America,* Vol. 51, No. 3, Part 2 (1972), 1063.

9. *Ibid.,* 1064.

10. *Ibid.,* 1065.

11. "The Age of Innocence and War in Oceanography," *Oceans Magazine,* Vol. I, No. 3 (March 1969), 7.

12. "Frequency Dependence of Reverberation in the Ocean," *Journal of Acoustical Society of America,* Vol. 41, No. 6 (June 1967), 1467.

13. *Ibid.,* 1474.

14. MPL Quarterly Report, 1 April - 30 June 1949, 1.

15. "Explosion Seismology at Sea," *Transactions of the American Geophysical Union,* Vol. 48, No. 2 (June 1967), 416.

16. SIO Annual Report, 1970, 15.

17. Charles Corry, Carl Dubois, and Victor Vacquier, "Instrument for Measuring Terrestrial Heat Flow Through the Ocean Floor," *Journal of Marine Research,* Vol. 26, No. 2 (1968), 168.

18. "The Flow of Heat Through the Floor of the Ocean," *The Sea,* Vol. 3 (New York and London: Interscience Publishers, 1963), 229.

19. "A Magnetic Survey off the West Coast of the United States Between Latitudes 32° and 36° N and Longitudes 121° and 128° W," *Geophysical Journal of the Royal Astronomical Society,* Vol. 1 (1958), 320.

20. *Ibid.,* 321.

21. "The Magnetic Field over the Oceans," *The Sea,* Vol. 3 (New York and London: Interscience Publishers, 1963), 194.

22. Memorandum to Vacquier, 16 January 1964.

23. Edward Bullard, "The Emergence of Plate Tectonics: A Personal View," *Annual Review of Earth and Planetary Sciences,* Vol. 3 (1975), 19.

24. V. C. Anderson, D. K. Gibson, and O. H. Kirsten, "Rum II – Remote Underwater Manipulator (A Progress Report)," Marine Technology Society, *Sixth Annual Preprints,* Vol. 1 (1970), 2.

25. *Ibid.,* 4.

5. Seeing the Light: The Visibility Laboratory

During 1948 a research group organized and directed by optical physicist Seibert Quimby Duntley at the Massachusetts Institute of Technology began a program of research on the penetration of daylight into oceans and lakes and on the visual sighting of underwater objects by swimmers and aviators. This group, which Duntley named the Visibility Laboratory, began its work under a contract from the U.S. Navy's new Office of Naval Research, which had replaced its wartime Office of Scientific Research and Inventions.

The Visibility Laboratory began its research in optical oceanography at sea off Key West, Florida, and in Lake Winnepesaukee, New Hampshire. That pattern of experimentation has continued without interruption to the present time, although the site of the ocean experiments was transferred from Key West to San Diego in 1951 with the beginning of a collaboration with Walter Munk of the Scripps Institution of Oceanography. Visibility Laboratory research at Lake Winnepesaukee continued through 1966, but Crater Lake in Oregon, Lake Pend Oreille in Idaho, and various sites in California also came into use for certain fundamental optical studies that could be performed more economically in lakes than in the oceans.

In the winter of 1951-52, the Visibility Laboratory was invited to become part of Scripps Institution. Space was not available in La Jolla, but the laboratory was able to obtain the use of a group of unoccupied Navy barracks buildings situated directly across McClellan Road from the Field Annex that Scripps then operated on Point Loma. The Visibility Laboratory's principal operation is still at that site, although it also has facilities and personnel in Sverdrup Hall and in UCSD's Muir College.

Funds for moving the laboratory to San Diego and for equipping and refurbishing the newly rented buildings were provided by the U.S. Navy. Duntley, while continuing his teaching at M.I.T., prepared the necessary Scripps contract proposal for outfitting the Point Loma facilities and conducting research in them. The resulting contract called for equal funding by the U.S. Navy and the U.S. Air Force.

The interior of the buildings on Point Loma was designed by Duntley in the winter of 1951-52 and construction began late in April. That month, Duntley employed a longtime friend from college days, John E. Tyler, to represent him at Point Loma while he continued to teach and operate the Visibility Laboratory at M.I.T. Daily telephone conversations with Tyler enabled Duntley to supervise the renovation of the leased buildings and to direct the procurement of the necessary furnishings and equipment with contract funds.

The new facility on Point Loma was completed in September of 1952 when Duntley arrived from M.I.T. with three graduate students and several staff members, to begin operating the Visibility Laboratory as a part of Scripps.

The Visibility Laboratory immediately began coordinated parallel measurements of the optical properties of the ocean and the atmosphere. Scripps ships and an Air

The Visibility Laboratory

Force B-29 aircraft were used to explore the limitations imposed by those media on the ability of man to see. Ancillary research included the quantitative properties of human vision and the capabilities of visual aids, such as optical telescopes, photographic cameras, and electro-optical devices such as television. One goal was to improve and extend the laboratory's capability to predict the limiting distances at which any specified object can be sighted and identified, day or night in all kinds of weather, through the atmosphere or through ocean water. That required a worldwide data collection activity as well as basic research of many kinds.

The Visibility Laboratory has always been interested in specific applications, including some needs of the military, as well as a wide variety of interests for the National Aeronautics and Space Administration in connection with manned space flights, interests of the National Oceanic and Atmospheric Administration in the remote sensing of marine resources, and concerns of the U.S. Coast Guard with the optimization of search-and-rescue operations and with the lighting of harbors and waterways at night. The Visibility Laboratory also has studied air collision avoidance for the Department of Transportation, the Department of Justice, and the Airline Pilots Association. In the realm of basic research, the laboratory's continuing research in optical oceanography has long been supported by the National Science Foundation.

The research program of the Visibility Laboratory did not originate in 1948 or in 1952. The program actually began in the autumn of 1940 when Dr. Duntley, then in his fourth year of teaching at M.I.T., became a member of a committee of scientists from M.I.T. and Harvard, chaired by President Karl T. Compton of M.I.T., that was concerned with the technical aspects of defending the continental United States against possible air raids like those that

Scripps Institution of Oceanography: Probing the Oceans

Germany was inflicting upon Great Britain. It was believed that, following the conquest of the British Isles, Germany might establish naval and air bases in the western hemisphere from which to bomb the eastern and central United States. Because radar had not yet been developed, both the bombing and the defense against it would depend upon direct human vision.

In the early deliberations of the Compton Committee, Duntley suggested that the ability of pilots and bombardiers to see specific objects on the ground should be predictable on the basis of optical data concerning the objects and their lighting, the clarity of the atmosphere, and measurable threshold properties of human vision. Such predictions would enable defenses to be planned accurately and operated optimally. That suggestion resulted in the establishment of a government-sponsored civilian research organization staffed by Harvard and M.I.T. personnel and called the Passive Defense Laboratory. That organization had three divisions devoted, respectively, to camouflage structures, camouflage materials, and visibility. The latter division was under the direction of Duntley.

The Passive Defense Laboratory existed for only one year. With the advent of war in December 1941, the divisions concerned with structures and materials were taken over by the United States Army and moved to Ft. Belvoir, Virginia. The Visibility division became a section of the Optics Division of the National Defense Research Committee of the Office of Scientific Research and Development in President Roosevelt's wartime Office for Emergency Management. Duntley was part of that section. The concept that he had initiated in the Passive Defense Laboratory became the central core of the wartime visibility program.

After the war, Duntley wrote a book about the wartime researches in visibility, and he participated in occasional committee activities concerned with peacetime applications of the scientific advances that had been made during the years of conflict. One of these applications concerned visual search for shallowly submerged objects, such as hazards to navigation as seen from low-flying aircraft. As a result of committee deliberations on that problem, based in part on a series of experiments Duntley had made during 1944 from ships and aircraft along the coast of Florida, he was urged to undertake a program of fundamental research on the visibility of submerged objects. That research began in 1948 under a contract between M.I.T. and the newly formed Office of Naval Research. For that work Duntley organized a small research group which he called the Visibility Laboratory since it was, in fact, a continuation of one aspect of the wartime program in visibility. The laboratory established a field station on Diamond Island in nearby Lake Winnepesaukee, New Hampshire, and carried out supplementary experiments from ships off Key West, Florida.

The visibility of submerged objects is greatly affected by water waves. In seeking information on waves, Duntley met Walter Munk of the Scripps Institution of Oceanography and became a frequent visitor to La Jolla. Together they continued electrical measurements of water wave slopes that had been begun by Duntley in fresh water at Diamond Island. The experiments at sea were conducted from the Scripps vessel *Paolina-T.* Professor Duntley's growing interest in optical oceanography and his friendship with persons at Scripps culminated in moving the Visibility Laboratory to San Diego in 1952.

A purse seiner of San Diego's tuna fleet. Photo by Joanne Silberner.

6. A Potpourri: The Institute of Marine Resources

In the early 1950s another fish story began. This one, like the statewide CalCOFI project (see chapter 3), also began at the urging of representatives of commercial fisheries; it became the Institute of Marine Resources (IMR) of the University of California, headquartered at Scripps.

The new project was founded much more slowly than had been the sardine project. In the spring of 1951 Roger Revelle, at the urging of members of the Marine Research Committee, pointed out to University President Robert G. Sproul that marine fisheries required long-term research:

> The Committee has been concerned with the fact that the Marine Life Research program, because it was tacitly assumed to be of a temporary nature, has been inadequately staffed with mature scientists of outstanding competence. It is now quite obvious to all concerned that the problems being studied can only be solved by a concerted attack over a period of many years led by experienced and creative scientists. The need for such scientists is particularly great in the biological aspects of the Marine Life program. We

have been unable to enlist such men on temporary research appointments, but there is convincing evidence for the belief that they could be attracted by the prestige and security of Faculty positions.[1]

In enlisting the support of Emil Mrak, then chairman of the Division of Food Technology of the Berkeley College of Agriculture (later Chancellor at the Davis campus), Revelle commented: "I am convinced that a very large percentage of the potential resources of the sea are not now being used for anyone's benefit and that such beneficial utilization could be greatly improved by scientific and technological development." He went on to note that "the problem is an extremely complex one, however, and involves far more than oceanography."[2]

The solution, Revelle felt, would be to establish a university-wide institute, which he recommended be based at Scripps Institution and should draw upon the expertise of other campuses as well.* At that time, such applied research was outside the usual role of the university, except in the field of agriculture, so a certain amount of soul-searching was necessary within the university administration.

Among the early supporters of the institute was Wilbert M. ("Wib") Chapman, then director of research for the American Tunaboat Association, and earlier an active founder of CalCOFI. Chapman proposed a university Institute of Marine Fisheries to the State Assembly Interim Committee of Fish and Game, which met in San Diego in October 1951 to consider legislation that might help the fisheries industry, especially tuna fishing. Chapman's proposal was quickly translated into a college of fisheries by legislators and reporters.

---

*An interdepartmental or an intercampus approach is implicit in the University of California's definition of an institute (but not of an institution).

Institute of Marine Resources

In the early 1950s the California fishing industry was the largest in the United States. Tuna fishing had begun in San Diego before World War I, but only for the albacore tuna; it expanded into the fishery for the wide-ranging yellowfin tuna in the late 1920s. Specialized vessels and improved freezing methods brought a further expansion after World War II. By 1947 San Diego's sixty-million-dollar business made it the "tuna capitol of the world."

President Sproul was cognizant of these points when he wrote to Chapman early in 1952:

> I should like to reassure you, and through you the members of the fishing industry as a whole, that the University of California is well aware of the importance of the industry. . . . If it sometimes appears that progress is made at too slow a pace, it is only because the University is unwilling to slight a careful study of the complex problems of coordination which always accompany new ventures.[3]

Sproul had appointed a special committee in October 1951, under the chairmanship of Baldwin M. Woods of UCLA, to look into the proposed new unit, already being called the Institute of Marine Resources. That committee soon reported that it was "of the opinion that an Institute of the type mentioned should be established. . . . The field of investigation proposed by Director Revelle and other members of the Committee is of University type and it is the belief of the Committee that no other agency in the State is so well qualified as the University to undertake the task."[4]

The function of the institute, as defined by Woods's committee, would be "to foster research, education, and public services by the University of California in the development of fisheries and other resources of the sea for the benefit of the people of California." The scope of the original idea was thus considerably enlarged, because "resources," as defined in the report, were "stipulated to include: plants and animals; minerals, both those dissolved in the water and those concentrated in bottom deposits; beaches and bays; highways of commerce over the surface; capacity for waste disposal; and the water itself."

In March 1952, the University Committee on Educational Policy of the Academic Senate met in La Jolla to consider the institute, and it also reported favorably. In the 1953-54 university budget, funds were allotted to launch the Institute of Marine Resources in January 1954. The allocation was $22,572 — nowhere near what Revelle had recommended as a minimum, or "core," budget of $120,000. This was in spite of an earlier statement by the committee chaired by Woods that "the Committee doubts the wisdom of initiating the Institute unless a sum approximately of this magnitude can be had."

Revelle was optimistic, however, and predicted that "the Institute of Marine Resources may become larger than Scripps Institution in the next five years."[5]

IMR began and has continued as an interdisciplinary and intercampus organization. Beyond the obvious interest of Scripps in marine-oriented research, the colleges of engineering at Berkeley and at UCLA and members of the Division of Food Technology at Berkeley (now at Davis) were included from the beginning. Projects already established on the Scripps campus were moved into the fledgling institute, for administrative purposes. (MLR was not one of these, in spite of a tacit assumption in the early years of IMR that logically that program would be incorporated.)

An executive committee was established, consisting of the director of Scripps; the director of IMR, who was to be a faculty member at Scripps, appointed by the regents

Institute of Marine Resources

of the university; the deans of the colleges of engineering at Berkeley and at Los Angeles in rotation every two years; and five faculty members, to be appointed by the president of the university for two-year terms. An advisory council was also established, whose fifteen members, not connected with the university, were appointed by the university president.

In January 1954, Rear Admiral Charles D. Wheelock (Ret.) became the acting director of IMR. Wheelock had joined the Scripps staff five months earlier as a research engineer, after some thirty years of Navy career as an engineering duty officer and, just before his retirement, as the inspector general for the Bureau of Ships. He was appointed director of IMR in 1958, and held the post until his retirement in 1961. The following year Milner B. ("Bennie") Schaefer became the institute's director.

An international fisheries expert, Schaefer concerned himself for many years especially with population dynamics of fisheries, and the economics of the industry. He also did research on marine pollution and on the disposal of radioactive wastes, and he served on committees concerned with the use of ocean resources and with the application of science and technology to economic development. From July 1967 until February 1969, Schaefer served as science adviser to Secretary of the Interior Stewart L. Udall, then returned to continue as director of IMR until his death in July 1970. In April 1971, John D. Isaacs became the director of IMR.

The first two faculty appointments in IMR were Harold S. Olcott in food technology at Berkeley, and H. William Menard in marine geology at Scripps, both in 1955. Menard's studies, long supported under IMR, are described in this narrative in chapter 12.

An umbrella as broad as IMR inevitably takes in projects that began in other ways and places. Its structure is diffuse, as the many projects included are supported by contracts and grants from various agencies. Approximately one-quarter of IMR's funds came from the university until 1971, when the federal Sea Grant program was incorporated into IMR and at Scripps became a major part of the total budget. Faculty members in IMR also hold appointments in an academic department on their own campuses, mostly at San Diego, Berkeley, and Davis. A number of graduate student fellowships and assistantships are provided by IMR funds.

Over the years the philosophy of the institute has inevitably changed somewhat, although always keeping in mind the dictum phrased by Schaefer: "What distinguishes the study of marine resources from other branches of marine science is the fact that the word 'resource' implies economic and social considerations."[6] Those considerations in California have carried IMR researchers to the beach, the market place, and the sewer; to the high seas and the floor of the ocean; and to pondering the law of the sea and the law of the land.

KELP STUDIES

One of the first projects moved into IMR was the study of giant kelp. The verdant forests of this overgrown alga, *Macrocystis pyrifera,* are found scattered along the Pacific coast (in the northern hemisphere) from Sitka, Alaska, to Magdalena Bay in Baja California, Mexico. The dense growths — rare jungles in the sparsely vegetated ocean — attract many fishes into their protection and provide food and shelter for crustaceans, mollusks, and other invertebrates. Sea lions glide among the stipes after unwary fish, and, in a few places, sea otters loll in the surface fronds or seek their food among the holdfasts.

Kelp itself, on the west coast, was first used as fertilizer, as animal food supplement, and for iodine, inorganic salts, and organic solvents. As early as 1911 the Bureau of Soils of the U.S. Department of Agriculture supported a study by W. C. Crandall, then secretary of the Marine Biological Association (which founded Scripps Institution), to estimate the quantity of kelp off southern California. During World War I, Crandall advised the California Fish and Game Commission on regulating the harvesting season.

In 1929 Kelco Company in San Diego began gathering the great fronds of *Macrocystis* for algin, a colloid used as an emulsifier and suspender to stabilize, smooth, and thicken many products such as ice cream, pharmaceuticals, paints, and the foam of beer. The distinctive kelp cutters mow swaths through the kelp beds, mechanically cutting about four feet below the surface and gathering the fronds amidships with mechanized loaders. Because of the plant's rapid growth — determined by Kenneth A. Clendenning to be the fastest of any plant on land or in the sea — an area can usually be harvested again in four months.

In 1948 diver-biologist and graduate student Conrad Limbaugh began a study of the kelp beds on his own, under what he called "primitive" methods of investigation. He was greatly boosted the next year by a fellowship grant from Kelco Company to determine whether the cutting of kelp did, as claimed by sportfishermen, reduce the numbers of fish. (Kelco had provided a fellowship for the kelp studies of J. Frederick Wohnus at Scripps from 1941 to 1943.) Limbaugh studied kelp beds at many points along the coast and islands from Monterey down to the San Benito Islands in Baja California, but he concentrated on the thick La Jolla kelp beds. He was among the first to use self-contained underwater breathing apparatus (Scuba) for scientific investigations, and he was an early Scripps developer of the techniques of underwater photography. On the kelp study he spent "thousands of hours . . . above and below the surface observing the organisms in their own enviroment."[7] He collected, identified, and photographed the richly varied life of the kelp forests from the lush surface canopy to the holdfasts as much as one hundred feet below, and he concluded that kelp harvesting "has no seriously detrimental effects on fishing." His own method of fishing was chiefly by hand, and he added many specimens to the fish collection and to the aquarium displays.

But the beds of giant kelp were declining all along the California coast, to the equal dismay of sportfishermen and harvesters. Controversy swirled over the causes of the decline, so the California Department of Fish and Game, which regulated harvesting, funded a five-year study, beginning in 1956, through IMR. The broad project, directed by Wheeler North from 1958, was intended to determine the reasons for the dwindling of the kelp beds, the effect of pollution on them, the effects of various methods of harvesting, the problems of frond litter on beaches, the possibilities of culturing other strains of kelp, and, again, the effects of harvesting on fishing. Kelco Company continued its support, with emphasis on habitat improvement; the State Water Control Board from 1957 financed a study of the effects of sewage discharge on kelp; and from 1960 the National Science Foundation supported a study on food-chain intermediates in the kelp beds. The kelp program lasted until 1963, through the unusually warm years of 1957 to 1960, which further reduced the vast beds, for they thrive only in cooler temperatures.

Limbaugh, founder of the scientific diving program at Scripps, and among the keenest advocates of safety in diving, drowned in March 1960, while exploring an underwater cavern in southern France. Others of the staff and students went on with the diverse diving project to understand the ecology of the remarkable underwater forests.

Michael Neushul checking the growth rate of *Macrocystis*, during the kelp project in the 1950s.

Scripps Institution of Oceanography: Probing the Oceans

The smallest vessel of the Scripps fleet, the 22-foot *Macrocystis*, was a common sight bobbing its way to and from the kelp beds southwest of Scripps as the diver-biologists tested harvesting techniques, measured growth rates, transplanted Mexican kelp plants adapted to warmer water, and determined that increasing numbers of sea urchins were overgrazing the kelp beds. The urchins had once been held in check by the marine mammal brought almost to extinction for its magnificent fur, the sea otter, long since gone from its former haunts along southern California. David Leighton suggested that quicklime might be effective against sea urchins, and indeed it proved to be, even in doses small enough not to injure abalones and other desirable kelp-dwellers. Quicklime and hammers have now brought the sea urchins under control in some kelp beds, so that the young plants are again becoming established; those beds are returning to their former size.* Some researchers suggested harvesting sea urchins, considered a delicacy in the Orient, as another way of controlling these kelp predators, and this industry is developing.

In spite of Limbaugh's conclusions, the argument between sportfishermen and kelp harvesters continued, so other marine botanists went on with studies of the kelp environment. Harvesting records of the 45 kelp beds of southern California had been kept since harvesting began in 1916. Kenneth A. Clendenning compared those records and sportfishing returns from specific kelp beds — some cut heavily, others lightly or not at all — and found that the catch of kelp bass, a favorite with sportfishermen, was actually higher in beds harvested regularly. He noted that the populations of kelp bass fluctuated from year to year, but irrespective of kelp harvesting. Numbers of sportfishermen

---

*This kelp research continues under the direction of Wheeler North, now at the California Institute of Technology.

increased steadily every year, also irrespective of kelp harvesting.

SCUBA

Although Conrad Limbaugh began the Scuba scientific diving program at Scripps, he was not the first diver at the institution. Ahead of him were two helmeted divers: C. K. Tseng, who collected the seaweed *Gelidium* in 1944 for cultivation for agar, and Frank Haymaker, who made direct observations of the Scripps submarine canyon for Francis P. Shepard in 1947 (see chapter 12).

At UCLA, the invertebrate section of the zoology department began a program of observations by divers about 1948, first with hand-operated air pumps and helmets, under the guidance and participation of Theodore H. Bullock. That group soon began using the Gagnan-Cousteau Aqua-lung, the self-contained unit that first appeared on the market in 1947. Scripps alumnus Boyd Walker of the UCLA ichthyology laboratory was another early user, with some of his students, of the Aqua-lung. By 1951 the UCLA researchers had established guidelines for using the new diving equipment. "It is our view," wrote Bullock, "that this powerful research tool for the study of biological and physical marine science should be treated as a potentially hazardous operation. . . . With caution, with use confined to individuals of cool judgment and gradually acquired experience, it is an eminently safe and useable technic."[8]

Already the self-contained underwater breathing apparatus (Scuba) was coming into use by enthusiastic swimmers and amateurs, a point that worried Bullock. He established safety rules and instructions for university students and personnel under his jurisdiction, and discussed the situation with Boyd Walker, with people at Scripps, and with safety supervisors.

Scripps people were naturally beginning to use Scuba to study their domain. Limbaugh was carrying out his kelp studies, using the institution's first Aqua-lung, purchased in 1950, and some of the geologists were making firsthand observations using Scuba of sea-floor structure and the movements of sea-floor sediments. James Snodgrass, then with the Special Developments Division, circulated a memorandum on his observations:

> It has been necessary for the Special Developments Division to repair the Aqua Lungs used by various Scripps personnel. This brings to light what we believe to be a very serious situation. The Aqua Lungs as manufactured are produced with many defective and unsatisfactory components. Some of these components are in critical positions in which their failure may well cause complete failure of the air supply.
>
> ... We feel that most of the users of the Aqua Lungs and the departments concerned do not properly appreciate the seriousness of the situation and that the narrow margin of safety under which they have been operating has been practically non-existent in several cases....[9]

Limbaugh was equally concerned for diving safety, and he began teaching classes in safe Scuba diving in 1951. The next year Scripps participated with the Navy in an Underwater Swimmer Panel, at which the use of Scuba for military purposes and the requisite training programs were intensively discussed. When, in 1952, two students from the Santa Barbara and Berkeley campuses of the university drowned while using Scuba, a statewide university

A motley crew of divers on Guadalupe Island, November 1954. From left: Earl A. Murray, T. U. Tiess, John Carter, Charles Fleming, Andreas B. Rechnitzer, James R. Stewart. The occasion was a research study of Rechnitzer's. Photo by Robert B. Haines.

committee was formed to establish training and certification procedures, and diving was authorized only at the Scripps campus. Limbaugh and others prepared a list of rules and regulations, first published in 1954. Diving programs were established on other University of California campuses during the 1960s, under the initial guidance of Scripps.

The Scripps diving classes were placed under IMR auspices in 1956. The diving locker was established for servicing the compressed air tanks and other equipment of the growing number of staff and student divers. About a dozen classes are now taught each year for Scripps staff and students, for state and national park naturalists, and for other eligible marine researchers. It is a rigorous program, one that takes into account techniques, hazards, and the hostile environment. The preliminary testing separates the casual from the serious swimmer, for the applicants are required to demonstrate their ability to swim 1,000 feet in a pool in less than ten minutes; to swim around the 1,000-foot Scripps pier in less than fifteen minutes; to swim under water for at least 75 feet; to swim under water for 150 feet, surfacing no more than four times; to demonstrate their ability to use mask, fins, and snorkel; to surface dive to a depth of 18 feet at the end of the pier, and to carry a struggling swimmer 75 feet.

The course lectures and films describe the physiology of diving, the nature and cure of divers' disorders, the varying conditions of waves, currents, and bottom topography, and the hazardous creatures that divers may encounter. The students are instructed on the maintenance of the Scuba gear. In the water and on paper they have to prove their competence before they are certified. Those who have completed the course may qualify for gradually deeper dives, under the supervision of other certified divers.

Swimmers who enter the blue world — the depths below

which the red component of daylight has been absorbed — tend to rhapsodize:

> At depths of 60 to 100 feet the diver enters a realm of stillness [wrote Wheeler North]. No water motion is ordinarily perceived. . . . The lower edges of the kelp beds lie in this range and the sea floor in these areas is scalloped into fantastic shapes by the activities of mollusks and echinoderms. The diver can now get well down into the upper reaches of the canyons. . . . He explores steep gorges, hundred-foot cliffs, caves, pinnacles, and the like, and he may feast his aesthetic senses upon gardens of fluorescent anemones or canyon walls covered with lace-like Gorgonian corals.

Working in deeper water is another story:

> It requires considerable fortitude [continued North] as well as a thick, foam rubber suit to do productive scientific work in the chilling depths below 100 feet. Illumination is reduced to twilight levels, hazards arising from decompression sickness must be avoided, and mental activities may be retarded by the effects of nitrogen narcosis. Such dives must be carefully planned to avoid accidents and to assure that the required work be accomplished.[10]

James R. Stewart, who began skin-diving in 1943 — well before Scuba gear — has been the chief diving officer at Scripps since Limbaugh's death. Over the years he has used all the common types of diving apparatus: closed circuit, semi-closed, open circuit Scuba, shallow-water mask, hookah, heavy dress, and saturation-diving techniques. He has dived in the Arctic, Antarctic, Atlantic, and

Scripps Institution of Oceanography: Probing the Oceans

Pacific oceans, the Gulf of Mexico, and the Mediterranean Sea. Stewart continues to emphasize safety, in his classes and in public talks; he helps to review and revise the statewide guideline for diving safety for the university campuses, and he has written several sections of the NOAA Diving Manual. Stewart's responsibilities, as well as that of the Scripps Diving Control Board, extend to about 150 certified divers on the UCSD campus, each of whom must make at least twelve dives – with a buddy – each year to retain his certification. Their record is noteworthy: over the years Scripps divers have carried out almost one hundred thousand dives without accident. The tragic death of Limbaugh, under unexplained circumstances in a Mediterranean cavern, has been the only Scripps fatality while using Scuba.

The early Scripps Scuba divers were inventors and innovators. George Harvey, for instance, devised what he called a Superman suit and described how he made it in a letter to Arthur Flechsig:

> ... It was as follows:
> 1. Purchased surplus army drawers and shirt, sewed zipper in drawers. Shirt was snug pull over.
> 2. Purchased rubber latex from Atlas Chemical Supply Co. in San Diego. ($7.50 per gal.) This can be obtained from any large chem. supply house. It comes in several grades, usually preserved with ammonia. The grade containing the largest amount of rubber is what I used. If a small amount of acetic acid is added the rubber will precipitate from the colloid suspension. Tetramethylthiuram disulfide, $(CH_3)_2$-N-C-S-S·S-CSN-$(CH_3)_2$ accelerates the setting of the rubber and causes some vulcanization. I didn't use this, but mixed a *small* amount of acetic acid and some H_2O_2 with about a pint of latex before using.

3. Put on the suit (drawers and shirt separately) and painted on the latex, saturating the cloth as thoroughly as possible, using a medium large paint brush.
4. Wore the damned thing till dry. This is very uncomfortable, but gives a very good form fitting job. During drying care is necessary not to allow adjacent surfaces to touch, or they will stick together. Also all wrinkles should be in the right places before drying. The H_2O_2 is supposed to give a tougher rubber. It would be wise to call a local latex products manufacturer to get his formula, since one suit I made got quite sticky and "runny" after a couple of months.

There is available 1/4" and 1/8" thick *cellular* rubber and also cellular plastic sheet that would make a much warmer outfit. My latex-cloth suit was quick and easy to make, and quite strong, but it only allowed me to stay out about 2 hours, and would have been much better with some of the cellular sheet cemented on. I think the wet type suit is fine if you have plenty of insulation, 1/8" to 1/4" of cellular material, and if the fit is snug enough to prevent water going in and out. . . . [11]

A major contributor to the present wetsuit used by divers (and surfers) was Hugh Bradner, who has been at Scripps since 1961. While he was at the University of California Radiation Laboratory in 1951, he devised a form-fitting underwater protective suit made of elastic foam rubber and provided with "a plurality of slide fasteners so located as to facilitate getting into and out of the suit." Bradner and his diving acquaintances made the first suits of this kind in 1952; commercial diving companies were very quickly marketing wetsuits of similar design.

Scripps Institution of Oceanography: Probing the Oceans

Scripps divers are in all branches of oceanography. On Capricorn Expedition in 1952 (see chapter 15) they pioneered in exploring submerged Falcon Shoal in the Tonga Islands. They have participated in the kelp studies, have made geological and biological observations, have studied sand transport and water motion, have explored coral reefs, submarine canyons, and sewer outfalls. They have inspected ship bottoms, fastened and removed equipment thereon; they have investigated pier pilings; and they have developed and tested new underwater equipment and techniques. They have advised on underwater parks. They have helped in underwater rescues and have investigated ocean tragedies. Stewart estimates that since 1965 at least 30 Ph.D. dissertations at Scripps have been completed in which diving was a necessary research tool.

In the early 1950s Scuba divers began finding Indian artifacts along the southern California coast. An especially rich source was the shallow zone in front of the La Jolla Beach and Tennis Club about a mile south of Scripps. Divers, including Scrippsians, soon gathered many hundreds of small mortars from that area, and smaller numbers at other sites along the coast. As Carr Tuthill and A. A. Allanson noted: "There must have been a grinding complex almost amounting to a compulsion [among the southern California aborigine culture] to account for the large number of mortars found just in the La Jolla area in underwater sites."[12] Those authors suggested that the mortars were found in now-drowned early aboriginal sites, and archaeologist and Scripps-draftsman James R. Moriarty later advocated that the sites had been drowned by the steady advance of sea level. Radiocarbon dating established that adjacent landward middens were in use by aborigines 5,000 to 6,000 years before the present, when sea level in the area was probably 40 to 50 feet lower than now.[13]

Several Scripps divers — Earl A. Murray, Arthur O.

Institute of Marine Resources

Flechsig, and graduate students Thomas A. Clarke, Morgan Wells, and Richard W. Grigg — were participants in Sealab II in 1965, the project of living on the sea floor sponsored by the Office of Naval Research. (Sealab I the previous year was off Bermuda.) For this experiment three teams of ten men each lived for two weeks at a time in a steel cylinder located half a mile off the Scripps pier in a gently sloping valley 210 feet below the surface. The living quarters were aerated with a mixture of helium, nitrogen, and oxygen. Communications to shore were supplied by the Benthic Laboratory developed by the Marine Physical Laboratory (see chapter 4). From their home base the divers could swim out daily to make observations and to carry out underwater experiments.

The biologists among them had a field day, for the presence of the steel cylinder on the sea floor attracted underwater life in great numbers, estimated to be 35 times the usual density of the sand-bottom locality. Zooplankton swarmed at the lighted viewing ports every night, drawing such numbers of fishes that they cut off the view as they inhaled the tiny zooplankton. Octopi snuggled into hidden corners all over the cylinder, scorpionfish piled up in the entryway (several divers were stung by them), and sea lions became regular evening visitors, startling off the fishes as soon as they appeared.

Other Scripps divers have participated in the later, shallower experiments in sea-floor living, the Tektite series of the University of Miami, which is particularly distinguished by having included women in its roster, three of them from Scripps. James Stewart participated in the Westinghouse Project 600, during which the first dives were made by scientists to a depth of 600 feet on the continental shelf.

The Scripps diving program itself does not include submersible vehicles, but some of the Scripps divers — and a number of other staff members — have ridden in various of the saucers and submersibles whenever opportunities have arisen.

TUNA RESEARCH

It was, after all, the tuna fishermen especially who had urged the founding of IMR. They gained benefits from the organization, in the form of sea-floor charts that provided the locations of seamounts and other submarine features that attract fish. That project, which began under the direction of H. William Menard, evolved into the Geological Data Center and is described in chapter 12.

In 1963 IMR spread its umbrella over another existing group on campus, the Scripps Tuna Oceanography Research program (STOR). This unit had begun in 1957 with support from the U.S. Bureau of Commercial Fisheries and in cooperation with several other campus groups, for the purpose of determining the distribution, abundance, and availability to capture of the fast, sleek, far-ranging tunas that constituted San Diego's major fishery. Maurice Blackburn directed STOR until 1971, and then joined the Coastal Upwelling Ecosystems Analysis (a project supported by the International Decade of Ocean Exploration). STOR ended in 1973.

This program concerned itself with all the tunas — "large, active, predatory, pelagic fishes that inhabit the upper layers of the World Ocean in tropical and temperate latitudes." Under intensive study by STOR were the yellowfin and the skipjack — the worldwide favorites when packed in small flat cans. The bluefin, caught, as are the yellowfin and the skipjack, chiefly by purse seiners, and the albacore, caught by line from coastal small vessels, were less intensively studied by STOR, but have been under scrutiny by the National Marine Fisheries Service and others.

Blackburn said that "STOR had its most productive period from 1962 to 1969, when its effort both on tuna ecology and pure oceanography was high. Research included effects of temperature and food on tuna distribution, seasonal change in biological oceanographic properties of the tropical ocean, vertical distribution of zooplankton, and very detailed descriptive and heat budget studies in physical oceanography. A method was developed for measuring sea surface chlorophyll continuously from an underway ship, which has been very widely used. . . . There was also a laboratory-experimental project in which phytoplankton cultures were grown under various nutrient conditions. It helped to identify nitrogen as the principal element limiting oceanic phytoplankton growth."[14]

STOR's program of studying the environment of the tunas was correlated with the Inter-American Tropical Tuna Commission (IATTC), which was established in 1949 by a treaty between the United States and Costa Rica. The objective of IATTC, as defined in its founding convention, is to maintain "the populations of yellowfin and skipjack tuna and of other kinds of fish taken by tuna fishing vessels in the eastern Pacific Ocean . . . at a level which will permit maximum sustained catches year after year." Other fishing nations — Canada, France, Japan, Mexico, Nicaragua, and Panama — have become members of IATTC, which recommends appropriate conservation measures for and carries out an extensive research program on the tunas and the bait fishes and other species taken by the tuna vessels. The commission offices were first located at the Scripps Field Annex on Point Loma, and were moved into the Fishery Oceanography Center on the Scripps campus when the building was completed in 1964. Milner B. Schaefer, who, in the interest of international fisheries research and cooperation, ranged the world even more widely than do the tunas, headed IATTC from 1951 until 1962, when he became director of IMR.

Both STOR and IATTC were deeply involved in the program called Eastropac, a series of three surveys in the eastern tropical Pacific in 1967 and 1968. As grand in scope as Eastropac was, it was but a portion of the study originally proposed in 1961, which was to have been a cooperative eight-year survey of the eastern tropical Pacific, including studies in meteorology, geology, geophysics, and physical, chemical, and biological oceanography, including fisheries. But it soon appeared that the costs were prohibitive and "some parts of the program had not attracted the scientists who would be required to carry them out."

The geological and geophysical work was omitted, and the meteorological program was reduced. Warren S. Wooster was appointed coordinator in 1966, under the Bureau of Commercial Fisheries, and he began the complex job of enlisting agencies and groups and nations into the program. Alan R. Longhurst assumed the post the following year. Chile, Ecuador, Peru, and Mexico all provided ships and personnel, as did several United States organizations and universities.

From Scripps the *Argo* and the *Thomas Washington* sailed separately for Eastropac. The *Argo,* which sailed on 24 January 1967, "returned to San Diego on 6 March after a successful cruise of 340 oceanographic stations, the routine having been broken only by an emergency rendezvous with a commercial vessel in the early part of the cruise so that the surgeon carried by ARGO could perform an appendectomy at sea on a crew member of the commercial vessel."[15] Among the researchers on *Argo* were observers provided through the Smithsonian Institution to take all-daylight observations on surface fishes, on cetaceans, and

Institute of Marine Resources

on birds, including flocks feeding on schools of fish.

The chief goals of Eastropac were to gather data to help define the relationship of fishery yields to ocean conditions and to search out unexploited skipjack tuna resources. A distinctive feature of the survey was its adherence to the original plan. "All participating ships undertook to perform a standard and basic suite of physical and biological observations, using standardized gear. . . . "[16] This simplified reducing the data to a consistently usable form.

Eastropac carried out three oceanographic surveys, six months apart, of the area from 20° N. to 20° S. latitudes and from the coast westward to 119° W. longitude — six million square miles of ocean. Survey cruises and monitor cruises by 16 ships gathered meteorological observations, water samples and analyses, and biological collections. The sorting of the extensive zooplankton and micronekton samples was handled at the Fishery Oceanography Center, aided by STOR and IATTC participants. The Data Collection and Processing Group of the Marine Life Research program assisted in the processing of the basic oceanographic data.

It quickly became apparent that "the data derived from the expeditions were so numerous as to render classical data reports impractical" — so the data were archived on magnetic tape at the National Oceanographic Data Center, which had been established in 1961, and were processed by computer. The results, compiled chiefly by Bruce A. Taft, are being published as a series of eleven Eastropac Atlas volumes, the first of which appeared in 1972.

A full program of research on tuna has continued through the years, particularly on the yellowfin tuna. Schaefer concluded in 1967: "Due to the very intensive and closely programmed research of the IATTC . . . and the results of the research of other agencies, the yellowfin tuna population of the eastern Pacific, concerning which very little of importance was known in 1951, has, during only fifteen years, become one of the best understood commercial fish populations in the world."[17]

FOOD CHAIN STUDIES

The Food Chain Research Group began at Scripps in June 1963, with the advent of British-born John D. H. Strickland from Nanaimo, British Columbia, originally a chemist, who was invited by M. B. Schaefer to establish the group within IMR. The goal of the research group, simply stated, has been to predict the formation and transfer of nutrients through the full cycle of life in the ocean.

> For the good of the fishing industry and for the guidance of those charged with the disposal of atomic wastes [wrote Strickland], we need to know what makes the aquatic environment 'tick.' . . . The basic stumbling block is one of sampling. Even in small ponds or lakes, where boundary conditions are generally well-defined and, in some circumstances, populations are relatively simple compared with field or forest, the problems are great. The medium is alien. . . . Much of the sampling has still to be done from top side, with the worker unable to see what he is doing, and studying only a minute fraction of the whole. When the area under investigation becomes an ocean, or even a fairly small fraction of an ocean, the size of samples relative to that of the environment becomes ludicrously small and, to compound our troubles, the environment will not keep still. Imagine the troubles of a field ecologist if the field he investigates one week moves by the next week into the adjacent county and splits into several segments in the process.[18]

Institute of Marine Resources

Tall, red-bearded John Strickland, "a man of boundless energy, . . . set a challenging pace and delved with enthusiasm into the myriad facets involved in the comprehensive study of plankton communities."[19] He died in November 1970, after a heroic and stoic battle of several years with kidney disease (he was the first patient of the UCSD hospital to be trained in home use of an artificial kidney machine). Since then the directing of the group has been handled by rotation among its senior staff, which includes John R. Beers, Angelo F. Carlucci, Richard W. Eppley, Osmund Holm-Hansen, Michael M. Mullin, and Peter M. Williams.

The problem for the Food Chain Research Group to solve, felt Strickland, was "a serious imbalance between observation and explanation" in understanding oceanic productivity. He went on to note: "Our information about the mechanisms responsible for what we observe is so inadequate that a crisis has developed."[20]

So, the researchers scrutinized bacteria, phytoplankton, zooplankton, and the constituents of sea water. Diurnal rhythms of diatoms and other minute plants were tampered with, aided by a searchlight bright enough to bring complaints from residents on Mount Soledad, four miles away. Minute organisms were kept alive in laboratory jars, but at first they were annoyingly uncooperative in reproducing. Plankton communities were placed in a 70,000-liter tank for controlled-environment studies. An Autoanalyzer was developed that "revolutionized the procedures for performing chemical analyses aboard ship."

The planktologists also went to sea. As soon as their laboratory was equipped, Strickland announced, in the true spirit of oceanography: "We plan to spend a week on a shake-down cruise this fall when as many people in the group as possible will attempt to measure as many things as possible at two or three stations in deep and

Michael M. Mullin, Jonathan Sharp, and Ralph A. Lewin — none of them ichthyologists — examining a fish collected on Southtow Expedition, 1972. Photo by Elizabeth Venrick.

shallow water off Baja California."[21] And a number of them did so.

It was but a beginning. Members of the group took to the *T-441* for weekly monitoring of the waters off La Jolla (shades of Erik Moberg on the *Scripps* in 1936), for studies of the cycle of "red tide" organisms, for investigations of the ecology of sewer outfalls, and for photographs of plankton *in situ*. They boarded larger ships to sample the California Current, the nutrient-rich waters of Peru, and the also-rich, cold waters of the Antarctic. Members of the Food Chain Research Group have been analyzing the structure of the microplankton population in the central gyre region of the North Pacific Ocean and also the structure of the population near the San Diego sewage outfall. Some have been investigating the effects of chlorine on phytoplankton, and others are in studies of microorganisms.

In 1973 the group participated in the multi-institutional project CEPEX (Controlled Ecosystem Pollution Experiment) — the "big bag experiment." This study was designed to help forecast long-term effects of pollutants on marine life. The "laboratory" is a group of large underwater enclosures (2.5 meters by 15 meters) located in Saanich Inlet off Vancouver Island, British Columbia. In these the natural phytoplankton communities are enclosed and sampled over a long period for comparisons with the surrounding water so that effects of measured amounts of introduced pollutants may be determined.

The program of the Food Chain Research Group is "open-ended," the members feel, "with no immediate end in sight to the work that needs to be done before the population dynamics and trophodynamics of the plankton is understood to a degree that will enable man to exercise satisfactory control of the environment and make useful predictions."[22]

OTHER IMR PROJECTS

One of the very extensive projects of the Institute of Marine Resources since its inception has been the food technology researches by Harold S. Olcott and his colleagues. This program, concerned especially with the handling of marine food resources and with the changes that occur in them during processing and storage, is skipped over here because it is other-campus; from 1955 to 1970 the researchers were located at Berkeley, since 1970, at Davis.

Some ocean engineering studies, in the department of civil engineering at Berkeley, have been under IMR. API Project 51 at Scripps, described in chapter 12, was administered through IMR from 1954 until the end of the project in 1962. Douglas L. Inman's work on nearshore processes (also in chapter 12) was partly supported by IMR for some years. Theodore R. Folsom's study of the Hyperion sewage-treatment plant (see chapter 13) was handled through IMR. Various studies on the economics of marine resources have been supported through IMR, which for a few years administered funds for studies by Carl L. Hubbs on the hydrographic history of southern California and for studies by Harmon Craig on water-vapor and carbon-dioxide analyses.

The Center for Marine Affairs, established in 1970 with Warren S. Wooster as director, was located in IMR from 1972. It was set up "to involve specialists from the social sciences, government, and other fields outside of the natural sciences in consideration of the scientific aspect of marine affairs."

From 1971 IMR has been the managing agency for the University of California's Sea Grant Program, part of an all-encompassing effort under NOAA to apply research to everyday use of marine resources of all kinds. Scripps established a Sea Grant program in 1968, and two years

Institute of Marine Resources

later joined with San Diego State University in an institutional program. This was enlarged to a statewide program, and the University of California received Sea Grant college status in 1973. It has supported a large number of projects at several University of California campuses, at state university campuses, at the University of San Diego, and at Moss Landing Marine Laboratory.

One of those projects, worked out by John Isaacs and Richard Seymour, is a dynamic breakwater that uses tethered floats. Waves passing through the rows of floats cause them to oscillate out of phase with the water motion, and the drag from these oscillations removes energy from the waves, thus reducing their height. In 1974 an experimental model, consisting of 20 floats fabricated from foam-filled scrap tires tethered to a floatable platform, was installed off San Clemente Island. In 1975 a tethered-float breakwater 150 feet long and 20 feet wide was tested in San Diego Bay.

The experience and expertise of the Institute of Marine Resources has been tapped at times by the state of California. In 1965, under the leadership of M. B. Schaefer, IMR prepared a significant report entitled "California and the Use of the Ocean"[23] for the State Office of Planning. This report, edited by Roger Revelle, which drew upon a large number of University of California researchers, summarized a study "of the problems and opportunities in the utilization and development of the resources of the sea, and the ways in which the sea and the utilization of its resources interact with the growing population of the State."

NOTES

1. Letter of 7 May 1951.

2. Letter of 8 June 1951.

3. Letter of 8 January 1952.

4. Draft of committee report, 16 November 1951.

5. *San Diego Union,* 20 July 1953.

6. "On the Policies and Program of the Institute of Marine Resources," statement of September 1962.

7. "Preliminary Résumé' of Results of Investigations Conducted from 1948 to 1953 Concerning Fish Life in the Kelp Beds and Effects of Kelp Harvesting on the Fish Populations," SIO Reference 53-41 (1953).

8. Letter to University Safety Supervisor T. E. Haley, 31 October 1951.

9. Letter to Revelle, 10 October 1951.

10. "Scientific Diving," *Yale Scientific Magazine,* Vol. 31, No. 3 (December 1956).

11. Letter of 15 March 1953.

12. "Ocean-bottom Artifacts," *Masterkey,* Vol. 28, No. 6 (1955), 226.

13. James R. Moriarty, "Submarine Archeology," *Science of Man,* Vol. 1, No. 4 (June 1961), 134.

14. Manuscript in SIO Archives.

15. Eastropac Information Paper 5, 19 June 1967.

16. Eastropac Atlas 1, 1972, xii.

17. "Fishery Dynamics and Present Status of the Yellowfin Tuna Population of the Eastern Pacific Ocean," IATTC Bulletin, Vol. 12, No. 3 (1967), 91.

18. "Between Beakers and Bays," *New Scientist* (2 February 1967), 276.

19. IMR annual report, 1972, 2.

20. *Ibid.,* 1963, 27.

21. *Ibid.,* 1964, 31.

22. *Ibid.,* 1968, 16.

23. IMR Reference 65-21 (1965).

"Spring Stirring," by Donal Hord.

7. Watching Waves in Land and Sea: The Institute of Geophysics and Planetary Physics

The redwood laboratory that houses the Institute of Geophysics and Planetary Physics on the Scripps campus — and the variety of projects there carried on by researchers — are the result of the persistence of one enthusiastic man, Walter H. Munk. He was described by Chancellor Herbert F. York in 1963 as having been "unceasing in his drive and efforts to bring about the realization of his dream."[1]

As noted in chapter 2, Munk had been a prewar and postwar student of Harald U. Sverdrup. After receiving his Ph.D. in 1947, Munk became an assistant professor at Scripps. When he became professor of geophysics in 1954, Munk also became associated with the intercampus Institute of Geophysics.

That institute, which had been established in 1946, was headquartered at UCLA. From 1947 to 1961 it was directed "with vision and energy" by Louis B. Slichter, who always had a keen interest in Scripps projects, and in fact had provided support for the Scripps Institution's first major expedition, Midpac, in 1950. Upon Slichter's retirement, Nobel Laureate Willard F. Libby became state-wide director. The institute has carried out a vigorous research program within the broad field of geophysics, with such diversity as the mechanisms of flow and fracture of rocks under pressure, the mechanism of earthquakes, magnetic fields in space, cosmic rays, meteorites, and more. Libby defined geophysics as "an extremely comprehensive discipline, embracing the basic chemistry and physics of the earth," and declared that "its potential in scientific research and technological development impinges directly upon the economy of the state and the health and well-being of its citizens."[2]

On the Scripps campus during the latter 1950s, the institute's sole representative, Walter Munk, was anticipating an expansion in geophysics. He was also considering an offer from Harvard. Roger Revelle asked, "What is it that you really want to do? Why is it that you could not do this better here than at Harvard?" So Munk consulted with Slichter, with Revelle and others, and in June 1959 he presented a proposal for the university's consideration:

> The tremendous publicity of the [International Geophysical Year] and the increasing activity in rocketry as a means of investigating physical conditions between the planets for eventual space travel have made geophysics and planetary physics fashionable sciences. . . . Because of the small number of outstanding workers, the competition for good talent will be fierce until more investigators and teachers are produced. In this difficult transitional period it is essential that the University of California maintain its favorable position in geophysics by keeping and attracting first rate men and by providing them with facilities and especially with the environment in which they would be most productive.
>
> Our university's position is favorable because it already has a nucleus of capable geophysicists and an

administrational framework which has served them effectively. But we need to provide room for new blood. . . . We propose that as part of the expansion of geophysics at the University of California a branch of the Institute of Geophysics on the UCLJ [= UCSD*] campus be established.

The Institute of Geophysics at UCLA (IGUCLA) has demonstrated how an intimate group of first-rate scientists can raise the standard of an entire campus community and be successful in attracting leading young people. The principal cause of its success is of course the quality of its senior members. However, the effectiveness of most of them is increased several fold by daily association with one another, which is difficult to bring about in too large and too diverse a group. To accommodate the diversity of interests in geophysics we propose to take advantage of the statewide aspect of the Institute of Geophysics. By creating a branch of the Institute on the La Jolla campus, the Institute of Geophysics can grow without suffering from elephantiasis.

. . . The establishment [of a branch on the La Jolla campus] will benefit the expanded campus at La Jolla in the same way as it has benefited UCLA, by providing opportunities for graduate and postdoctoral research for outstanding young people.

. . . Emphasis is to be placed on appointments of young men. Large project-type research activities are to be avoided. The field of research is, of course, the concern of each senior investigator. Initially the appointments might reflect the activities of the senior investigators available for the formation of its nucleus.

---

*For a short time UCSD was known as the University of California at La Jolla.

> We suggest that it be planetary physics with initial emphasis on the Earth-Moon system.
>
> ... We thus envision an institute located near the Scripps Institution ... which by 1964 might consist of a dozen senior investigators and be about 10 per cent the size of Scripps. The combined Los Angeles and La Jolla branches of the state-wide institute would possess a faculty in Earth (and planetary) science second to none in the country.[3]

Munk's proposal, favorably received by university officials, led to the establishment of the La Jolla branch of the Institute of Geophysics in 1960. As a modest beginning, Slichter provided $2,500 for a student fellowship and $1,000 for "unrestricted use in meeting minor needs" at the new facility in its first year. Munk was appointed director of the La Jolla branch and associate director of the statewide institute, which, also in 1960, added "and Planetary Physics" to its original name. Since then the Institute of Geophysics and Planetary Physics has been referred to usually as IGPP.

Munk stated his objectives:

> We plan to study the planet Earth, its atmosphere, oceans and interior, using the methods of experimental and mathematical physics. ... We propose to form a group that is at the same time *small* and *non-specialized*, and it is this unusual combination that will make our institute distinctive. ... By our insistence to remain small we shall form a closely knit (though heterogeneous) group, requiring a minimum of administration and permitting all of us, including myself, to devote our time to teaching and research. ... Dr. Revelle has suggested twelve senior investigators as the ideal *ultimate* size.[4]

Institute of Geophysics and Planetary Physics

Within its first year the new branch of IGPP had results to show. George E. Backus joined the staff and began mathematical approaches to the problem of rotational line splitting. Richard A. Haubrich arrived, and set up a precise seismic station on Miramar Ranch, once the home of E. W. Scripps. Munk and Gordon MacDonald (of the UCLA branch) undertook a study of atmospheric tides. The direction of arrival of ocean waves was also under study, and from the records Munk was able to identify long swell that had been generated halfway around the world in the Indian Ocean. Records from earthquake-generated sea waves, tsunamis, were scrutinized; the resonant oscillations along the California coast were recorded. Visitors came, some for a few days and some for many weeks. The first of these was the already frequent visitor to Scripps, Sir Edward Bullard, who began developing a generalized program for the analysis of geophysical time series. That computer program was named BOMM for its several devisers: Bullard, Florence Oglebay, Munk, and Gaylord Miller.

The earliest members of the IGPP staff at La Jolla did not easily have "daily association with one another," as they were scattered throughout the Scripps Institution buildings. In 1960 the institution was serving as the staging area for the School of Science and Engineering and for the incipient UCSD, and everyone had office space problems. At the end of the first year of the new organization, Munk's complaint was that "we have no home." He set out to find one.

Estimates indicated that a fully equipped laboratory of appropriate size would cost about one million dollars. Because the institute appointments were to be joint ones with teaching departments, the regents of the university agreed to match funds acquired from outside sources. Munk's efforts brought in contributions of $200,000 from the National Science Foundation, $20,000 from the

Westinghouse Corporation, $20,000 from the Research Corporation, and $170,000 for scientific equipment from the Air Force Office of Scientific Research.

Meanwhile, a scenic site was found for the new building overlooking the sea, near the wooden cottages on the upper slopes of the Scripps campus. University architects commented that a building on that site could someday slide into the sea (an admittedly awkward disaster for geophysicists, who should know better), and advised against redwood, as such material would last only a century. Munk and his wife Judith, an architectural designer and sculptor, insisted on the site and the material — and won, chiefly on the leverage of the funds Munk had obtained from outside sources. Judith Munk advised on the building's design throughout the planning sessions. When the split-level building, stepping up the slope, was completed in 1963, the Munks were gratified that it had been built at the lowest cost per square foot of assignable space ($20.90) for a university laboratory in many years. The redwood building contains laboratories, offices, a machine shop, a library and reading room, and a conference or lecture room within its four levels.

To enhance the grounds of the new building, philanthropists Cecil H. and Ida Green offered to buy a sculpture that Judith Munk had long admired: "Spring Stirring," by San Diegan Donal Hord, under whom Judith Munk had studied sculpture. The massive work, carved from black diorite from Escondido, and mounted on a matching base, was installed in the patio of the building in the fall of 1963. It represents a huddled figure, partly shrouded and stirring from sleep; at its feet, as though pushing from the earth, are small sprouts. Cecil Green, equipment manufacturer and founder of Texas Instruments, noted that "this gift arises out of our very high regard for the unique ability and valuable contributions being made by Professor Munk to

The redwood laboratory of La Jolla's Institute of Geophysics and Planetary Physics, in 1969.

advanced learning and research in earth sciences."[5] The sculpture was on prominent display at the dedication of the building on 26 February 1964, at which Leland Haworth, director of the contributing National Science Foundation, was the principal speaker. In 1972, Mr. and Mrs. Green established a foundation to promote scientific and educational projects at IGPP, the funds from which have been used each year to bring a visiting scientist to the laboratory as a lecturer-researcher.

Within its first few years the new laboratory reached the size advocated by Revelle and Munk, about twelve faculty appointments. As stipulated by the university regents for an institute, faculty appointments in IGPP are joint ones with a teaching department of the university. Because of the proximity and similarity of interests, many of the appointments at the La Jolla branch of IGPP are jointly with Scripps Institution. The remainder are with the upper-campus teaching departments of applied mechanics and engineering sciences (AMES) and physics. In addition, several Scripps researchers not officially part of IGPP, but whose studies are related, have offices in the IGPP building. Carl Eckart was one of these for several years prior to his death in 1973. A number of graduate students, working under the faculty members in their respective departments, have office and laboratory space there also.

To all intents and purposes, the La Jolla branch of IGPP operates as a laboratory within the Scripps Institution. The geophysical studies carried out in the redwood laboratory are selected by the senior scientists. Under the theme that "land geophysics and ocean geophysics are part of the same subject," these studies have been aimed at both the restless ocean and the quivering earth. With ships and equipment, ashore and afloat, for fifteen years this group of geophysicists has been seeking answers to puzzles from the surface of the sea to the core of the earth.

Institute of Geophysics and Planetary Physics

Walter Munk's own studies have been chiefly concerned with the ocean in motion: its waves and its tides. As noted in chapter 2, during World War II he had devised the means of forecasting sea and swell for amphibious landings. After the war he continued with studies of ocean waves. Using records from wave meters installed outward from the pier, he worked out the relationships for computing the travel time and distance of a storm. He also developed equations for estimating the forces and motions exerted by waves on vertical beams. Munk's colleague, Frank E. Snodgrass, an ingenious engineer, designed sensitive pressure meters to track wave trains across the ocean. In 1963 a series of six tide stations was set up from Alaska to New Zealand, five of them on islands and the sixth on *Flip,* on station northeast of Hawaii. Munk and his family lived for three months in a native village in American Samoa, tending one of the stations. From the array of stations Munk could trace waves generated by storms in southern waters as they moved northward, to reach the northernmost station two weeks after their origin.

For studies of tides in the deep ocean, Snodgrass in the early 1960s designed an instrument capsule in the form of a pair of aluminum spheres that could be dropped free-fall to the sea floor, where it could be left for as long as several weeks and could be recalled by an acoustic command that causes the release of a link to the weighting storage batteries. On magnetic tape in the capsules could be recorded pressure fluctuations as small as the equivalent of one one-hundredth of an inch in sea level at a depth of three miles, temperatures to a resolution of a few millionths of a degree, and tidal currents as slow as ten feet per hour. The first tide capsule, launched in 1965, was christened Judith, at the suggestion of her husband; each succeeding one was christened by one of the project's participants for his wife. Despite many early problems with the capsules and their release systems, only one was lost: Dottie, named for Snodgrass's wife, which vanished after its launching late in 1965. Over a period of several years the tide capsules were placed on the sea floor at approximately fifty locations off the continental shelf of California to determine the pattern of tidal variations in the northeast Pacific Ocean and to define the properties of the boundary layer at the bottom of the deep ocean. For the launching and retrieval of the units, the *Ellen B. Scripps,* with her capacity for handling equipment vans, proved especially useful.

In 1969 Snodgrass boarded the National Science Foundation research ship *Eltanin* to set out three tide capsules at depths of 12,000 to 18,000 feet in the Antarctic Ocean, an area in which the tidal configuration was unknown. Munk joined Snodgrass during the Antarctic summer to participate in the successful recovery of the instrument packages. Analysis of the Antarctic data indicated that the tides generated in that ocean did not create as high tides in mid-ocean as had been previously guessed.

An international survey of deep-sea tides was proposed by Munk in 1967, but it was indefinitely postponed because the complex instrument capability was not that nearly ready. Munk's comment to science writer Daniel Behrman was: "You know, there are times I think we may have started too soon and gone too quickly on the deep-sea tide survey. But we have time. At the rate that tidal friction is changing the configuration of the solar system, we have a billion years to finish our measurements before something happens."[6]

The deep-sea tide measurements by Munk and Snodgrass ended in 1971, and the capsules were modified for studies of internal waves in the ocean. One more capsule was built then and placed on the sea floor for a year for a long-term record of low-frequency fluctuations of pressure, temperature, and water velocity.

Four of the free-fall capsules used during the latter 1960s by the Institute of Geophysics and Planetary Physics for deep-sea tide measurements.

Scripps Institution of Oceanography: Probing the Oceans

Also concerned with the characteristics of moving water was John W. Miles, who joined the La Jolla IGPP in 1963 from UCLA, and turned to making waves. By means of hot-wire anemometers, he measured the disturbances in the airstream above waves generated in the wind-water tunnel in the Hydraulics Laboratory, in order to determine the transfer of energy from wind to waves. These measurements were compared satisfactorily with theoretical predictions. Miles also undertook calculations on rotating flows of liquids in which Coriolis forces are dominant, and on the resonant response of harbors and bays to earthquake-generated tsunamis.

Microseisms — the minute, continuous quiverings of the earth — have been recorded on seismographs and pondered by geophysicists for years. At the Scripps IGPP Hugh Bradner in the mid-1960s developed a free-fall seismometer to measure microseisms directly on the sea floor. Also interested in these tiny tremors has been Richard A. Haubrich, who in 1970 determined that some microseisms are caused by traveling storms at sea.

Much larger earth motion — that is, earthquakes — became a major study at the La Jolla branch of IGPP during the late 1960s. In 1969 and 1970, following the installation of the seismic station by Haubrich and others at Miramar Ranch, a station was set up on Navy land at Camp Elliott, twelve miles inland from the institute, by Ralph Lovberg and Jonathan Berger. At this recording station a laser-interferometer strain meter capable of measuring the compression in rocks down to one part in a billion was installed, to record strain, tilt, and vertical movement caused by earthquakes, earth tides, ocean and atmospheric loading, or nuclear explosions. In the early 1970s another complete geophysical station was installed, under Berger's direction, at Piñon Flat in San Bernardino National Forest, near the San Jacinto and San Andreas

Institute of Geophysics and Planetary Physics

fault systems. This observatory includes a three-component laser strain meter, three gravimeters (superconducting, La Coste, and quartz-fibre), an array of eight tiltmeters, and three-component long-period seismometers. Recorders installed in 1973 in the IGPP laboratory provide a visual record of earthquakes recorded at the remote stations.

Seismologist James N. Brune, who joined the Scripps staff from Caltech in 1969, was provided with an office in the IGPP building, and he became a participant in the earth-motion studies. Brune also became an associate director of IGPP in 1973. His interest has been in the total system of earth strain in the general southern California region, from the San Andreas fault southward through the Gulf of California to the East Pacific Rise.

The chain of undersea mountains known as the East Pacific Rise, extending from Antarctica to the edge of the North American continent, is one of the active spreading centers of the world. From its motion, the peninsula of Baja California is being pushed away and northward from the mainland of Mexico. As Brune said, "What happens in the Gulf of California determines what happens in Southern California." Brune enlisted the cooperation of the University of Mexico in setting up a network of seismograph stations around the Gulf of California, to try to determine the total input of energy into the fault system by the relative movements of the North American and Pacific tectonic plates.

Brune and his associates also prepared an array of portable seismic recorders to transport to the site of earthquakes to record the aftershocks. Brune, Hugh Bradner, and William A. Prothero set up a system of recording earthquakes on the floor of the ocean by means of ocean-bottom seismometers and sonobuoys. These have been deployed both in the Gulf of California and on the crest of the East Pacific Rise about 150 miles south of the tip of Baja California.

Other researchers at the Scripps IGPP have turned their attention to the nature of the solid earth itself, its internal constitution and its oscillations. George Backus, the first appointee to the branch institute, has pondered the geophysical inverse problem: Given the frequencies of the earth's normal modes, what can be inferred about the interior distribution of density and the elastic contents? Freeman Gilbert, who joined the IGPP group in 1961, applied the inverse theory to a study of the earth's normal modes, the mechanical structure of the earth, and the source mechanism of deep earthquakes. Barry Block and Robert D. Moore developed a low-frequency accelerometer to record the normal modes of the earth and to determine the threshold at which surface waves can be detected.

Richard A. Haubrich measured the tilt of the earth by ocean loading and investigated possible causes of the wobble of the earth. He could find no correlation of the wobble with major earthquakes, as had been proposed at times.

Robert L. Parker, who arrived in 1967, set out to determine the electrical conductivity deep within the earth from measurements of the slow variations of the geomagnetic field. Also, working in cooperation with the Deep Tow group of the Marine Physical Laboratory, he investigated the direction of magnetization of seamounts to determine their drag by tectonic movement along the sea floor. To speed up the time required for calculating great quantities of data on magnetism from intensive surveys, Parker developed a method of using Fourier transforms for handling the data. His general computer program, called Supermap, for plotting worldwide geophysical data using any conceivable projection, has proved useful for establishing the relative motion between two crustal plates.

Parker got drawn into an extracurricular project in 1967 while sharing bachelor quarters with a vigorous coffee-stirrer. Continued observations and experiments resulted

Institute of Geophysics and Planetary Physics

in a paper by W. E. Farrell, D. P. McKenzie, and Parker, entitled "On the Note Emitted from a Mug While Mixing Instant Coffee."[7] The group observed that "If the bottom of the mug was tapped repeatedly with the spoon as the powder was stirred into the water, the note emitted could be heard to rise in pitch by over an octave . . . in a matter of seconds." They concluded that bubbles trapped on the powder and released into the water created the change in tone. The manuscript was first submitted to *Nature,* which rejected it, and returned the pages — marred by coffee-stained rings.

A regular visitor to the La Jolla IGPP since its establishment has been Sir Edward ("Teddy") Bullard, whose name, said Willard F. Libby, is one "to conjure with throughout the world of geophysics."[8] Bullard entered that world in its infancy when the few participants built, hauled, and repaired their own sometimes hazardous recording devices. "I migrated from physics to geophysics in 1931," he wrote, "and spent some years learning the techniques of applying physics to the earth and trying to understand the modes of thought of geology. My initial idea was that geophysics should be used to solve specific geological problems, the paradigm being the applications to prospecting for oil. Gradually I realized that, important as such applications were, I was more interested in major problems of earth structure and history."[9]

In 1949, on a visit to Scripps, Bullard developed the heat-flow probe with Arthur E. Maxwell (see chapter 15). A graduate of Cambridge University, Bullard was assistant director of Naval Operational Research for his native England in the latter years of World War II. He was professor of physics at the University of Toronto in 1948-49, and then director of the National Physical Laboratory in England until 1956, when he became assistant director of research at the Department of Geodesy and Geophysics at Cambridge University and in 1964, professor, until his retirement in 1974. For some years, Bullard has customarily spent three months a year at the La Jolla IGPP — where he has an office whose door is adorned with a remarkably informal photograph of himself. Throughout the years, whatever his location, he has "played a part in the transformation of a backwater into a bandwagon" — and has enlivened geophysical discussions with ingenious ideas and spicy anecdotes.

NOTES

1. Letter to Cecil H. Green, 27 November 1963.

2. Director's Report of IGPP, 1965-70, 3.

3. Original proposal, 2 June 1959.

4. Proposal to Max C. Fleischmann Foundation of Nevada, January 1960, 1.

5. Letter to University Vice President Thomas J. Cunningham, 14 November 1963.

6. Daniel Behrman, *The New World of the Oceans* (Boston and Toronto: Little, Brown, & Co., 1969), 138.

7. *Proceedings of the Cambridge Philosophical Society,* Vol. 65 (1969), 365-67.

8. Letter to University President Clark Kerr, 30 September 1964.

9. "The Emergence of Plate Tectonics: A Personal View," *Annual Review of Earth and Planetary Sciences,* Vol. 3 (1975), 8.

Harbor seal in one of the research pools of the Physiological Research Laboratory.

8. A Different Point of View: The Physiological Research Laboratory

Per F. ("Pete") Scholander brought an idea when he arrived at Scripps Institution in 1958 as a professor of physiology:

> For several years I was able to watch the progress of oceanographic biology at Woods Hole and other places, and I always had the strong feeling that this discipline is greatly in need of a vigorous experimental approach through physiology and biochemistry.[1]

To Scholander the "vigorous experimental approach" required a laboratory and a ship. The ship, which to him was to be a seagoing laboratory, must have "space and sophisticated equipment that would permit the study of physiological phenomena in live specimens at or in their natural habitats."

In 1962 the National Science Foundation agreed to provide money for a building, for adjacent pool facilities, and for a laboratory ship. NSF guaranteed the ship-operating costs, and the National Institute of Health guaranteed funds for operating the laboratory for seven years.

With these generous funds promised, the Physiological Research Laboratory was established in 1963, with Scholander as director. Its four chief projects were to be: "cardiovascular and respiratory research on large marine vertebrates and 'aquatic' man; studies of various transport mechanisms on cellular and organistic levels; neurophysiological studies, in collaboration with UCLA; and behavioral studies of marine mammals, also in collaboration with UCLA."[2] The new laboratory was expected to form a bridge between physiologists at Scripps and biologists at the new School of Science and Engineering of UCSD and the medical school to be established at UCSD.

"Because of the extent, uniqueness and international scope of the operation," said Roger Revelle, "I am appointing an advisory board [named the National Advisory Board at the request of NSF] of distinguished scientists from outside the University to advise Professor Scholander and me on the scientific program and optimum use of the facilities."[3] A. Baird Hastings was appointed the first chairman of the advisory board; a professor emeritus from Harvard, Hastings was then with the Scripps Clinic and Research Foundation in La Jolla and also a Research Associate in the UCSD Medical School with an office at Scripps Institution. It was not only a wise appointment, it was also just, for Hastings had advocated the need for an advisory board to Revelle on the basis that one person – Scholander – could not be both ashore and afloat simultaneously, attending to everything. The other distinguished members of the National Advisory Board were Eric G. Ball (Harvard), Lawrence R. Blinks (Stanford), Theodore H. Bullock (then UCLA), Wallace O. Fenn (University of Rochester), Knut Schmidt-Nielsen (Duke University), and H. Burr Steinbach (University of Chicago).

To staff the new research unit, PRL, Scholander wanted "persons with proven accomplishments and perspective in the basic science of physiology, who see in the marine environment a unique opportunity to penetrate further into basic biology."[4] He felt that five academic members would be the minimum necessary to create an "intellectually sustainable" group. The first research appointments to PRL, besides Scholander, were Theodore Enns, Robert Elsner, Dean Franklin, and Edvard Hemmingsen. It was expected that the group would quickly double in size, but it has not done so.

Scholander had found early support for his goals in the staff of the Brain Research Institute at UCLA. In fact, they obtained the funds that made possible a third story on the proposed building and additional pool facilities. The title of Marine Neurobiology Facility was given to the UCLA laboratories that occupied the third floor. Susumu Hagiwara from UCLA transferred into these quarters as soon as the building was completed, to be in charge of the Marine Neurobiology Facility and to be Associate Director of PRL. He was recruited as a full professor on the faculty and was the first appointee under the unique arrangement whereby the UCSD Medical School supports certain Scripps faculty members in marine biomedicine. The Marine Neurobiology Facility has operated from the beginning as a joint facility between Scripps and the Brain Research Institute; it is directed by a Scripps faculty member who is also a member of the Brain Research Institute. Since 1969 Theodore H. Bullock has headed the unit.

The Physiological Research Laboratory building (sometimes jokingly called "Per's Porpoise Palace") was completed in June 1965, just seaward of the Aquarium-Museum. Alongside it was built a doughnut-shaped pool, ten meters in diameter, equipped with a trolley for instruments and a center island containing a wet laboratory and a dry laboratory. A round pool and a rectangular one for holding aquatic animals were also constructed, as well as a machine shop. The pools have intermittently housed sea lions, seals, and even penguins. For non-marine physiological studies, the Comparative Cardiovascular Physiology Facility — more often called just "the farm" — was also set up in 1965 in Seaweed Canyon, just below radio station WWD. The Scripps Clinic and Research Foundation contributed $5,000 toward this facility, which has housed sheep, dogs, and horses for various studies.*

Finding a suitable research ship was its own problem. Scholander first explored the possibilities of converting an available Navy yard freighter into a laboratory vessel. While on a visit to Norway, he concluded that a Norwegian trawler was the type of ship most suitable for his purposes. He brought back ship plans, and then found that American shipyards could not build the vessel as inexpensively as the home country. An American version of the ship was designed by Lawrence M. Glosten and Associates, and J. M. Martinac Shipbuilding Company of Tacoma, Washington, was selected as the builder.

Some referred to the proposed ship as "Schoboat." Scholander almost named her "Caprice," then "Baluga," then "Proteus," but before the launching he had settled on *Alpha Helix* in honor of the helical configuration of protein molecules.

The launching of the *Alpha Helix* was unique. Directly afterward, Baird Hastings, who had discovered on arrival that he was chairman of the launching ceremony, recounted the tale:

*The highly regarded Scripps Clinic in La Jolla and the widely known Scripps Institution have often been confused with each other. One story relates that a doctor arrived at the San Diego airport and asked a taxi driver to take him to "Scripps." He was delivered to the front door of old (and deteriorating) Ritter Hall on the campus. After walking past a musty aquarium room, down a dingy hallway, and through a lingering odor of formalin, he stopped short and exclaimed: "This *can't* be a hospital!"

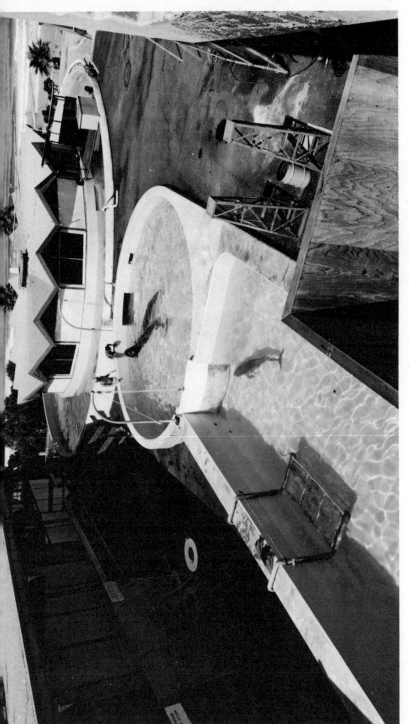

The pools used for studies for marine mammals by the Physiological Research Laboratory.

She was conceived in 1959, she was designed in 1962, she was funded in 1963, she was constructed between December, 1964 and June, 1965, she was scheduled to be launched and christened the Research Ship R/S "Alpha Helix" at 6:30 P.M., June 29 — but at 5:30 P.M. she took matters into her own hands and launched herself — practically unobserved. She did it expertly, safely and expeditiously.

At 5:25, Professor Per ("Pete") Scholander, Mrs. Susan Scholander, Mrs. Grey Dimond, Isp (a Spitz) and I arrived in Scholander's car at the gates of the J. M. Martinac Shipbuilding Company in Tacoma. I stopped to take a picture of the two ladies while Pete went to find a way into the shipyard. As we entered through the front gate, there was a resounding noise — part crash, part thump — from the huge shed that housed our unlaunched lady, a new type of sea-going laboratory.

Almost immediately Mr. Glosten, the designer, appeared running from the shed toward the office, blanched and shouting, "It's a catastrophe!" We ran toward the shed. I scurried down its east side to the waterfront just in time to see the "Alpha Helix" slide quietly, on even keel, with unpardonable pride and unperturbed dignity, across the river toward the other bank.

A small tug, which had been steaming by, had responded to Mrs. Glosten's timely call to "Get a line aboard her!" and took the lady in tow. With some coaxing, he persuaded her to come with him, and with gentle prods and pulls he brought her back to our side of the river. There, with the help of professional hands, who had arrived from nowhere, and two larger tugs, the as-yet-unchristened, newest lady of the seas was secured alongside a more sedately launched sister ship of the Martinac yard.

Meanwhile, her shattered cradle — something akin to her placenta — had been recoverèd from the river by three men in an oversized skiff. They had come running, jumped into the boat and were at their job of clearing the river of flotsam in what seemed one motion. I couldn't hear their words, but from their expression and relieved laughter, I could tell that they were thankful that they had not been working under her when she took it into her head to be born.

It was now six o'clock, still half an hour before the scheduled launching time, but Mr. Martinac, many of his staff and yard personnel, as well as early arrivals for the ceremony had arrived. Though all were surprised and somewhat shaken by the turn of events, the mood was one of gratitude for the absence of injuries or damage which had accompanied the self-executed launching. With efficiency, the program was revised from "a Christening and Launching of the Alpha Helix" to a program of "Christening, after Launching, of the Alpha Helix."

Speeches were shortened and revised to fit the new circumstances and Susan Scholander, the sponsor, with the platform party, was conducted to the bow of a large tug, which was maneuvered to the bow of the now quiet and high-floating lady. There, Susan with "I do christen thee Alpha Helix" anointed the well-reinforced prow and herself with champagne from an expertly shattered bottle. Pete and I managed to get a few drips into our hands and thence to our mouths so that we could henceforth say that we shared her christening carbon dioxide.

Thus ended an unpredicted incident in the birth of the "Alpha Helix." Since she thus began her marine living in the manner of the Prince of Serendip, may her voyages be in like spirit — unexpected adventures and happy endings.[5]

Scholander consoled a much-upset Mr. Martinac with: "Everybody expects a few irregularities in matters in which I am involved, and we all think the launching was charmingly distinctive."[6]

The ubiquitous Captain James L. Faughn took on the task of supervising the ship's completion at the shipyard, from seeing that the lab floors were level to checking on the painting and the engines, and the installation of laboratory equipment and ship's gear. Finally, on 26 February 1966, Faughn and a Scripps crew, accompanied by shipyard personnel, took the ship to sea for trials:

> About five miles seaward of Tatoosh Island [Washington, wrote Faughn in his daily log] slowed to steerage way on cse. due west, then together with Mr. Skewis [chief engineer of Martinac Shipbuilding Company] and Joe Martinac [president of the company] went over documents for signature and those to be presented to the vessel by the builders. At 1547 zone +8 time and with final signatures on transfer papers, took delivery of the vessel from the builder in Latitude 48°26.7′ N Long. 124°57.3′ W, depth of water $\sim$110 ftm, about nine miles W X N of Tatoosh Island.*
>
> . . . Because of a slight attack of Mal de Mer on part of Mr. Martinac no ceremonies were held at time of delivery. On way back in to port, Joe turned in on my bunk and Ed turned in on Ch. Scientists bunk.

---

*The transfer at sea, outside territorial waters, was arranged in order to avoid the requirement ashore of paying Washington state sales tax.

Physiological Research Laboratory

Scholander called the vessel a floating laboratory. Its purpose was to provide laboratory facilities for physiological studies where the animals themselves were, as previous experience had shown that most marine animals could not be carried alive for very long. The 133-foot ship was — is — equipped with a main laboratory, 24 by 26 feet, on the main deck, an electrophysiological laboratory below deck, a freeze laboratory, photographic darkroom, and a machine shop. The vessel can carry a scientific party of ten, members of which are rotated by air to and from the laboratory's location. The ship is designated a national facility, to be used by scientists throughout the country and abroad, when their projects have been approved by the Alpha Helix Review Committee, successor to the National Advisory Board.

On 11 March 1966, the jaunty *Alpha Helix* shared dedication ceremonies with the *Thomas Washington* and the Chester W. Nimitz Marine Facilities. Eight days later the *Alpha Helix* departed on Billabong Expedition on Australia's Great Barrier Reef. There, alongside a bit of land dubbed "Botany Spit" in the Flinders Islands group, she served as laboratory and support base for six months for 44 scientists who came from around the world to analyze and study the corals and the mangroves, the ghost crabs and mud crabs, the python and the skink, mudskippers and pufferfish, giant clams, dugongs, crocodiles, and many more denizens and growths of the stunning reef region. On shore the researchers set up a small village, in the form of two six-man tents, a tent messhall and storage area, and laboratory facilities. In spite of the land comforts, the ship and her crew represented life to the expedition participants, Captain Faughn concluded, "and a ready source of general assistance which even our most rugged staff are reluctant to let out of sight."[7] The ship provided fresh water from its evaporator, and electrical power for the camp facilities. Her

Scripps Institution of Oceanography: Probing the Oceans

smallboats *Waltzing Matilda* and *Serendip* transported the scientists up and down the nearby coast and mudflats for collecting forays. The ship's laboratories were also extensively used, and the fantail was quickly littered with aquariums, from which such occupants as moray eels occasionally escaped. Captain Faughn spent many an evening dipnetting squid and salps and small fishes and other creatures from alongside the ship, so expertly with a borrowed fine-mesh dipnet that it was finally ceremoniously presented to him. The gastronomic fare was as diverse as the subjects of experiments, for many of the creatures under study went into the pot or barbecue pit when physiological studies were completed. Crabs and assorted fish were special favorites, certainly more appreciated than the roast python. Hunters in the group occasionally added wild pigs, wild goats, and ducks to the menus, especially for festive occasions, which were also enlivened by songfests.

The *Alpha Helix* was built for all climes, and she has visited the extremes and the in-betweens. After Billabong Expedition in 1966, she sailed the next year to Brazil and twelve hundred miles up the Amazon River to a jungle base camp for seven months. Among the subjects scrutinized there by 86 scientists from 12 countries were boa constrictors, electric eels, piranhas, cicadas, sloths, freshwater dolphins, tropical plants, and more mangroves. En route home the ship paused at the Galápagos Islands for collecting and field work, especially on fishes.

The 1968 trip was a six-month voyage to the Bering Sea, where, accompanied by the Coast Guard icebreaker *Northwind,* the *Alpha Helix* worked her way through light and heavy ice, amidst seals, walruses, and sea birds until firm ice forced her back to open water. "There," wrote Scholander and Laurence Irving (his father-in-law, another noted physiologist), "we were removed from the concentration of life prevalent within the Bering Sea ice, and the

The *Alpha Helix* on station in the Antarctic in 1971. Photo by Edvard A. Hemmingsen.

ship's lively motion precluded analytical procedures."[8] During the Bering Sea Expedition, various participants studied sea birds, walruses, seals, sea otters, reindeer, eel pouts, crabs, musk ox, salmon, and more.

In 1969, *Alpha Helix* turned to the tropics again, this time to New Guinea, "for broad investigations into the comparative physiology and behavior of a wide variety of mammals, birds, and fishes; the comparative biochemistry of proteins in vertebrates; and the phenomenon of bioluminescence, the heatless light produced by fireflies, fungi, certain fish, and other organisms."[9] Alongside the ship a comfortable shore camp was established at Maiwara Mission, ten miles north of Madang.

Shortage of funds for operating the *Alpha Helix* and its program for visiting scientists necessitated cutting short the New Guinea expedition, and the ship returned home. She was used for local trips until the latter part of 1970, when she sailed for Central America and again, the Galápagos. In 1971 the sturdy ship headed south, to the Palmer Peninsula in Antarctica for two months of icy summer.

The National Science Foundation in 1972 changed the funding system for the *Alpha Helix* to one of single grants, to be selected by the advisory board and put together into a unified expedition. A trawl winch and an A-frame were added to the ship in 1970 for additional oceanographic studies, which allowed for a greater variety of programs. In 1973, for example, the *Alpha Helix* again went north to the Bering and Chukchi seas, for investigations on marine mammals, then turned south to Hawaii for work on deep-sea fishes, and en route home was used for plankton studies.

Since early 1973 the ship has been under the auspices of the University-National Oceanographic Laboratory System (UNOLS), and the National Advisory Board was reconstituted as a UNOLS Review Committee. The complex coordination of the *Alpha Helix* expeditions became

Physiological Research Laboratory

the task of program manager Walter F. Garey at PRL for several years.

The researches carried out in connection with the *Alpha Helix* have involved hundreds of scientists from the United States and from many other countries. The work is almost entirely biological, with the emphasis on physiology. Among the demands on an *Alpha Helix* scientist is that he must present an abstract of his research accomplishments before leaving the ship. This has been obeyed faithfully, and the printed outcome is impressive.

> A large and important scientific literature has resulted from the research conducted aboard the Alpha Helix [noted Garey]. Not directly reflected by the publications, however, are the significant concomitant personal learnings and appreciations gained through the inquiry and cross-fertilization processes; as participants, commonly representing different disciplines, work and live together in a wholly stimulating scientific and intellectual environment free from the intrusions of telephone calls and administrative concerns.[10]

PRL is unique. The spark for the unit has always been Per Scholander, a man whose interests and enthusiasm are boundless. Among other accomplishments, he "sings Scandinavian drinking songs with verve and plays the violin with sensitivity."[11] Born in Örebro, Sweden, on 29 November 1905 to a Norwegian mother and a Swedish father, he earned both an M.D. (1932) and a Ph.D. (1934, in botany) at the University of Oslo, where he continued as a research fellow until 1939. Scholander then came to the United States as a Rockefeller Fellow at Swarthmore College. After wartime service in the Air Corps (from captain to major) and service as an aviation physiologist until 1946, he returned to Swarthmore. From 1949 to 1951 he was a research fellow, with A. Baird Hastings, at the Harvard Medical School, from 1952 to 1955 physiologist at Woods Hole Oceanographic Institution, and from 1955 to 1958 professor of physiology at the University of Oslo. From there he moved to Scripps Institution.

Scholander's tendency is to study "situations where life hangs by a thread."[12] He began with lichens in Greenland, went on to the respiration of diving mammals and birds, then to that of sloths and man. Recording and analyzing adaptations to adverse conditions have long appealed to Scholander, and have led him from the tropics to the poles, from the Andes to the beach. While at Swarthmore he developed an inexpensive gas analyzer that became standard throughout the country; at Harvard he developed the Scholander respiration method, a simple, accurate, and economical means for measuring respiration in small samples; the following year he devised a means of measuring the oxygen consumption of single cells during division — a feat never previously accomplished. This device was dubbed Scholander's Bubbleometer.

Scholander's interests are catholic. "During the year 1959," for example, Hastings noted, Scholander "worked on such diverse problems as dating the age of Greenland's glacial ice, the bradycardia that accompanies diving, the reason that fish in subzero sea water do not freeze, and how the blood pigment hemoglobin facilitates the passage of oxygen through membranes."[13] At that time Scholander found his office at Scripps "so full of secretaries [two] and alligators [one large one, in a bathtub] that he had no room to do his work," so he availed himself of space in Hastings's new laboratory at the Scripps Clinic and Research Foundation for his hemoglobin project. At the same time he and Hastings had the opportunity to discuss the laboratory that Scholander envisioned.

Physiological Research Laboratory

Plants and animals have received equal time from Scholander. One of his longterm studies has been on the mechanism of osmosis in mangrove trees as they thrive knee-deep in salt water. His chief interests are in how living systems function, and his philosophy is that "there is much to be learned *from* the ocean, not only *about* the ocean." He has yearned to devise a means of getting to the ocean depths himself to determine how abyssal creatures manage to survive there.

The work of other staff members at PRL has also ranged throughout physiology. Robert Elsner for some years concentrated on diving physiology and asphyxia in marine mammals, man, and pregnant sheep. Gerald Kooyman has also worked on respiratory adaptations of deep-diving animals — birds, sea turtles, sea snakes, and various mammals, particularly the Weddell seal of Antarctica. Harold T. Hammel has studied regulation of body temperatures in various vertebrates — seals, penguins, dogs, and turtles — and has studied osmosis as well. Edvard A. Hemmingsen has investigated the respiratory and circulatory physiology of fishes, and the properties of gases in solution, in an effort to understand the mechanism of the formation of bubbles during certain conditions of decompression. A. Aristides ("Art") Yayanos has concerned himself with the effects of temperature and pressure on amino acid molecules; he devised a collecting device that can maintain specimens in their native pressure and temperature, including bacteria from 7,000-meter depth. Arthur L. DeVries has studied the physiology of certain fishes, especially those of the Antarctic, which are capable of resisting freezing in waters below the freezing point.

Human subjects for PRL's studies are volunteers, who are persuaded that their metabolic reactions in time of stress are of significance to science. Thus, graduate students held their faces under water long enough for PRL researchers to determine that the heart rate slowed and the blood circulated chiefly from heart to brain and less to the extremities.

Early in 1970 a committee was appointed to consider PRL's role at Scripps, and to find a replacement for Scholander, who was about to leave on sabbatical and wanted to be relieved as director. Andrew A. Benson was selected as the new director of PRL. The committee learned that some at Scripps felt that PRL's work was not sufficiently marine-oriented and that there was a lack of communication between the laboratory and the rest of the institution. They also found the Physiological Research Laboratory to be "a real asset . . . both to S.I.O. and to the world of science"; they complimented the laboratory on the high quality and the impressive number of its publications. PRL's physiological studies and its rapport with the UCSD Medical School have continued apace.

NOTES

1. Memorandum to Acting Director Fred N. Spiess, 17 October 1962.

2. Announcement to Staff Council by Roger Revelle, 1 August 1963.

3. *Ibid.*

4. Memorandum to Acting Director Fred N. Spiess, 17 October 1962.

5. Memorandum of 2 July 1965.

6. Letter of 9 July 1965.

7. Journal entry of 14 August 1966.

8. "Introduction," R/V Alpha Helix Bering Sea Expedition, 1969.

9. SIO Annual Report, 1970, 17.

10. "The *Alpha Helix* Kona Expedition," *Comparative Biochemistry and Physiology,* Vol. 52B (15 September 1975), 5.

11. Mary Harrington Hall, "Inside UCSD. Part II. Scripps: Where the Four Winds Blow," *San Diego Magazine,* Vol. 16, No. 7 (May 1964), 46.

12. Daniel Behrman, *The New World of the Oceans* (Boston and Toronto: Little, Brown, & Co., 1969), 25.

13. Manuscript in SIO Archives.

Lionfish on display in the Scripps Aquarium — and those spines are poisonous.

9. "Let's Visit Scripps": The Thomas Wayland Vaughan Aquarium-Museum

A public aquarium has been a feature of the Scripps Institution since before it was an institution, because the members of William E. Ritter's summer study sessions, which began in 1892, always had a few display tanks for interested visitors. When the Marine Biological Association of San Diego set out to establish Scripps Institution in 1903, they listed as one of their objectives: "to build and maintain a public aquarium and museum."[1]

In 1905, in the "little green laboratory behind the bath house" in La Jolla Cove park, a few shelves were set aside for a museum display, and a central counter held open containers of live specimens (some of which vanished with visitors). Five years later the public aquarium was located on the ground floor of the institution's first permanent building, the George H. Scripps Laboratory.

In 1915 a separate public aquarium was built, a wooden building 24 feet by 48 feet, just north of Scripps Laboratory; it held 19 aquarium tanks, in capacities from 96 to 228 gallons. The following year the museum, previously located on the second floor of Scripps Laboratory, was moved into the ground floor of the newly completed library building. The primary purpose of the museum was to "exhibit as large a part as possible of the local fauna,"[2] and, if funds could be secured, it was intended that exhibits of ocean research would be displayed. The curator of both aquarium and museum was Percy S. Barnhart, who had joined Scripps in 1914 from the Venice (California) aquarium operated by the University of Southern California.

Barnhart's usual technique of acquiring exhibits for the aquarium was by fishing from the pier. In the 1930s some specimens were gathered from the *Scripps* and the *E. W. Scripps,* and, during World War II when the institution had no ship, Barnhart kept a trap set off the end of the pier for gathering new material.

As early as 1925 the curator was complaining, in his annual reports, that the aquarium tanks were gradually disintegrating, and that "bad water, cracked glasses and broken tanks [were] a constant source of worry and aggravation." He pointed out that the wooden structure had been intended only as a temporary site for the aquarium. In 1931 he added that the museum location would soon be required by the library.

So — while repairing tanks, making plaster casts of some large fishes and skin mounts of others, storing the biological collections, walking through the aquarium at nine each night and at five each morning to check for leaks and other problems, building new shelving, writing a book on the fishes of southern California, and answering visitors' questions — Barnhart also dreamed of the ideal aquarium-museum. His vision was a building in which brightly lighted display tanks would form a periphery around a central museum room. That vision became the design theme of the new aquarium-museum.

Barnhart retired in 1946, while his dream was still only in the planning stage, and Sam Hinton, who had joined the staff a few months earlier, became the next curator.

Thomas Wayland Vaughan Aquarium-Museum

Hinton is a versatile person: folk singer, illustrator and artist, reptile fancier, curator (versatile there too, for one post was with a desert museum and the other with the marine aquarium), and helpful adviser to high school students. His first job in San Diego was as editor and illustrator at the University of California Division of War Research (see chapter 2). In his 18 years at Scripps, besides handling the aquarium duties, Hinton answered the endless questions of visitors, served as the public information office for news media, and designed the lighted exhibit explanations above the aquarium tanks. He also drew many a certificate for a Scripps expedition, equator crossing, or any other special occasion. On outside time he became a nationally renowned folk singer, and he has livened many local occasions with folk songs from around the world and with an occasional ditty of his own, all cleverly presented.

Hinton began at his new post in 1946 with planning the new building. This led him into "meetings with the University architects and engineers, extensive correspondence with museum people in all parts of the country, research in the literature on the subject, and direct observation of the behavior of visitors in the present museum and aquarium."[3] Hinton also collected specimens from the *E. W. Scripps,* and he obtained others from local commercial fishermen. He and Claude Palmer, the only other aquarium employee then, gathered sand crabs and red worms to supplement the fish purchased to feed to the aquarium inhabitants. When the Pacific Division of the American Association for the Advancement of Science met in San Diego in 1947, Hinton gathered enough fresh fish for a fish fry, and he directed a grunion hunt for the group afterward.

The new aquarium-museum building was completed and occupied in October 1950. Its dedication was hailed as a major occasion and was set for the University's Charter

Scripps Institution of Oceanography: Probing the Oceans

Day celebration in March 1951. Denis L. Fox pronounced the light green concrete building "a model of sound construction, beauty and dignity," in a letter to T. Wayland Vaughan, second director of the institution, in whose honor the building was named. Vaughan, who had retired to Washington, D.C., in 1936, was unable to attend the ceremony, but he sent a recorded message to be played during the program. University President Robert G. Sproul spoke in praise of Scripps Institution's wartime contributions in oceanography and the accomplishments of its sardine studies, the two-year-old Marine Life Research program. Detlev Bronk, president of Johns Hopkins University, of the National Academy of Sciences, and of the American Association for the Advancement of Science, reminded the largest crowd ever gathered on the Scripps campus until that time that "the ocean is a natural, if not a unique, focus for many fields of learning."[4] George F. McEwen, Denis L. Fox, Claude E. ZoBell, Martin W. Johnson, and Acting Director Roger Revelle dedicated the attractive structure to their former colleague, Vaughan. Barnhart came for a tour of inspection on dedication day; he gave his stamp of approval and the comment: "E. W. Scripps promised me that building thirty years ago."

Hinton, who had put considerable thought into the planning, later qualified the result:

> . . . The staff of this Aquarium-Museum is very well satisfied with our aquarium, but there are a few things that we should do differently, given the opportunity. Perhaps one of the strongest restraining factors in the design of aquariums is the fact that so few people ever have the chance to design two of them![5]

The new building was three times the size of the previous space for the aquarium and museum, but not all of

The old aquarium as it appeared in 1933, when it was, to Percy S. Barnhart, "a constant source of worry and aggravation."

it was available to them. The second floor was preempted for the director's offices for some years (and somehow acquired the nickname of "green zoo"), and the space has continued to be used for non-aquarium offices. Also, the institution's growing collection of preserved fishes was stored in the basement of the building until the second addition to Ritter Hall was completed in 1960.

As the Aquarium-Museum has never been supported by research funds, attempts to make it self-supporting have periodically led to the suggestion of charging admission. Hinton wrote eloquently against the idea in 1954:

> . . . Many La Jollans have a proud sense of proprietorship in the Scripps Institution, and enjoy bringing their visitors to see "their" aquarium and museum. For example, last summer when we had trouble with our water supply, and were losing fish steadily, a number of local people made daily visits as if to a sickroom, inquiring anxiously as [to] the welfare of our creatures. Many local youngsters start nearly every summer day with a routine tour of inspection of the aquarium; these boys and girls are always delighted with new specimens and new exhibits, and frequently conduct their parents on guided tours on weekends. Lots of groups of families organize beach picnics and parties with the Aquarium as a meeting place; a tour of the place before and after is usually the order of things. These examples of the public attitude are individually small, but they add up to the fact that we enjoy a position of high prestige, and are considered by the community as a whole as part of the civic family. It would be most regrettable if this standing were to be lowered, as I feel it would be if each person were required to pay for admittance. It is surprising to realize that a considerable amount of

Thomas Wayland Vaughan Aquarium-Museum

bitterness still exists because of our having closed the pier to public fishing, nearly fourteen years ago*; the exacting of a toll would arouse even more ill-feeling.[6]

The issue of charging admission was shelved at that time. Some proceeds were derived from the sale of books, shells, and postcards that Hinton instituted in 1953.

Carr Tuthill joined the staff as a museum preparator in 1952 and was later put in charge of the aquarium exhibits. Richard H. Rosenblatt became overall director of the Aquarium-Museum in 1961, and Sam Hinton continued in charge of the museum displays until 1964, when he transferred to the UCSD office of relations with schools. In January 1965, Donald W. Wilkie became director of the Aquarium-Museum.

A native of Vancouver, British Columbia, Wilkie had taught mathematics, science, and physical education in his native province before earning a B.S. in zoology and an M.S. in ichthyology at the University of British Columbia. He had served as assistant curator at the Vancouver Public Aquarium and curator in charge of mammals and fishes at the Philadelphia Aquarama before becoming director of the Scripps Aquarium-Museum. In his new post Wilkie soon established a laboratory for researches on fish ailments, with microscopes and "behind-the-scenes" aquariums for ailing or for new specimens. The aquarium staff has made significant contributions in diagnosing and curing the ailments of marine creatures and in the handling of them from sea to shore.

In addition to researches on ailments of aquarium fishes,

---

*The Scripps pier was a favorite public fishing spot from the day it was built in 1915 until 10 December 1941, when wartime studies (and pier disrepair) put a stop to public fishing. An occasional oldtime La Jollan used to present his "lifetime" permit to fish from the pier, but these too were turned away.

especially on protozoan diseases, Wilkie has investigated the development of differences in skin pigments of certain fishes of the intertidal and subtidal zones. He has also worked with Denis L. Fox in analyzing the pigments of other marine animals. Like Barnhart, around his many other obligations, Wilkie for several years has been planning a new — much larger — aquarium-museum for Scripps.

Sea-water aquariums are much more difficult to maintain than fresh-water ones. Corrosion of pipes, growths of barnacles and other fouling creatures inside the piping, toxic "red tide" blooms, compatibility of various creatures, and ailments and diets of the aquarium inhabitants create constant problems. Under Wilkie's direction, records have been kept of the food preferences and intake of the animals in every tank, and the water has been monitored daily. The use of sub-sand filters has provided more flexibility in the sea-water circulation, which helps to control red tide problems and makes it possible to maintain warm-water fishes throughout the year.

The score of illuminated display tanks, in several sizes up to 2,000 gallons, have presented a colorful array of inhabitants through the years. Recently the displays have been composed of natural habitat groupings, chiefly of the San Diego area and of the Gulf of California region, where a much more tropical fauna dwells. Much admired by visitors are the orange garibaldis, dubbed the "La Jolla goldfish," and their blue-spotted young. An especially popular aquarium personage for some years was Harvey, a 100-pound grouper, who majestically circled a large corner tank from 1956 until his death in 1973. Morays, as they stretch open their mouths to display an array of needle-sharp teeth, draw awed gasps from viewers. At times sea turtles have swum the rounds of the larger tanks, and at least briefly one tank displayed highly venomous

Thomas Wayland Vaughan Aquarium-Museum

sea snakes, carried home from a tropical expedition.

Visitors sometimes fail to find the well-camouflaged flatfishes and skates against the matching sand. Certain sea anemones provide stunning color in some of the displays, and gracefully undulating nudibranchs — so unlike their relative, the garden slug — add natural artistry. Local fishermen like to compare notes on some of the species that they regularly catch. Whenever an octopus is among the inhabitants on display, it brings pause to most of the visitors, especially when it flashes through remarkable changes in color.

The sea-water line was renovated in 1964, to increase the capacity and improve the filtration system. Sea water for Scripps needs, including the aquarium, is drawn in from the far end of the pier, by means of two vacuum-assisted pumps that lift the water to a wooden trough that slopes to the landward end of the pier. It passes through a sand-bed filter and is then stored in two tanks near the aquarium building. From one the water flows by gravity through the Aquarium-Museum, the Experimental Aquarium, the Physiological Research Laboratory, Ritter Hall, and Scripps Building. From the other storage tank sea water is pumped up to the Hydraulics Laboratory and the Southwest Fisheries Center. At Wilkie's suggestion, a duplicate polyvinylchloride (PVC) line was installed on each section; these are used alternately, to hold the growth of barnacles, mussels, and other clogging creatures to a minimum. Charles J. Farwell, who joined the aquarium staff in 1969, has, among his other duties, carried out studies on water quality and its control.

As a convenience to the public, a sea-water tap was installed at the landward end of the pier in 1972. Home aquarists stop there regularly to fill jerry cans, bottles, and buckets. One woman takes home sea water to use in cooking — and one man drives to Scripps twice or oftener each year from Arizona to fill a tank truck with sea water to sell to desert aquarists.

An especially useful service of the aquarium personnel for many years has been the supplying of specimens to other Scripps researchers — some of whom are "softies and laboratory bound," according to one of them, but dependent upon "a steady supply of live material for daily experiments." Since 1966 Robert S. Kiwala has been the official collector, and he is often aided by aquarium helpers and other Scripps assistants in the gathering of several thousand marine specimens annually. This group can tell many tales of marine creatures that, for one good reason or another, they did not catch. Collecting trips by the aquarium staff have been as far afield as the Gulf of California, and sometimes collectors have joined more distant expeditions of other researchers. Much of the collecting has been carried out with the aid of Scuba gear in recent years. Rare specimens are sometimes exhibited and observed in the display tanks before they are preserved for the fish collection.

The Aquarium-Museum is the institution's door to the public. In 1950, while the present building was under construction, 50,000 visitors filed through the old wooden structure. One staff member estimated that at least two groups of school children each month were conducted through the aquarium. Within the first month after the new building opened, visitors from all 48 states and from Hawaii, Alaska, and the nation's capital had signed the visitor register. In fiscal year 1975-76 the numbers had reached 411,914, plus 61,364 students in conducted school groups, from San Diego, from Tijuana and other points south, from Los Angeles and other points north and east.

When the issue of charging admission was again raised in the late 1960s, Wilkie suggested that a voluntary donation box be tried instead. This has proved satisfactory — indeed, quite profitable. The sales desk has also been expanded, under the management of Bernice K. King since 1967, and does a lively business in books, sea shells, and marine-oriented items.

Wilkie has emphasized the educational role of the Aquarium-Museum. Soon after his arrival, he found himself inundated with 800 school children on one frantic morning. The solution, he felt, was a docent program, which he instituted — with one docent — in 1966. The docent group has expanded to number sixty or more each year. Since 1970 Patricia A. Kampmann, who was one of the first docents, has supervised the group. These enthusiastic volunteers primarily conduct school groups, but some visit classrooms and hospitals with displays, and others help with exhibit preparation, collecting, the sales desk, and even patrolling the Scripps Shoreline Reserve to advise visitors not to remove invertebrates from the protected area. The docents and the aquarium staff have prepared lesson packages for visiting school groups, which include study material to be used before, during, and after the group's tour of the aquarium and the museum.

In cooperation with area schools, Wilkie set up a career-training program for high school students interested in aquarium or marine biology careers. Several "graduates" of this program have become student employees at the aquarium, and some have gone on to advanced degrees in biology. The aquarium staff also offers short summer courses for several age groups of school children and organizes symposia on marine subjects for teachers.

The very popular Junior Oceanographers Corps (JOC), for students from fourth to twelfth grade, is also under the auspices of the Aquarium-Museum. Roger Revelle and Sam Hinton began it, originally as a means of allowing enthusiastic young fishermen to fish from the pier and as a source of specimens for the aquarium tanks. Hinton

Scripps Institution of Oceanography: Probing the Oceans

organized JOC in March 1959, as a monthly lecture program; members had the privilege of fishing from the pier on weekends, but with the requirement that the catch must be offered first to the aquarium. Finding adult supervisors proved difficult, and over the years boats and equipment on the pier were occasionally damaged, so in 1964 all pier fishing was again forbidden. JOC has continued as a combined lecture and field trip program, usually under the direction of a graduate student. Two of these JOC leaders — Leighton R. Taylor, Jr., and John E. McCosker — have gone on to careers at other aquariums.

The central museum area was given a major face-lifting in 1968, when, thanks to several generous donations from local citizens and businesses, it was possible to set up a large number of new exhibits in the rearranged central room. San Diego's Mayor Frank Curran presided at the ribbon-cutting ceremonies in August of that year.

To represent the variety of researches within oceanography, new exhibits are prepared frequently for the museum. Many oldtimers recall the large oarfish and other hand-painted specimens prepared by Percy S. Barnhart. In more recent years exhibits have illustrated and explained scientific concepts such as sea-floor spreading and ocean circulation, the undersea vehicles of the Marine Physical Laboratory, the programs of the Deep Sea Drilling Project and of Sea Grant, and such processes as wave motion and action (by means of a small wave tank). An unusually large manganese nodule has been displayed for many years. Models of the offshore bathymetry and of the submarine canyon that heads just beyond the Scripps pier have also been museum features for some time.

In the spring of 1975 the Aquarium-Museum opened its outdoor tide pool exhibit, which was based on a design by Wilkie and supervised by Farwell. The tide pool "rocks" were formed by pouring concrete into latex molds taken

Looking down one corridor of the Vaughan Aquarium-Museum, early 1970s.

from natural rock formations on the Scripps beach, under the direction of landscape artisan Julian George of Los Angeles. The display features a unique tidal cycle: a three-foot tide that rises and falls every four hours. Periodically waves wash through the pool to simulate natural conditions; the waves are created by a vacuum-chamber generator designed by John D. Powell of the Hydraulics Laboratory. Local marine denizens from nearshore areas inhabit this display, in a hide-and-seek fashion that is typical of tide pools of the San Diego area.

Aquarium tenders learn to avoid the poisonous stonefish and lionfish, and the threatening gape of the morays (only Ben Cox regularly petted those, during his many years of feeding the fishes). They have found the octopus to be the most troublesome aquarium inhabitant, as it is inclined to wander. Several times octopi have been found on the floor by startled janitors or staff. Late one night a restless — or hungry — octopus crawled from his own tank into his neighbors', inadvertently dragging along his probably life-saving refrigeration unit. The neighbors were crabs, which the octopus was eating when discovered. A hastily called aquarium crew wrestled for twenty minutes with the 35-pound octopus, prying loose the 2,000 suckers of his eight arms. As they lifted him out of the tank, he reached back to snatch one last crab.

NOTES

1. Helen Raitt and Beatrice Moulton, *Scripps Institution of Oceanography: First Fifty Years* (Los Angeles: Ward Ritchie Press, 1967), 21.

2. Report of the Scripps Institution for Biological Research, 1 July 1916, *in* Annual Report of the President of the University of California for 1915, 222.

3. Aquarium-Museum report to the director, 1946-47.

4. La Jolla Light, 29 March 1951.

5. Brochure of the Vaughan Aquarium-Museum (May 1954), 1.

6. Memorandum to business manager C. Earle Short, 25 March 1954.

Part III
The Disciplines

10. Life in the Sea: Studies in Marine Biology

For approximately its first quarter of a century, the laboratory by the sea was the Scripps Institution for Biological Research. Biologist-Director William E. Ritter considered chemists and physical oceanographers to be chiefly contributors to the total biological picture. Even dynamic oceanographer Harald U. Sverdrup said, "The major duty of a physical oceanographer is to provide a background for biologists."[1]

Times have changed, as the physics, chemistry, and geology of the sea have become disciplines of their own, rather more than "background for biologists," at Scripps and at other marine institutions. Biology at Scripps now involves approximately one-third of the staff researchers, scattered through several administrative units. Among the groups that deal primarily with biology but have been covered separately in this book are the Marine Life Research Program, the Physiological Research Laboratory, and portions of the Institute of Marine Resources.

Within the very broad field known as biology, the major contributions of Scripps have been in systematics, distribution, and ecology of fishes, of innumerable ocean invertebrates, plants, and microorganisms, and in biochemistry, developmental biology, and physiology.

In 1953, soon after the great expansion of oceanography immediately following World War II, Carl L. Hubbs noted:

> In general the physical sciences have been supported more generously than the biological by contract grants from the Department of Defense. The biological sciences at Scripps have profited to a limited extent directly, and to a larger measure indirectly from such support. A more auspicious balance would result if major support could be obtained for biological work at Scripps.[2]

Major support for biology was already being sought, and in 1954 resulted in the awarding of one million dollars by the Rockefeller Foundation "as an outright grant to the University of California for the development of a research program in marine biology at the Scripps Institution of Oceanography." The Rockefeller grant was in response to a formal proposal prepared in 1953: "Proposed development of marine biology at the Scripps Institution of Oceanography," which opened:

> The ocean and life within it are among the last frontiers of exploration — not a mere fringe but by far the largest habitat on the earth. . . . We must concede that the mere description of the ocean and of its inhabitants is in its infancy. The ocean still keeps a stupendous wealth of mysteries well guarded in its abysses, its vastness, and in the surprising biologies of its creatures.
>
>   . . . In spite of inherent obstacles, however, our insight into the ocean's movements, chemical

constitution, bottom structure, and interactions with the atmosphere has deepened during the last few years.

By contrast, marine biological knowledge has not kept pace with the progress of physical oceanography, nor has it taken full advantage of the experimental approach, which has brought about sweeping advances in other fields of biology. Marine organisms have provided tools for important advances in experimental biology, but marine biology as a whole, with few exceptions, has remained at a descriptive level. . . . At present, thanks to the greater familiarity with the sea, to the development of new tools and theoretical approaches, and to the deeper insight into general biological problems obtained by biophysicists, biochemists, geneticists and microbiologists the time is ripe for a planned, broad, frontal attack on the problems of marine biology.

Such an attack will have to be commenced on many fronts, in various waters, in several laboratories, taking advantage of the international cooperation, traditional for this field of biology. But a nucleus of activation of this field is needed. This function, we think, could be well performed by the Scripps Institution of Oceanography.

The Rockefeller funds, which were intended to be used over an eight-year period, established four new professorships in biology, one visiting professorship, several graduate and post-doctoral fellowships, and were also used for laboratory equipment and to add to the resources of biology staff members. At the end of the eight-year period, by previous agreement, the university assumed responsibility for the salaries of the new faculty appointments. These four were researchers in widely varying aspects of biology: E. W. Fager in ecology, Per F. Scholander in physiology,

Benjamin E. Volcani in microbial biochemistry, and Ralph A. Lewin in experimental phycology.

The Rockefeller Foundation funds also contributed to an international symposium, "Perspectives in Marine Biology," held in the "swank bungalow-type" La Jollan Hotel in March 1956, under the auspices of the International Union of Biological Sciences and sponsored jointly by Scripps and the Office of Naval Research. Adriano A. Buzzati-Traverso "organized the program, took the brunt of intricate diplomatic problems, and directed the large staff in managing a complex function."[3] More than 170 scientists from 14 nations arrived for the eight-day meeting, to present and discuss papers, and to talk shop in small "idea groups." Four invited Soviet scientists who had accepted the invitation failed to appear. Their leader, L. Zenkevich, had previously suggested that a biological program should be added to the forthcoming International Geophysical Year, and his suggestion was formally proposed to the meeting by Roger Revelle, whereupon the members of the symposium enthusiastically endorsed the idea. "The sense of the meeting," summarized Joel W. Hedgpeth, "as I got it from my own inevitably biased viewpoint, was that marine biology, if it can be said to have one perspective rather than several, should be directed toward the study of life in the sea as an organically interrelated complex, as an ecological unit."[4]

Let us look at some of the people and projects in biology at Scripps Institution, starting in 1936 when Harald Sverdrup became director.

MARINE VERTEBRATES

Francis B. Sumner was the senior biologist at that time. He had begun his professional career with fishes, and he ended it with fishes, although his reputation was chiefly established by population studies on deermice. In the latter 1930s and until his retirement in 1944, Sumner undertook studies on the colors of fishes and the role of coloration in providing them protection. This gentle man, who had emerged from what he called a "morbidly bashful" childhood, was a longtime tie with the very early days of the institution and an especially helpful colleague to younger staff members.

Carl L. Hubbs was invited to Scripps from the University of Michigan to replace Sumner in 1944. The appointment elicited the comment from University President Robert G. Sproul: "[Hubbs] is an exceptionally prolific writer, having published more than three hundred papers since 1915. His fertility in producing sound ideas is as amazing as is the energy he brings to his work."[5]

Born in Williams, Arizona, Hubbs had spent most of his childhood in southern California, and had in fact lived in San Diego from the age of two months until he was thirteen years old. As a boy he relished hunting horned toads on the hills around the small city of those days, and he paddled a "sneakboat" through the unspoiled bird-filled marshlands of San Diego Bay and False (now Mission) Bay. He attended Stanford University, and has always been proud to have co-authored papers there with David Starr Jordan. From 1917 to 1920 Hubbs was assistant curator of ichthyology and herpetology at the Field Museum of Natural History in Chicago, and he then became instructor of zoology and curator of fishes at the University of Michigan, which conferred upon him a Ph.D. in 1927 (unsought, and without requiring of him a dissertation, on the record of his publications). In 1940 Hubbs became a professor of biology at Michigan.

Hubbs's major contributions have been in the taxonomy of fishes, a subject that represents about one-half of his published works (a total of 690 papers to the end of 1976). He established the standard procedure for measuring fish specimens. He also has devoted considerable effort to fisheries and to the distribution of fishes. At the University of Michigan he set up the organization of the fish collection and increased tremendously its holdings in freshwater fishes. When he joined the staff of Scripps in 1944, he established the same system for the small collection of fishes then under the care of Percy S. Barnhart. Through Hubbs's initial enthusiasm and through additions from the wide-ranging Scripps expeditions — and occasionally through helpful fishermen — that collection has been greatly increased. Under Hubbs's direction it was at first maintained successively by graduate students Kenneth S. Norris and Arthur O. Flechsig, until Richard H. Rosenblatt became curator of fishes in 1958. The marine vertebrates collection contains more than one million specimens, representing at least 2,500 species of marine fishes.

At Scripps Hubbs also became interested in marine mammals. Soon after his arrival, he began taking notes on the annual migration of gray whales, which were slowly recovering from near-extinction after intensive whaling. In 1948 he enlisted the support of actor Errol Flynn (son of marine biologist T. Thomson Flynn) for a plane and helicopter flight to the whale breeding grounds in Baja California. That was a near-disaster, for the helicopter had three forced landings from engine troubles and the Piper Cub tangled with a chubasco. Undaunted, Hubbs continued his studies of the gray whale. He lined up volunteers for the annual census of whales as they passed Scripps southward to the breeding lagoons each December and January; for this the chief watching post was the "whale loft" atop (old) Ritter Hall, a small room equipped with a pair of powerful binoculars. From 1952 to 1964 Hubbs and his wife Laura flew to the Baja California lagoons almost every February to

Carl L. and Laura C. Hubbs on a research trip to Baja California, 1954.

count the whales and their calves, usually piloted by colleague Gifford C. Ewing. Hubbs helped persuade the Mexican government to set aside the major breeding lagoon, Scammon's, as a wildlife preserve.

He also rediscovered the Guadalupe fur seal, thought to have been exterminated in the early 1930s, at least until George A. Bartholomew (of UCLA) tentatively identified one on San Nicolas Island in 1949. In 1954, on a trip to Guadalupe Island, Hubbs finally found a small harem of the elusive fur seal among the rugged rocks on the eastern side of the island, and he greeted the defiant roar of the bull with his own equally loud roar of discovery. The fur seal survivors have proved prolific: the herd, protected by Mexican legislation, has increased to more than 500 individuals.

Guadalupe Island and Baja California have been favorite study areas of Hubbs for many years. To determine the temperature profile of the water along the shore, and its relation to upwelling, he carried out monthly temperature runs from 1948 for some years, from San Diego on the spine-jarring road down the Baja California peninsula to Punta Baja, just southwest of El Rosario, stopping at 61 stations to record ocean temperatures along the coast. Allan J. Stover, who often went along on those trips, took over the grueling task in the later years.

Facets of natural history other than temperatures also appealed to Hubbs on the monthly trips: birds, mammals, cacti and other peninsular plants, and evidence of early human occupation. That led him into a long archaeological study of the middens of La Jolla and Diegueño Indians, the early inhabitants of southern California and northern Baja California. When Hans E. Suess established the La Jolla Radiocarbon Laboratory in 1957 (see chapter 13), Hubbs was able to acquire, through precise analyses by that laboratory, the dates of occupancy of many of the middens. He

Marine Biology

also determined past ocean temperatures from the midden shells by means of oxygen-18 measurements, and established evidence for several periods of warmer and cooler temperatures through past ages.

Other stamping grounds for Hubbs have been the California deserts and the western basin-and-range domain. These interests began in his college days under John Otterbein Snyder, with whom he carried out a survey of the fishes of Utah in 1915. Over the years Hubbs often returned, with his family, to the Great Basin and desert areas, to work out the patterns of stream flow that in Pleistocene time formed a vast network of waterways, and, as the climate changed, dried to trickles and remnants in which a few species of fish survived. Hubbs persuaded the federal government to set aside Devils Hole National Monument to save the last handful of the Devil's Hole pupfish, *Cyprinodon diabolis,* and he has persisted in efforts to save other endangered species and subspecies of relict fishes throughout the western United States and in Mexico.

A Ph.D. project by one of Hubbs's students created surely more voluntary cooperation than has any other student project at Scripps. This was Boyd W. Walker's study of the habits of the silvery scaly grunion, the coastal fish that chooses to spawn in the moist sand left by a retreating high tide, during several nights after full and new moons from February to August. During his three-year study Walker enlisted the aid of more than 250 people, many of them from Scripps, who cheerfully paced the beaches late at night to look for, gather, count, sex, and tag grunion. On one April night more than one hundred volunteers watched for grunion on the beaches from San Francisco Bay to San Quintín in Baja California; and on one busy weekend Walker and thirty-odd assistants gathered 5,000 grunion from the Scripps beach by hand and surf-net, tagged them by fin-clipping, and returned them to the sea.

This is science? Volunteers coping with a good grunion run, for Boyd Walker's dissertation research in 1949. Photo by Lamar Boren.

In recognition of valued services, each volunteer received a certificate of merit, drawn by Sam Hinton, which read:

> Neither surf nor sand nor wee small hours
> have held terror for this brave worker in the
> cause of science; slippery fishes, scalding
> coffee, sharp-edged buckets, lofty sea walls —
> all have been met, and fought, and conquered.
> In recognition of these activities, so far be-
> yond the call of duty, it is hereby ordained that
>
> [volunteer]
>
> is a Life Member of the
> Society for the Investigation of Non-Gastronomical
> Characteristics of the Grunion
>
> and is hereinafter privileged to spawn on the beach
> at high tide.

Walker's study very precisely established the spawning pattern of the grunion.

One of Boyd Walker's students from UCLA was Richard H. Rosenblatt, who joined the Scripps staff in 1958, and simultaneously, as mentioned, became curator of the growing fish collection. Rosenblatt's own studies have been directed to the taxonomy and distribution of a great variety of marine fishes, especially of the eastern Pacific, from nearshore to midwater and deep ocean areas. He has been a leader or participant in a number of expeditions, including the attempt on Antipode Expedition in 1971 to photograph and capture a live coelacanth near the Comoro Islands. Although the trip did not attain that goal, it did add a number of significant specimens of other fishes to the collection. In 1975 Scripps alumnus John E. McCosker acquired in the

Comoros a frozen coelacanth that has been preserved and put on display in the Aquarium-Museum.

Robert L. Wisner, who began work at Scripps in 1947 for Francis P. Shepard, soon returned to his earlier interest — fishes — as the Marine Life Research program grew. He turned to the large unstudied collection of the usually mesopelagic lanternfishes (family Myctophidae) gathered at Scripps chiefly with the Isaacs-Kidd midwater trawl. Among fishes, these twinkling, small creatures — one-half inch up to twelve inches long — are exceeded in numbers in the oceans probably only by the cyclothones, and their taxonomy is complex.

In 1973 Walter F. Heiligenberg, who received his Ph.D at the University of Munich, joined the Scripps staff, and he has been pursuing studies on electrolocation in fishes, with emphasis on problems of signal detection and information processing in the nervous system.

Others who conducted studies on marine vertebrates at Scripps for shorter periods of time were Theodore J. Walker, who in the 1960s made observations on gray whales in the breeding lagoons and along the migration route, and earlier investigated the role of the lateral line of fishes; and Grace L. Orton, a specialist in amphibian larvae, who in the 1950s and early 1960s studied the eggs and early stages of various marine fishes. Also, Adriano A. Buzzati-Traverso, a population geneticist who was at Scripps from 1953 to 1962, set out to establish a field of marine genetics through the use of the brine shrimp *Tigriopus*. In a project for the Marine Life Research program, he also adopted the technique of paper partition chromatography in an effort to distinguish individual populations of sardines. Samples from various specimens of the same species, he found, in collaboration with Andreas B. Rechnitzer, were "remarkably constant, irrespective of the size or age of the fish."

Marine Biology

The researchers on fishes and fisheries at Scripps have for many years maintained a close working relationship with their colleagues in the National Marine Fisheries Service, a rapport that was nurtured by the California Cooperative Oceanic Fisheries Investigation (see chapter 3). Several of the senior researchers of the federal laboratory have served as lecturers and adjunct professors at Scripps. The fisheries laboratory maintains an extensive collection of eggs and larval fishes, to which Scripps has often contributed specimens. Researchers from each organization have often participated in the other's expeditions.

MARINE INVERTEBRATES

For some years, the one who devoted the greatest time at Scripps to studies on invertebrates was Wesley R. Coe. New Englander Coe had been a faculty member in biology at Yale University for 42 years before his retirement in 1938, when he became a visiting professor at Scripps. His field of research was the growth and reproduction of various marine invertebrates, especially the nemerteans and various species of mollusks.

As a visitor at Scripps during the academic year of 1926-27, Coe began a study of the attachment and growth of organisms on surfaces introduced into the ocean. "During the first year the program as outlined met with many vicissitudes," noted Coe, "owing mainly to the writer's inexperience with the powerful force of the breaking surf on an exposed sea-shore."[6] All the introduced blocks were carried away by storm waves except those set in deep water at the end of the pier. Replacements were installed, and Winfred E. Allen continued the observations after Coe's return to Yale; the two researchers published on the long project together. "During the nine years from October, 1926, to

Scripps Institution of Oceanography: Probing the Oceans

October, 1935, series of wooden and cement blocks, cement plates, and glass plates [were] placed in the water at intervals of two and four weeks in order to obtain information concerning the season of attachment, the rates of growth, the periods of reproduction, ecological conditions, and life histories of some of the numerous species of organisms which attached themselves to the exposed surfaces."[7] Throughout the years the species and organisms were nearly uniform, but the abundance of each varied a great deal from season to season. Coe and Allen recorded the seasonal distribution for a large number of marine invertebrates common to the La Jolla area.

For a study of *Mytilus,* the California mussel, Coe and his younger colleague Denis L. Fox gathered more than a thousand mussels of various sizes, placed them in screened boxes, and suspended them below the water surface at the end of the pier. The length of each mussel was measured every month until waves from a late December storm snapped the supporting boom and ended that project in its twelfth month. A second group of mussels was gathered to start again, and from the two years of data the rate of growth of mussels of various ages was meticulously tabulated.

That younger colleague, Denis L. Fox, by 1975 had gained the distinction of having been the longest-term living staff member at Scripps Institution. His career at the laboratory by the sea began in 1931, when as "a sharp young graduate from Berkeley and Stanford with fresh experience in industry and degrees in Biology and Chemistry, [he] was buttonholed by T. Wayland Vaughan."[8] Fox was born near Rye, England, in 1901 and in 1905 moved to the United States with his parents, to a farm near Napa, California. After earning his B.A. in chemistry at Berkeley, he worked in chemical research for Standard Oil Company for four years before entering Stanford for a Ph.D. in biochemistry. In June of 1931 Director Vaughan offered him an instructorship at Scripps, which Fox began that fall as soon as he had completed his dissertation. He was advanced to professor in 1948 and became professor emeritus in 1969. With the exception of a few weeks after his arrival, he has occupied the same suite of offices on the second floor of then-new, now-old Ritter Hall.*

Fox later said: "The colleague who had the greatest influence upon my professional growth at Scripps was Francis B. Sumner. . . . It was he who introduced me to the whole general subject of animal pigments, and with whom I conducted joint researches on the subject for the first several years."[9] These studies were particularly on the carotenoid pigments of fishes, and the research led Fox into further work on other animal pigments, to the extent that an admiring colleague could say: "Armed with new methods and uncanny ingenuity he wrenched the colorful secrets from anemones and ascidians, from gorgonians and garibaldis, from Metridium and medusae."[10]

All these and others too, for Fox's studies have been principally aimed at unraveling the metabolic fractionation of pigment molecules. His subjects have also included the orange-bodied California mussel, nudibranchs, barnacles, octopi, squid, primitive plants, and a number of fishes — and "the largest brilliantly coloured birds of the New World," the flamingos. In a Gorgonian coral and in two hydrocorals, Fox and his colleagues in 1970 described the first known examples of chemical binding of brightly colored carotenoid acids in calcareous skeletons.

For Fox's long study of flamingos he worked closely with the San Diego Zoo. Flamingos, as zookeepers had long known, tend to lose their bright coloring in captivity. When Fox began his study in 1954, the flock of flamingos at the

*This book includes events through the end of December 1976, at which time the Marine Biology Building was not completed.

Scripps Institution of Oceanography: Probing the Oceans

San Diego Zoo were fading in color, so finely ground lobster shell was added to their food, and within a few months the birds' color "exhibited a conspicuous return of pink to salmon or vermilion colours."[11] Fox determined that with supplemented diet, notably including β-carotene, the flamingos developed rich stores of carotenoid pigments — canthaxanthin and astaxanthin — which provided striking pink or red color in the feathers and in the naked skin of their webbed toes, tarsals, heel-joints, tibiae, and lower bill-mandibles.

In 1953, Fox published the first edition of his summary treatise, *Animal Biochromes and Structural Colours*,[12] written, according to one reviewer, "with a zest that betrays his own delight in his subject." A revised second edition of the text by Fox was published by the University of California Press in 1976.

The most intensive ecological study of nearshore creatures at Scripps was carried out by Edward W. ("Bill") Fager. He had received his Ph.D. in organic chemistry at Yale in 1942, had participated in the wartime Manhattan Project, and had received a Ph.D. in zoology at Oxford before joining the Scripps staff in 1956 to pursue ecological studies. A longterm project of his was analyzing the community structure of a shallow-water area off the Scripps beach, often called "Fager's Half-acre." The area selected was the sand plain between the two heads of the La Jolla submarine canyon, and the study method was direct observation by Scuba diving. The water depth of five to ten fathoms was "chosen because it allowed 50-60 min working time underwater per dive and avoided the complications of surf in shallower water and the problems of decompression introduced by prolonged work at greater depths."[13] Fager, with Arthur O. Flechsig as his most frequent partner, dived repeatedly in the area over a three-year period (and less often for another three years), to monitor the most common invertebrates and their locations. Among his early conclusions was that "it is often misleading to label some species in a community 'important' and others 'unimportant,' especially if this means that the latter are ignored in studies of community structure and dynamics."[14]

Fager and Flechsig next set out four artificial "rocks" — one-meter cubes of asbestos board on iron frames — on the underwater sandy area, to determine the development of a community in a new environment. These were monitored, "weather permitting," every two to three weeks for several years. As encrusting invertebrates and algae became established on the new homesites, fishes, starfishes, crabs, lobsters, and octopi moved in for food and shelter. The new communities had both similarities and differences, which led Fager to conclude that "chance plays a much larger role in the establishment and maintenance of communities than had been thought."[15]

The ecological studies by Fager and his diving assistants and colleagues were combined with computer simulation studies to try to understand the dynamics of simple communities and to predict the effects of various types of perturbation on them. Fager concluded in 1971: "All of the studies on the computer that have been even moderately successful in modeling the observed events have had a prominent random component in them."[16]

These studies were in progress when Fager was incapacitated by pneumococcal meningitis early in 1973 (which led to his death in 1976). Paul K. Dayton began supervising some of the projects under way. He had joined the Scripps staff in 1970, and among other studies turned to an ecological survey of kelp communities that had been started off Del Mar by Westinghouse Corporation in 1967. This project by Dayton's group "involves studies of the individual growth rates, recruitment, and mortality patterns of several kelp species, the foraging patterns and effect of their

E. W. Fager maneuvering a bongo net on Piquero Expedition, 1969. Photo by Elizabeth Venrick.

herbivores, and the community implications of several carnivores."[17] Dayton has also been directing a study of Antarctic sea-floor invertebrates.

One of the many students of Fager's was James T. Enright, who, after receiving his Ph.D., was at the Max-Planck Institute in Germany from 1961 to 1963, then at UCLA until 1966, when he became a staff member at Scripps. He began his researches on circadian rhythms, using first a beach-dwelling amphipod (*Synchelidium*), which customarily migrates up and down the beach with the tide. Enright found that under laboratory conditions the tidal rhythm of the amphipod persisted for several days. He also determined that *Synchelidium* can perceive pressure changes of less than 0.01 atmosphere, that another amphipod, *Orchestoidea corniculata*, orients itself to moonlight, and he has continued with researches on the parameters that establish rhythmicity in various animals, including the house finch. Enright's titles are often eye-catchers: i.e, "The Internal Clock of Drunken Isopods," and "When the Beachhopper Looks at the Moon: The Moon-Compass Hypothesis."

Elizabeth Kampa moved to Scripps Institution at the same time as Carl L. Hubbs in 1944, having been his assistant at the University of Michigan after receiving her B.A. there. She earned her Ph.D. with work done at Scripps in 1950, simultaneously with fellow-student Brian P. Boden; they had been married the previous year. Brian Boden had begun his researches on diatoms, but soon both Bodens were investigating the ups and downs of the deep scattering layer. They determined that an important component in the migrating layers was the abundant group of shrimplike crustaceans known as euphausiids. Brian Boden, with Martin W. Johnson and Edward Brinton, prepared a taxonomic study of the Euphausiacea, and he measured the bioluminescence within the deep scattering layer. Elizabeth Kampa

Scripps Institution of Oceanography: Probing the Oceans

Boden pursued studies on the light sensitivity of various euphausiids and found that the eyes of those within the scattering layer were more nearly alike, even in different families, than like their relatives that did not live within the layer. With sensitive light meters she has measured the intensities of light within ocean depths and has found these to vary much more than expected; scattering-layer organisms, she concluded, react to minute variations in particular intensities and migrate upward or downward to find the deep shade that they prefer. In 1973 she carried her equipment aboard the Texas Maritime Academy ship *Texas Clipper* to observe the effect on the organisms in the deep scattering layer of a total solar eclipse. During totality the animals rose rapidly upward but were unable to reach their optimum light conditions, probably because they could not swim fast enough. As sunlight reappeared, they resumed their usual haunts.

Joel W. Hedgpeth was at Scripps from 1950 to 1957, during which time he compiled and edited the *Treatise on Marine Ecology and Paleoecology,* a classic work in the subject, published by the Geological Society of America in 1957. Besides his broad interests in ecology and in the history of marine biology, his outrage at the distribution of debris by humans across the landscape and seascape, and his fondness for Puckish humor and the Celtic language, Hedgpeth is an expert on pycnogonids — marine spiderlike arthropods that typically have four pairs of legs. He doesn't always take them seriously, however. After identifying the one pycnogonid gathered in a dredge haul on Transpac Expedition in 1953, he could not resist comparison, in a brief classic paper, with the 1948 Swedish Deep Sea Expedition: one pycnogonid in three dredge hauls on the Scripps trip and one pycnogonid in nine dredge hauls on the Swedish trip. "The only justifiable conclusion," said Hedgpeth, "and one that cannot offend any national sensibilities, is that the total pycnogonid population of the overall world ocean has increased threefold since 1948. . . . It is estimated that the pycnogonids may support a major fishery sometime in the next millenium."[18]

William A. Newman's researches on invertebrates are both biological and geological, in the classical tradition of Charles Darwin and former Scripps director T. Wayland Vaughan — among the few who have combined studies of living and fossil organisms. Newman joined the Scripps staff in 1962 for a year, spent two years at Harvard, and returned to Scripps in 1965. He has served since then as curator of the Scripps collection of benthic invertebrates. Newman's research has been especially on the taxonomy, distribution, and evolution of barnacles throughout the world from tropic isles to the Antarctic. The distribution of these shallow-water creatures has been of interest to geologists concerned with the fluctuations in sea level and with the sinking of guyots. Newman has also dated the levels of terracing of coral formations to determine former still-stands of sea level on various islands and shores.

On Styx Expedition in 1968, Newman, along with Richard H. Rosenblatt and Edwin C. Allison (then a San Diego State University geology professor), discovered another seamount in the Mid-Pacific Mountains west of Hawaii. This proved to be a drowned atoll which had been at the surface in Cretaceous time. In honor of the originator of the accepted explanation of atoll subsidence, the Styx researchers named the feature Darwin Seamount.

Andrew A. Benson joined the Scripps staff in 1962, from UCLA. He turned to studies of the waxes and fats stored by marine animals, especially in the copepods, "small marine crustaceans that are enormously abundant; indeed, in most oceanic areas they are the largest single component of the zooplankton."[19] Copepods are also a major link in the food chain from phytoplankton to the higher animals, including species of fishes of commercial importance. The minute, colorful copepods store waxes apparently as a reserve for times of starvation. Benson and colleagues have compared the wax content of copepods from various depths in the subtropical and temperate regions of the Pacific Ocean with specimens from the Arctic Ocean; they found that larger amounts of wax are found in copepods in colder and deeper waters. The researchers have also determined that waxes play a role in the food chain of tropical coral reefs. The corals exude a slimy mucus that is avidly eaten by various reef fishes; this mucus was found to contain a wax similar to that of copepods. The coral-destroying crown-of-thorns starfish *(Acanthaster planci)* exudes digestive enzymes that are "highly efficient in their action on the wax components of coral. No other animal we have examined [wrote Benson and Richard F. Lee] has so thoroughly adapted its digestive system to wax nutrition."[20]

Other work on invertebrates begun at Scripps in recent years has been that of Robert R. Hessler, who joined the staff in 1969 from Woods Hole Oceanographic Institution, where his studies demonstrating the great diversity in the supposedly sparse deep-ocean fauna had attracted attention. Hessler, with colleagues, has been making an intensive survey of the benthic community under the center of the North Pacific gyre. From South-Tow Expedition in 1972 he acquired what he considered to be "one of the most complete samplings of deep-sea bottom communities in a single ocean location."

Also, Lanna Cheng (Mrs. Ralph A.) Lewin has since 1970 been gathering specimens of the only truly marine insect, the several species of the fragile sea-skater of the genus *Halobates,* for studies of their distribution and behavior.

A great deal of work on invertebrates – i.e., plankton organisms – has been done by the Marine Life Research program and is cited in chapter 3.

Marine Biology

Some of the marine invertebrates have been studied at Scripps by researchers who call themselves geologists and who are especially concerned with the dating of sediments by means of fossil organisms. The researches by Milton N. Bramlette on nannoplankton, by Fred B Phleger and colleagues on foraminifera, and by William R. Riedel on radiolarians are cited in chapter 12. Another researcher who devoted considerable time to invertebrates while at Scripps was Robert H. Parker, a biologist by background, who did extensive studies of the taxonomy and ecology of macroinvertebrates in the sediments of the Gulf of Mexico and the Gulf of California under API Project 51 (see chapter 12).

MARINE MICROBIOLOGY

Claude E. ZoBell was at Scripps as an assistant professor of marine microbiology in 1936, having joined the institution's staff in 1932. Born in Utah in 1904, he was raised in the upper Snake River Valley in Idaho; he received his B.S. and M.S. in bacteriology at Utah State University in Logan, and his Ph.D. at Berkeley. Most of his doctoral work was done at the Hooper Foundation for Medical Research in San Francisco, and ZoBell had a six-month appointment there before transferring to Scripps. For a number of years he served as an advisor on sanitation and public health matters at Scripps, along with his other duties.

The subject that ZoBell began at Scripps was large. As he pointed out in 1959: "Bacteria, yeasts, fungi, microflagellates, blue-green algae, and allied microbes are widely distributed in the sea. . . . Besides being concerned with the deep sea and open ocean, [the marine microbiologist] must also be cognisant of conditions of microbial life in the littoral zone, estuaries, inflowing rivers, and related lakes. Thus the domain of the marine microbiologist

Claude E. ZoBell attaching messenger to wire above a J-Z sampler, in the 1940s

Marine Biology

encompasses more than two-thirds of the earth's surface and 99 per cent of the hydrosphere."[21]

In the mid-1930s ZoBell determined that various organic compounds are adsorbed on solid surfaces submerged in the ocean in greater concentration than in the surrounding water. This encourages the attachment and growth of bacteria and other microbes. ZoBell defined the sequence of events in the fouling of submerged surfaces, and during World War II he participated in studies of the fouling problem for the U.S. Bureau of Ships.

In 1942, ZoBell became a participant, with Denis Fox and Roger Revelle, in the first American Petroleum Institute grant to Scripps, for fundamental research on the occurrence and recovery of petroleum. ZoBell's researches were aimed at determining the microbial modification of marine sediments, in part using the mud cores taken from the floor of the Gulf of California in 1939 and 1940 (see chapter 2). The petroleum institute was soon fascinated by ZoBell's discovery that some of the bacteria in those cores proved capable of freeing oil from sediments. The minute bacterium with the ambitious name of *Desulphovibrio halohydrocarbonoclasticus,* ZoBell found, caused oil to separate from sediments in several ways, and he also found that bacteria can free oil from tar sands and oil shales. Tests were carried out in Pennsylvania on the possibility of inoculating old oil fields with bacteria to release unrecovered oil. In 1947 ZoBell received a patent on the oil-releasing process, which he promptly assigned to API for free use to the public. The process was not used extensively in the United States, but was used for a slight increase in oil production in old wells in several European countries.

For various studies on the oil-forming process, the API grant was continued until 1952. ZoBell determined that bacteria tend to make organic matter more petroleum-like by removing oxygen, nitrogen, and certain other elements, and that bacterial activity tends to create reducing conditions, and so favors the formation and preservation of petroleum hydrocarbons.

ZoBell and his colleagues have also researched the manner in which oil on the sea surface is degraded by bacteria. Free oxygen proved to be the most important condition affecting the oxidation of oil. Temperature is also important, through its effects on the mixture of oil with water and the reproduction rate and metabolism of bacteria. Cold-acclimated Arctic bacteria were found capable of oxidizing oil at temperatures slightly below 0° C. In the breakdown of oil, ZoBell found that several species of bacteria acting together degrade petroleum more rapidly than does a single species.

ZoBell has also investigated the microorganisms of the surface and of the middle depths of the ocean. For the latter he devised the J-Z sampler (designed by ZoBell, constructed by buildings and grounds foreman Carl Johnson), a sealed sterilized bottle that is designed to draw in a water sample when lowered to the desired depth and triggered when the attached glass tube is broken by a messenger traveling down the wire. The samplers can be attached at intervals along the wire to collect at a number of depths on the same lowering. ZoBell has found that bacteria are abundant in the uppermost layers of the ocean where plankton are found, less common in mid-depths, and very abundant at the interface of the ocean water and the sediments of the ocean floor, where detritus accumulates.

On the *Galathea,* during the Danish Deep-Sea Round-the-World Expedition in 1951, ZoBell acquired microorganisms from the almost-greatest depths of the ocean and "brought 'em back alive." Hakon Mielche, who was also on the expedition, recalled the moment on 15 July of awaiting the return of the sampler from the floor of the

Marine Biology

Philippine Trench, 10,000 meters below*:

We had worked intensely and had not slept for 24 hours. We looked unshaven and red-eyed. The coffee-pot was perking incessantly in the [galley]. The whole ship was in a fever of excitement. Instruments had never been sent down to such great depths. Most scientists the world over claimed that organic life could not exist in the ice-cold, totally dark depths of the sea under a pressure of 1,000 atmospheres.

. . . The one-meter-long metal [corer] appeared above the water and was hauled aboard. We had barely time to take pictures and film of it, before ZoBell had a hold of it and disappeared down below to his little, private laboratory embracing the [encased core of mud] as if it were pure gold. In spite of the fact it only contained mud. But the mud was applied to glass [slides] for the microscope and then the rest was quickly divided and deposited in fifty [piston-stoppered vials], which immediately were put under pressure of 1,000 atmospheres and cooled to . . . 2 degrees Celsius, which temperature the thermometer had shown was found at the bottom of the ocean. There *was* organic life in the mud. Long chains of bacteria, creatures on the border of the animal- and plant-world — but living creatures.[22]

After the expedition, from time to time the vials, containing either the original mud samples or nutrient medium inoculated with such mud, were examined for evidence of living microorganisms. On a visit to ZoBell's laboratory at Scripps 15 years after the trench haul, Mielche found that

---

*ZoBell's student, Richard Y. Morita, on Midpac Expedition in 1950 had retrieved bacteria from the sea floor at 2,700 meters. See chapter 15.

four of the original 50 pressurized vials were still holding living bacteria from the Philippine Trench.

By 1967 ZoBell had "explored" — with corers on various expeditions — ten trenches deeper than 7,000 meters, and had found several different varieties of living bacteria in them. Over a number of years he has investigated the effects of pressure on deep-sea organisms and the rates of growth of microorganisms at various hydrostatic pressures. He and his colleagues have found that some microorganisms from surface-pressure environments survive deep-sea pressures at deep-sea temperatures, although existing under such conditions only in a dormant state. This has led to comparisons with deep-sea organisms, which are active at high pressures and low temperatures.

MARINE PLANTS

Plants of the ocean have been scrutinized by several Scrippsians. Some of the early studies on the biggest ocean plant — the giant kelp, *Macrocystis pyrifera* — have been described in chapter 6. Much smaller, but much more significant in both volume and in the food chain of the ocean, are the phytoplankton, of which the major representatives are the diatoms — microscopic algae with cell walls fortified by silica; and the dinoflagellates — plantlike protozoans with two flagellae for locomotion. In 1936, the chief researcher on these at Scripps was Winfred E. Allen, who had first spent a summer at the institution in 1917 and continued as a full-time staff member from 1919 until his retirement in 1946 (he died the following year). His special interest was in diatoms and their distribution. Allen designed a closing bottle for subsurface sampling in 1925, and he set up standard collecting methods that involved filtering or settling of the contents of water samples. He also inaugurated daily collections of plankton samples from the Scripps pier in 1919 and at Point Hueneme the following year. These collections were continued for twenty years at Point Hueneme and for much longer at Scripps. For two years Allen made twice-daily collections from the pier. At his request, Coast Guard officers took daily collections at the Scotch Cap Lighthouse in the Aleutian Islands from 1925 until 1933; daily collections were also made at the Farallon Islands and at Oceanside for five years and at Balboa, California, for three years. From the accumulated material, Scripps Institution's first woman alumna, Easter Ellen Cupp (Ph.D. 1934), compiled a monograph on the marine plankton diatoms of the eastern Pacific, published in 1939.

While making his collections over many years, Allen also logged his observations from the pier: of schools of mackerel and of the first hammerhead shark reported in the area; of myriads of sea birds — "about 500 pelicans (part count, part estimate)" one early morning, and "several hundred or even thousands of white bellied shearwaters" fishing north of the pier; of sea lions in groups of ten to twenty; of siphonophores; of red tide; and of double rainbows, and even one meteor.

In his analysis of twenty years of plankton observations, Allen found such abrupt variations in numbers of individual organisms at each station that he was led to the "unavoidable conclusion" that "attempts to coordinate and correlate chemical and physical data with records of changes in activity or productivity of diatoms are likely to be grossly misleading if based on observations made at intervals longer than one day."[23] In fact, he noted: "To those who may have thought of conditions in the ocean as being so nearly uniform that the routine of life must be rather monotonous, it may seem somewhat shocking to know that our records show no two years alike in the twenty, no two months alike, and no two weeks alike."[24]

In addition to the daily collections, Allen also gathered and received material from short cruises of the *Scripps,* from longer cruises of the *E. W. Scripps* for the U.S. Bureau of Fisheries from 1938 to 1941, from the expeditions in 1939 and 1940 to the Gulf of California, from collections by the U.S. Coast and Geodetic Survey in Alaskan waters, from collections by the U.S. Navy from California to Australia, and from collections by G. Allan Hancock and by Templeton Crocker. He published on the relationships of abundance of phytoplankton to the hydrographic conditions of the areas covered. The problem was big, as he observed: "I think that enough may have been said to indicate that no matter how busy we have been and how many problems we have attacked with variable success, there is still plenty to do."[25]

Half a century later, James T. Enright turned back to the data that Allen had accumulated during his two years of twice-daily sampling, which Enright called "one of the most extensive phytoplankton sampling and identification programs ever undertaken by a single investigator at a single location."[26] Allen had found "that the estimates of abundance of diatoms for the evening samples tended, in general, to be lower than those for the preceding or following mornings," and he suggested that the diatoms increased through reproductive activity at night. After computer analysis of Allen's material, Enright concluded that the data could be best explained by an increase through reproductive activity at night and extensive grazing by zooplankton in daytime.

Marston C. Sargent began studies on marine algae at Scripps in 1937, after receiving his Ph.D. at Caltech and serving there as a staff member for three years. His researches were on the photosynthesis and growth of kelp and other seaweeds and on organic productivity in the ocean. On Operation Crossroads in 1946, with Thomas S. Austin, he made the first measurements of organic

Winfred E. Allen, with his plankton collector, on the end of Scripps pier, about 1926.

production on a coral reef. From 1951 to 1955 Sargent was head of training at the Navy Electronics Laboratory, after which he became oceanographer for the Office of Naval Research; in that post he was responsible for relations with all west-coast oceanographic organizations that had contracts with ONR, and he was conveniently located at Scripps. When Sargent retired from ONR in 1970, he became coordinator for the California Cooperative Oceanic Fisheries Investigation (see chapter 3) until his second retirement in 1974.

Another Scripps researcher on minute marine plants has been Francis T. Haxo, who received his B.A. from the University of North Dakota and his Ph.D. at Stanford in 1947. He was an assistant professor at Johns Hopkins University before joining the Scripps staff in 1952. In 1963 he advanced to professor. Haxo's studies have specialized in photosynthesis and the pigments of algae and dinoflagellates. For example, on Billabong Expedition to the barrier reef of Australia in 1966, Haxo and colleagues determined from studies of living giant clams that the brown-colored dinoflagellates inside the clam's mantle tissue produce more than enough oxygen and glycerol for their own needs and so provide some to their host. On the Bering Sea Expedition by the *Alpha Helix* in 1968, Haxo measured the photosynthesis and respiration of Arctic cold-adapted seaweeds.

From 1947 to 1961 at Scripps, Beatrice Sweeney also conducted studies on phytoplankton, especially the culture and characteristics of dinoflagellates. She continued, among other projects, the observations begun by W. E. Allen on red tides — "when the sea turns the color of tomato soup" — from vast numbers of organisms, often dinoflagellates, containing orange or red pigments. She discovered and characterized the natural diurnal rhythms in the bioluminescence of *Gonyaulax polyedra,* a subject that has been pursued by a number of investigators ever since.

Marine Biology

Ralph A. Lewin's studies have been chiefly with various species of algae, especially on their biochemistry and their sensitivity to pollutants. Lewin was born in London and earned his B.A. and M.S. at Cambridge University, which also awarded him a D.Sc. in 1972. He earned his Ph.D. at Yale in 1950 and joined the Scripps staff in 1960 from the Marine Biological Laboratory at Woods Hole, Massachusetts.

Over several years Lewin and his coworkers analyzed the flexibacteria – brightly colored gliding microbes, some of which grow on organic matter, some of which are parasitic and others predatory. Lewin's group developed techniques for isolating and culturing flexibacteria from coastal waters, hot springs, and other habitats, which made it possible, with computer analysis, to classify six genera and 27 species of this difficult group. Lewin also isolated diatoms from brackish, marine, and highly saline waters, and has determined their tolerance to these varying salinities.

This enthusiastic man began the Sumnernoon film-or-slide programs that during the 1970s became a popular presentation each week – in Sumner Auditorium – during the academic year. Lewin has for some years also been a keen advocate of the international language, Esperanto.

OTHER BIOLOGICAL STUDIES

Various workers at Scripps have used marine organisms as study subjects in cell physiology, embryology, and development – traditional fields for marine biological laboratories.

Benjamin E. Volcani, for example, has pursued studies on the biochemistry of diatoms. Born in Ben-Shemen, which is now in Israel, Volcani received in 1941 the first Ph.D. in microbiology awarded by Hebrew University. He joined the Scripps staff in 1959. Volcani and his colleagues have investigated the role of silicon — long thought to be biologically inert — in the cell development of diatoms, which take silicic acid from water and deposit silica as an intricate shell, interlocked within an organic casing constituting the cell wall. Volcani's group also found that silicon is necessary to the diatom for the synthesis of DNA molecules, and the DNA polymerase and thymidylate kinase enzymes; thus, the diatom, which is absolutely dependent upon silicon, offers an experimental system par excellence for exploring the biological role of silicon.

In the late 1960s David Jensen carried out researches on the rudimentary heart of the hagfish and on cardiac control in various animals. Theodore Enns since the mid-1960s has carried out studies on the formation of urea in the horn shark and on the transport of water, gases, electrolytes, and other substances in various marine organisms. Nicholas D. Holland, since 1966 at Scripps, has studied, with the aid of electron microscopy, the mucous secretion in sea urchins and the detailed anatomical structure and spermatogenesis in sea lilies (crinoids). David Epel, on the Scripps staff since 1970, has pursued studies of fertilization and the early stages of development in the eggs of sea urchins and starfishes. George N. Somero, who also joined the staff in 1970, has been investigating the biochemical changes involved in adaptation to different temperatures in fishes.

In 1967 the Neurobiology Unit was formed, under the leadership of Theodore H. Bullock, to recognize the substantial group of neurophysiologists at the institution. These include some from the Marine Neurobiology Facility located in the Physiological Research Laboratory building and various other administrative units, with overlap to the UCSD Medical School.

ECOLOGICAL RESERVES

Scripps Institution biologists established the first two reserves belonging to the university, now incorporated into a statewide system called the University of California Land and Water Reserve System. The various parcels of land throughout the state — 22 of them in 1975 — have been set aside both for study purposes and to preserve natural habitats. Included are desert dunes, coastal marshes, mountain meadows, islands, chaparral, mountain and coastal streams — representing many of the remarkably diverse habitats of the large state.

Long before that system was established in 1965, the first reserve had been set aside: the shoreline and coastal area in front of Scripps Institution, now called the Scripps Shoreline-Underwater Reserve.

Percy S. Barnhart, the longtime curator of the Aquarium-Museum, started the shoreline preservation early in 1926, because he was concerned about the "constantly increasing number" of people "coming to our local rocky beach for the purpose of obtaining lobsters, abalones, and other animals of the shore for food purposes."[27]

Director Vaughan seconded Barnhart's memorandum and on 7 January 1926 forwarded it to University President W. W. Campbell. Other biological stations, especially Stanford's Hopkins Marine Station, had encountered similar depredations, and had tried as early as 1919, but unsuccessfully, to have legislative action taken to protect the area in front of the several marine stations, for study purposes of the researchers. Vaughan pursued the matter in 1926, and he enlisted the aid of the California Fish and Game Commission, the San Diego Natural History Museum, and members of the state legislature. A bill was presented to the California legislature early in 1927, specifically to set aside as a biological preserve the thousand feet of shoreline and portion

Scripps Institution of Oceanography: Probing the Oceans

of ocean adjacent to Scripps Institution (although inexplicably without including other marine stations). Rod-and-line fishing and recreational beach use were not prohibited. The bill passed both legislative houses but was pocket-vetoed by Governor Clement C. Young on the ground that its constitutionality was doubtful. In 1929 a similar bill was again approved by the legislature and that time was signed by Governor Young.

The waters adjacent to Scripps came under protection in 1948, under the enforcing jurisdiction of the Commandant of the Eleventh Naval District. The regulations prohibit vessels, other than federal or Scripps ships, from anchoring in front of the institution, or from dredging, dragging, seining, or otherwise fishing within the area. This provided protection for the various pieces of equipment installed by Scripps researchers along the pier and from the beach outward. In 1972 an underwater reserve was established by the city of San Diego and the state of California from La Jolla Cove up the coast for almost six miles to the northern limits of Torrey Pines State Reserve. Within the park is an ecological reserve, from which no marine life may be removed. At the dedication of the park, a plaque was unveiled in memory of Scripps diver Conrad Limbaugh (see chapter 6) and Harold Riley, one-time president of the San Diego Council of Diving Clubs and advocate of the underwater park, who drowned off Torrey Pines in 1970.

Scripps supervised another piece of land of ecological distinction for some years before it too was placed in the university reserve system. This was a parcel in an area that Ritter long before had advocated preserving: Mission Bay. After his directorship had ended, Ritter summarized to the chairman of the State Park Commission his views:

> . . . For years [False Bay or Mission Bay] has been one of the richest sources of certain kinds of marine

Marine Biology

life for research and study at the Scripps Institution. . . . During the years of my directorship of that institution we often considered the possibility of preserving at least some portion of that area as a permanent source of such forms of life. Unless measures are taken soon the whole area will be diverted to uses such that it will become almost useless in this way.[28]

Director Vaughan in 1928 tried to enlist the aid of the Fellows of the San Diego Society of Natural History in setting aside part of Mission Bay for scientific purposes.

The actual establishment of a Mission Bay marsh reserve came about through negotiations by Carl L. Hubbs with two owners of marsh property: Mrs. Oscar J. Kendall and the San Diego Beach Company, founded by A. H. Frost. These owners donated 20 acres of prime marshland on Mission Bay to the University of California in 1952,* Mrs. Kendall as a memorial to her husband and son, and the corporation as a memorial to its founder. The Kendall-Frost Reserve is one of the few remaining saltwater marsh areas in southern California and supports a large population of resident and migratory birds.

Hubbs was chairman of the UCSD unit of the University of California Land and Water Reserve System for a number of years, and was a major participant in the arrangements for the system to acquire two other parcels in the San Diego area: Dawson Los Monos Canyon Reserve, near Vista, and Elliott Chaparral Reserve, near Miramar Naval Air Station. Since 1973 Paul K. Dayton has been chairman of the UCSD reserve committee.

---

*Shortly after the city had threatened to condemn it to preempt as park land.

NOTES

1. *In* Brian Boden, "Bioluminescence in Sonic-Scattering Layers," *Proceedings of International Symposium on Biological Sound Scattering in the Ocean,* 1971, 65.

2. Memorandum to Roger Revelle, 21 July 1953.

3. Theodore Bullock, "Marine Biology," *Science,* Vol. 124, No. 3215 (10 August 1956), 281.

4. Mimeographed summary.

5. *University Clip Sheet,* 7 September 1944.

6. "Season of Attachment and Rate of Growth of Sedentary Marine Organisms at the Pier of the Scripps Institution of Oceanography, La Jolla, California," *Bulletin of Scripps Institution of Oceanography,* Technical Series, Vol. 3, No. 3 (1932), 39.

7. W. R. Coe and W. E. Allen, "Growth of Sedentary Marine Organisms on Experimental Blocks and Plates for Nine Successive Years at the Pier of the Scripps Institution of Oceanography," *Bulletin of Scripps Institution of Oceanography,* Technical Series, Vol. 4, No. 4 (1937), 101.

8. Tribute to Fox at SIO Staff Council meeting, 22 May 1969.

9. "Again the Scene," Manuscript in SIO Archives, 1975, 124.

10. Tribute to Fox at SIO Staff Council meeting, 22 May 1969.

11. D. L. Fox, "Metabolic Fractionation, Storage and Display of Carotenoid Pigments by Flamingoes," *Comparative Biochemistry and Physiology,* Vol. 6 (1962), 6.

12. London: Cambridge University Press.

13. Edward W. Fager, "A Sand-bottom Epifaunal Community of Invertebrates in Shallow Water," *Limnology and Oceanography,* Vol. 13, No. 3 (July 1968), 448.

14. *Ibid.*

15. SIO Annual Report, 1970, 26.

16. SIO Annual Report, 1971, 26.

17. SIO Annual Report, 1974, 38.

18. "Reports on the Dredging Results of the Scripps Institution of Oceanography Trans-Pacific Expedition, July-December 1953," *Systematic Zoology,* Vol. 3, No. 4 (December 1954), 147.

19. Andrew A. Benson and Richard F. Lee, "The Role of Wax in Oceanic Food Chains," *Scientific American,* Vol. 232, No. 3 (March 1975), 77.

20. *Ibid.,* 86.

21. "Introduction to Marine Microbiology," *Contributions to Marine Microbiology,* New Zealand Department of Scientific and Industrial Research Information Series, 22 (1959), 761.

22. "Meeting Microbes from the Deepest Sea," translated by Greta Ashdown from *Aalborg Stiftende,* 29 January 1967.

23. "Summary of Results of Twenty Years of Researches on Marine Phytoplankton," *Proceedings of Sixth Pacific Science Congress,* Vol. 3 (1941), 581.

24. *Ibid.,* 580.

25. *Ibid.,* 583.

26. "Zooplankton Grazing Rates Estimated Under Field Conditions," *Ecology,* Vol. 50, No. 6 (Autumn 1969), 1070.

27. Memorandum to Director Vaughan, undated.

28. Letter of 29 March 1928.

The Scripps pier, from which various oceanic measurements have been recorded for almost three-quarters of a century.

11. The Ocean in Motion: Studies in Physical Oceanography

"The particular satisfaction in having a physical laboratory operating in conjunction with the biological work lies in the fact that whenever a special biological question comes along requiring information from the physical side, the physicist can be appealed to *then and there.*"[1] So said Director William E. Ritter in 1908 in reference to Scripps Institution's first physical oceanographer, George F. McEwen, a graduate student in physics from Stanford. McEwen was often appealed to, for his association with Scripps continued until after his retirement in 1952; he had first worked at the institution during the summers until he finished the requirements for his Ph.D. from Stanford in 1911, then spent one year as instructor in mathematics at the University of Illinois, after which he joined the Scripps staff.

In the course of his long career, according to his colleagues, McEwen:

> applied both statistical and physical methods in studies of the variation in temperature and other properties, and sought relations between ocean changes and

weather and climate. These investigations stimulated him to make an early (1919) estimate of turbulent eddy transfer in the ocean surface layer and later (1938) to introduce an energy equation for computing values of evaporation over the eastern Pacific.

In 1946 several colleagues were involved in diffusion studies in Bikini Lagoon and at their request McEwen developed models of turbulent diffusion from radioactive source areas. The success with these applications led him to devise a model to explain the decay of the large horizontal eddies observed off the coast of Southern California. The paper on these results was published in 1948 as his contribution to the Sverdrup Anniversary Volume.[2]

McEwen installed a tide gauge on the Scripps pier as soon as that structure was completed in 1915, and shortly afterward he installed a complete weather station there also. For a number of years he prepared long-range forecasts of Pacific coast seasonal air temperatures and precipitation, based on correlation with ocean temperatures. These were provided to farm advisers, chambers of commerce, and many businesses, until the project was discontinued during World War II. McEwen also participated in studies by visiting Polish climatologist Wladyslaw Gorczynski in 1939, on a comparison of southern California climates with those of similar sunny regions in Europe and Africa — a project that "involved a large amount of compilation and computing and the preparation of many charts."[3]

Scripps had taken on another laborious task in 1935: supervising the compiling of oceanographic data gathered by "all ships in the Pacific" from 1904 through 1934. The project was under the auspices of the Work Projects Administration, and was located in three rooms in the Long Beach Municipal Auditorium building. As Sverdrup described it:

Physical Oceanography

After data are transferred from the coding sheets to punched cards, the cards are sent to the Washington Office [of the Navy Hydrographic Office] for mechanical tabulation. In accordance with arrangements made with the United States Hydrographic Office, the final coding sheets of data thus compiled are filed at the Scripps Institution for use in various studies of ocean-surface conditions, in particular for investigating variations in the Japan Current.[4]

When the project was ended in 1939, some 60,000 coding sheets from data in the Pacific Ocean were on file at the institution.

The advent of Harald U. Sverdrup to the institution in 1936 led to a considerable increase in physical — or dynamic — oceanography, Sverdrup's own specialty. In his first year he installed at the end of the pier "an electrically operated device for recording the highly localized [current] movements."[5] During World War II, he taught classes for military officers and participated in compiling current charts for life rafts and in the research by Walter H. Munk on forecasting sea and swell for amphibious landings (see chapter 2). In his years at Scripps, Sverdrup's publications covered practically all aspects of his field: upwelling and water masses, evaporation from the ocean, currents and circulation, geostrophic flow, lateral mixing, and oceanic turbulence. His interest influenced many studies and drew many students into physical oceanography at the institution. Prewar student Walter H. Munk returned to Scripps after the war to continue his studies, which became both dynamical and geophysical (see chapter 7).

The establishment of an extensive program in physical oceanography — and its usefulness to the Navy — attracted a number of early postwar graduate students to Scripps. In 1939, Scripps had 8 registered students; in 1947 there were

41, a large proportion of whom were in physical oceanography. Their resources at the institution included daily records of temperature from the pier recorder back to 1915; the records compiled by the WPA-sponsored prewar project; and records from steamship companies that routinely provided surface temperature measurements from their ships in the Pacific to Scripps. After World War II Scripps received the records from stationary weather ships that had been established under Navy auspices during the war, and had recorded every six hours the wind velocity, air temperature, and barometric pressure. Sverdrup persuaded the Navy to add the taking of data on swell condition and water temperatures down to 450 feet every two hours on each of the 20 weather ships stationed from California to Japan.

Among the early returning students in physical oceanography was Robert S. Arthur, who had been assigned to duty at Scripps by the Navy in 1944. After receiving his Ph.D. in 1950 he continued at the institution, becoming professor in 1963. Arthur's early studies were on the refraction and diffraction of waves, and his later work has focused on currents and upwelling. For example, he devised an improved method of predicting mean monthly anomalies of sea-surface temperatures; he has studied oscillations in sea temperature at the Scripps and Oceanside piers; he has researched the dynamics of rip currents; and he has investigated methods for calculating currents at the equator and for determining upwelling velocities along coasts. Arthur has long carried a heavy teaching program at the institution, and has always devoted a great deal of time to students.

BATHYTHERMOGRAPHS

In 1938 the bathythermograph (BT) came into use; this ingenious device for determining temperature below the

Physical Oceanography

ocean surface was the brainchild of Athelstan F. Spilhaus, then at Woods Hole Oceanographic Institution. The original device was modified during World War II especially to gather data on the temperature structure of the ocean for the Navy; it consists of a temperature-sensing element to which is attached a stylus that scratches a trace on a glass slide, recording temperature against pressure. The early slides were prepared "by rubbing a bit of skunk oil on with a finger and then wiping off with the soft side of one's hand," followed by smoking the slide over the flame of a Bunsen burner.[6] The mechanical BT is lowered by means of a small winch on the ship. The instrument drops nearly freely through the water, dragging out wire as it sinks; when the maximum depth is reached, a brake is applied, and the BT is drawn to the surface. Over the years BTs were developed for lowering to 900 feet. But Li'l Abner's Skunk Works was put out of business — at least for BT slides — by replacing skunk oil with an evaporated metal film.

The first BT slides handled by Scripps were a few sent to the institution by Woods Hole in late 1940; Eugene LaFond made prints of these — the first time this had been done — and returned them. During World War II LaFond was in charge of the BT center at the University of California Division of War Research and carried out the analysis of the slides on the Scripps campus. LaFond's group compiled sonar charts of the BT data for the Pacific and Indian oceans for use by Navy ships. After the war Scripps was designated the repository for the BT soundings in the Pacific Ocean that had been taken by Navy and Coast Guard ships — some 100,000 of them, accumulated especially for studies of underwater sound. When LaFond transferred to Navy Electronics Laboratory in 1947, graduate student Dale Leipper was put in charge of the group of Scripps technicians who converted the BT readings into depths, and he was followed by graduate students Wayne V. Burt and John Cochrane.

In the latter 1940s Margaret Robinson entered the scene, first as a draftsman in the BT section. She soon enrolled as a graduate student — in spite of Director Sverdrup, who told her that "women will never be accepted as oceanographers."[7] Mrs. Robinson received her M.S. in 1951, and in 1957 became supervisor of the Bathythermograph Temperature Data Analysis Section, a post that she filled until her retirement in 1974. With "energy, ingenuity and perseverance," Margaret Robinson and her assistants (all women) processed vast amounts of ocean temperature data, from BTs and from shore stations.

The early slides processed by the BT group were gathered laboriously. On Midpac Expedition in 1950, for example, according to watchstander Edward S. Barr:

> The BT winch ... was used every hour.... [It] was operated from the side of the ship. One would lower the recording device — looking like a rocket — over the side, and let it drop, free wheeling, to a predetermined depth. Then the brake would be applied, stopping its descent. Winching power was then applied to reel the device back to the surface and aboard. ... In any kind of rough weather, this BT position was frequently subject to waves making a clean sweep of the deck. In spite of breaking waves over the side, the operator had to hold his station, because the equipment was already over the side. One couldn't run for shelter as the brake and hoisting power were combined in a single hand lever. To let go of this lever would cause all the wire on the winch to unwind, sending the recording device and all its cable to the ocean bottom forever. It was not at all uncommon, from the protective position of the laboratory door, to look back and

see your watchmate at the BT winch completely disappear from sight as a wave would come crashing over the side. . . . We also took turns taking BT readings. It wasn't fair for only one person to get wet consistently.[8]

Crashing waves were not the only problem. As B. King Couper and Eugene C. LaFond noted of the wartime technique: "Probably the greatest hazard was a swinging BT after it left the water. A familiar saying was: 'sight, surface, oh that son of a gun,' or words to that effect, as the instrument swung in circles around the boom."[9] James M. Snodgrass noted that "Oftentimes, in somewhat heavy — and not so heavy — weather, [the BT] behaves as a tethered lethal missile. The bathythermograph cable with bathythermograph attached has occasionally been wrapped around the ship's funnel."[10]

Snodgrass, in fact, found himself "appalled at the primitive nature of the BT lowerings"; a physics graduate of Oberlin College, he had served at the University of California Division of War Research during World War II, and after two years in industry, he returned to Scripps at the invitation of Roger Revelle in the fall of 1948. He participated in the development of several of the new oceanographic tools of the early 1950s. By 1958 he was in frequent correspondence with B. King Couper of the Navy Bureau of Ships over the possibility of developing an easier method of taking BTs, specifically an expendable bathythermograph. Snodgrass's idea was to use a wire-connected device, instead of an acoustic signal. As he described it to Couper:

> Briefly, the unit would break down in two components, as follows: the ship to surface unit, and surface to expendable unit. I have in mind a package which could be jettisoned, either by the "Armstrong"

Louis Garrison taking a mechanical bathythermograph recording on Midpac Expedition, 1950.

method, or some simple mechanical device, which would at all times be connected to the surface vessel. The wire would be payed out from the surface ship and not from the surface float unit. The surface float would require a minimum of flotation and a small, very simple sea anchor. From this simple platform the expendable BT unit would sink as outlined for the acoustic unit. However, it would unwind as it goes a very fine thread of probably neutrally buoyant conductor terminating at the float unit, thence connected to the wire leading to the ship.[11]

A number of companies were entering the growing field of ocean industry, and Snodgrass urged them toward developing an expendable bathythermograph (quickly called an XBT), which was much desired by the Navy. He discussed with engineers from various companies his ideas on a feasible approach. In the early 1960s the Navy called for bids on an XBT, and three companies, including Sippican Corporation of Marion, Massachusetts, received contracts to provide a small number of units. Within a short period of time, Sippican became the sole supplier of XBTs.

Their unit, based on Snodgrass's original idea, consists of: a probe that falls freely at 20 feet per second in the ocean; a wire link; and a shipboard canister that remains in the launcher until the measurement is made. The wire link, which is wound on a spool in the probe and a second spool in the canister, is unreeled from both spools as the probe sinks and the ship moves away, so that the wire remains stationary. A thermistor temperature sensor in the probe is connected electrically to a chart recorder through the three-conductor fine wire and a cable from the launcher to the recorder. The sinking rate of the probe determines its depth and so yields a temperature-depth trace on the recorder.

Scripps Institution of Oceanography: Probing the Oceans

Over the years the quality of XBTs has been improved well beyond the accuracy of the laborious mechanical BTs, and at a cost per unit measurement that is considerably less. Also, the depth to which XBTs can be used is much greater.

In 1967, Jeffery D. Frautschy, Marston C. Sargent, and Phillip R. Mack, supervisor of the staff shop, designed and built the BT digitizer, a machine to digitize analog BT traces automatically. Computer programs were developed to speed up the data reduction and to conform to the National Oceanographic Data Center system. The programs could also be used to derive annual temperature distribution from 125 meters to the ocean bottom and to determine annual salinity distribution from surface to bottom.

The processing group handled BT data from many other institutions in several countries, as well as Scripps and Navy BTs. From the mass of material Margaret Robinson prepared "carefully compiled atlases of the temperature distribution of the oceans and seas of the world." The first one, on the North Pacific Ocean, was prepared in 1971 for the U.S. Naval Oceanographic Office; it provided monthly temperature distribution at the surface and at levels of 100, 200, 300, and 400 feet. Atlases for the Gulf of Mexico, the Caribbean, the Red Sea, the Mediterranean, and the Black Sea followed; these brought the comment to Mrs. Robinson, upon her retirement, from Chief of Naval Research Rear Admiral M. D. Van Orden that her compilations had "formed the basis for the Fleet's oceanographic forecasting capability, the source material for many scientific evaluations of the physical ocean, especially the North Pacific, and provided the framework for many oceanographic expeditions." By that time Margaret Robinson — and "her girls" — had processed 477,483 bathythermograph slides.

Couper and LaFond noted:

Physical Oceanography

Although the accuracy of [BT] data is not always as good as desired, they have proven extremely useful to both the Navy and others in understanding water structures and the physical, chemical and biological processes which occur in the upper layers of the sea. For instance, the correlation between time-lapse photographs of moving slicks and BT layer-depth information established the relationship of surface slicks to internal waves. Even in meteorology, heat transfer can be more accurately established with BT data.[12]

CURRENTS

At about the time that Sverdrup arrived at Scripps in 1936, McEwen wrote: "It is especially desirable to undertake a systematic investigation of the cold California Current, since virtually nothing is known as to the amount of water carried by this current or its seasonal variations or its changes from year to year."[13] Sverdrup agreed with this point, and encouraged a program of repeated cruises with the institution's new vessel, the *E. W. Scripps,* to study the oceanic region from San Diego to San Francisco out to several hundred miles (see chapter 2). In addition, Richard H. Fleming rode the Fish and Game Commission ship *Bluefin* twice and Erik Moberg rode it once in 1936, for various studies including the release of 6,000 drift bottles to try to trace currents. Within a few years the California Current and its fluctuations had been outlined.

The early exploratory expeditions by Scripps made current measurements well beyond the California Current. On Midpac Expedition, researchers using a borrowed geomagnetic electrokinetograph (GEK) found unexpected surface eddies near the equator. On Northern Holiday

Expedition, 305 observations were taken with the GEK. On Shellback Expedition a new device was tried: a current cross, originally devised by Donald W. Pritchard and Wayne V. Burt, and modified by John A. Knauss. "Techniques in handling the cross and in measuring wire angles were primitive," noted Wooster,[14] but the measurements were consistent with those from the GEK. Knauss commented ruefully: "The truth of the matter is that no completely satisfactory method of measuring currents has been devised."[15] Over the years improvements led to developing parachute drogues, current meters, and anchored buoys; the results were combined with calculations on geostrophic motion and analyses of salinity and dissolved-oxygen distribution, to determine the circulation of the waters of the world's oceans. Drift bottles have also been used, especially by the Marine Life Research program, for tracing nearshore surface currents.

Warren S. Wooster turned his expertise on the currents of the deep ocean toward international solutions to sticky problems, especially in fisheries and cooperative researches. During Navy service in World War II he determined to study the oceans, received his M.S. at Caltech, and then entered Scripps for his Ph.D. As a student he participated in CalCOFI cruises throughout the California Current, and he was chief scientist on parts of Northern Holiday and Shellback expeditions. Armed with his just-received Ph.D., he led Transpac Expedition to Japan in 1953. During 1957 and 1958, Wooster was introduced to international oceanography as director of investigations for the Council of Hydrobiological Investigations in Lima, Peru. There he became interested in the unusual El Niño current that intermittently destroys Peru's fishery and guano industries.

On his return to Scripps he was soon organizing another current-chasing expedition, Step-1, which in 1960 sailed to South America in cooperation with the Inter-American

Physical Oceanography

Tropical Tuna Commission and the Bureau of Commercial Fisheries. Studies of currents on the Scripps expeditions during the International Geophysical Year, led by John A. Knauss (see chapter 15), had left unsolved questions on the water movement beneath the equatorial countercurrent. The researchers on Step-1 found a new surface current 300 miles wide a few degrees south of the equator — which had been predicted by Joseph L. Reid, Jr., shortly before. They also found another predicted current beneath the Peru Current. In tracing the Peru Current and the Chilean Current, they found, between the two off northern Chile, a patch of relatively warm water which was rich in tuna. They still could not determine, however, where the waters of the equatorial undercurrent — the Cromwell Current — moved eastward of the Galápagos Islands.

On various other expeditions over the years Wooster pursued studies of ocean currents and of regions of coastal upwelling — for example, along the coast of Africa, a study that was based on measurements of temperature, wind, and ship drift collected by merchant vessels during the past century.

From 1961 to 1963, Wooster was the first Director of the Office of Oceanography of UNESCO, and secretary of the Intergovernmental Oceanographic Commission, located in Paris, and there he "quickly experienced the harsh realities of international science. The incessant demands of bureaucracy, the incessant digestion and production of paper and then the incessant and interminable meetings where participants behave less than scientific have all left him unperturbed while he has proposed some action which would be accepted by both the governmental and the scientific communities, and which would break down another barrier interfering with the progress of marine science."[16]

Wooster served as secretary to and, from 1968 to 1971, president of SCOR (Scientific Committee on Oceanic

Research), the advisory body to the oceanographic program of UNESCO. Other international committee appointments, concerned with fisheries and with freedom of scientific research in the oceans, came to Wooster as he continued as a professor at Scripps and as the first director of the Center for Marine Affairs. In 1973 he left Scripps to become director of the University of Miami's Rosenstiel School of Marine and Atmospheric Science, a post he held until 1976.

John A. Knauss began studies of currents as a Scripps graduate student, on Shellback Expedition in 1952. He became especially interested in the Cromwell Current, first observed by researchers who were longline-fishing for tuna at the equator on the U.S. Fish and Wildlife Service ship, *Hugh M. Smith,* in 1951. The longline gear drifted to the east while the surface drift of the ship was to the west. In 1952 Townsend Cromwell, then with the Fish and Wildlife Service in Hawaii (later at Scripps), with R. B. Montgomery and E. D. Stroup, made direct current measurements of the newly found current — the first large permanent subsurface current to be identified. They proposed the name Pacific Equatorial Undercurrent for it. Knauss and Cromwell traced it with free-floating parachute drogues on Eastropic Expedition in 1955, and Knauss continued the study on Dolphin Expedition of the International Geophysical Year in 1958. On 3 June, three days after the Dolphin measurements were completed, Townsend Cromwell was killed in a plane crash in Mexico en route to the *Spencer F. Baird* for tuna studies. Knauss proposed that the major subsurface flow be named the Cromwell Current. (Both names are presently used.)

The current was found to be a thin, swift flow of water moving eastward beneath the equator at speeds up to three knots. For several years the whereabouts of the current east of the Galápagos Islands was unknown. Knauss tried to trace it in that area on Swan Song Expedition of 1961 — his

Physical Oceanography

last for Scripps before leaving to become dean of the Narragansett Marine Laboratory of the University of Rhode Island. Finally, measurements made from an Ecuadorian ship, *Huayape,* on Eastropac Expedition in 1967-68 and from the *Thomas Washington* on Piquero Expedition in 1969 convinced physical oceanographers Merritt Stevenson (of the Inter-American Tropical Tuna Commission) and Bruce A. Taft (of Scripps) that the undercurrent continued as a high-salinity core beneath the southern edge of the Equatorial Front.

Joseph L. Reid, Jr., entered physical oceanography by way of expeditions into the California Current, and went on to studies of other surface currents, to the circulation of intermediate-depth waters, and to the movement of bottom currents. Texas-born Reid received a B.A. in mathematics from the University of Texas in 1942 and entered the Navy. After World War II he enrolled at Scripps and received his M.S. in 1950, and he continued at the institution in the Marine Life Research program, of which he became director in 1974. In 1955 he coordinated the extensive synoptic survey called Norpac (see chapter 3).

To determine what happens in the North Pacific in the winter, Reid set up Zetes Expedition in 1966 and was chief scientist on the second leg, from Kodiak to Hakodate in midwinter; that leg was named Boreas for the god of the North Wind. The Scripps ship *Argo* endured wind gusts up to 70 knots, plunged into waves up to forty feet high, and at times her decks were covered with ice. In the Sea of Okhotsk near the coast of the Soviet Union, *Argo* acquired a watchdog: a Soviet Navy ship that hovered nearby and watched every move of the American researchers. The Russians would not communicate (except for a few waves) and their ship pulled away when *Argo* put over a small boat in the hope of trading ice cream for vodka. From data gathered on Zetes Expedition, Reid demonstrated that "the characteristics of the low-salinity intermediate layers of middle latitudes are derived not directly from the sea surface in high latitudes but by vertical mixing which takes place beneath the mixed layer."[17]

Ready for warmer weather, Reid led two legs of Styx Expedition to Samoa in 1968. The aim was to determine the flow of the circulation of the bottom waters. The deep, dark river surrounding Hades for which the expedition was named proved to be — as expected — a submarine current that transported cold Antarctic bottom water through a gap 80 to 100 miles wide into the Pacific Ocean at a velocity of six inches per second, "by far the highest on record for an ocean bottom current," according to Reid.

From his analyses of the circulation of the bottom waters, Reid has concluded:

> From the Norwegian-Greenland Sea the cold and saline water is traced southward through the Denmark Strait, where vertical mixing raises both temperature and salinity to their maximum values in the central North Atlantic. From there the temperature and salinity decrease monotonically southward toward the Weddell Sea, partly by lateral mixing with the cold, low-salinity waters on this stratum where it lies near the sea surface in the Weddell Sea, and partly by vertical mixing with the underlying Antarctic bottom water. From the southern South Atlantic the high values of temperature and salinity (the stratum now lies close to a vertical maximum in salinity) extend eastward with the Antarctic Circumpolar Current into the Indian and Pacific oceans, with monotonically decreasing temperature and salinity as further vertical mixing erodes the maximum in salinity, until the

"Oceanography is fun." The North Pacific leg of Zetes Expedition, 1966.

salinity maximum is found at the bottom in the North Pacific Ocean.[18]

AIR-SEA INTERACTION

The many studies at Scripps that have begun with the ocean and turned toward the weather are an example of the institution's diversity and complexity. They also show that the organization of Scripps shifts as readily as do the sand grains in a rip current.

In 1967, for example, Director Nierenberg observed that "an inventory at the Institution shows that no less than eight groups are working on aspects of [air-sea interaction]."[19] Administratively, these groups were scattered among the Applied Oceanography Group, the divisions of earth sciences, of marine biology, and oceanic research, the Marine Life Research group, the Visibility Laboratory, and the Institute of Geophysics and Planetary Physics. Since then, the division names have been changed, AOG has expired, some of the people have turned to other studies, and others have changed administrative units. But the weather continues much the same — and continues to be studied.

The earliest attempt of Scripps to observe the area where the sea meets the sky was ill-fated. It began in August 1931 with an offer by Lt. Harold L. Kirby that he would gather oceanographic and meteorological data for the institution from a seaplane. Lt. Kirby's background and experience were impressive: U.S. Naval Air Force, public-spirited civic movements; research in meteorology and in flying; pilot for Pacific Marine Airways from Los Angeles to Catalina Island; U.S. Army Corps Reserve.

His request was for official support from the institution — and a month's salary — in order to arrange, through

Physical Oceanography

other sources that he knew of, for the purchase and operating expenses of an airplane, sources that wanted "to have him associated with a high-grade institution before actually paying over any funds."[20]

Several Scripps staff members were quickly persuaded of the value of gathering data from both sea and air by this means. They envisioned the airplane as capable of "carrying the gear necessary for all of our work at sea except the especially heavy equipment used in deep sea dredging"[21] — at no cost to the institution. The Naval Air Station at North Island offered to provide Douglass seaplanes to Kirby for a short series of flights that would take aerial measurements up to 1,000 feet and sea measurements down to 50 meters. After those flights Kirby hoped to design a seaplane especially for marine observations.

He was given an appointment as Associate in Meteorology for one month, and the enthusiasm for the project continued. But at the end of the month his report was not finished, and so the appointment was extended another month. By that time the special meteorological fund had been sadly depleted, and doubts were creeping in as to the possibility of the outside sources of funds. In addition, Vaughan noted that "Lieutenant Kirby's conduct around the Institution has been such that he has offended a number of people. Apparently he thinks that he may order certain people connected with the Institution in a way in which not even I myself would think of doing."[22] The appointment was not continued.

The next Scripps group to use an airplane for ocean research was more experienced: the Applied Oceanography Group (AOG), which was established in 1961 to study certain problems of particular interest to the Navy. Its area of study was especially the top few millimeters of the ocean surface. The laboratory was directed by Edward D. McAlister until 1968, when the last members of the group were transferred into the Marine Physical Laboratory as a separate unit. In its early years the laboratory included about twenty people and was located in Scripps Field Annex on Point Loma.

During World War II McAlister had pioneered in the development of proximity fuse for antiaircraft guns. He joined the Scripps staff in 1961 from the Naval Ordnance Division of Eastman Kodak and turned to studies of the mechanism of heat transfer in the topmost layer of the ocean. He designed an infrared radiometer, "a very ingenious gadget . . . which looks at the water through two different windows — one at around 2.5 microns and the other at around 3 microns." McAlister found that "the difference in absorption of water in these two bands is very great and, therefore, gives him simultaneous measurements of the radiant heat flux from two different depths right near the surface and, therefore, a very delicate measure of the thermal gradient due to evaporative cooling."[23]

In 1962 AOG leased a DC-3 airplane, later given to them, on which they mounted the infrared radiometer to measure heat flow from the ocean. With the airplane it was possible to survey 10,000 square miles of sea surface in 24 hours. The top 30 meters of the world's oceans store 10^{21} calories or 10^{15} kilowatt hours of solar energy in an average day. The release of this heat depends upon wind speed, cloud cover, air temperature, and other factors. Through various measurements, McAlister and colleagues concluded that different mechanisms of heat transfer to the surface dominate within different depth regions.

In other studies of the topmost layer of the ocean, AOG researchers measured the vertical motion in the water by means of Rhodamine-B dye tracer plus fluorometer, an extremely sensitive method aimed at determining the convective activities of the ocean above the thermocline, and they measured changes in the wave spectrum by means of telemetering buoys.

Gifford C. Ewing, already a staff member at Scripps, was one of the organizers of AOG, and he continued with that group until 1964, when he left for Woods Hole Oceanographic Institution. He had participated in Operation Crossroads in 1946 (see chapter 15), and then entered Scripps for graduate work. For his Ph.D. dissertation he set out to determine what causes band slicks — "calm streaks on a rippled sea." He photographed them, he measured temperatures in and alongside them, he watched the behavior of bits of paper dropped near them. Slicks, he concluded, "are formed by the ripple-damping action of a surface film of organic matter which occurs naturally on biologically productive waters." The compaction of the surface film can be caused in several ways; one of them, "typical of summer conditions on the California Coast, is a train of long internal waves in a shallow thermocline."[24]

Ewing was a proficient airplane pilot and, in his own airplane, he occasionally transported other Scripps researchers to remote areas of Baja California for special studies, such as coastal lagoon work by Fred B Phleger and gray whale censuses by Carl L. and Laura C. Hubbs. Ewing knew, he was sure, every possible landing strip for an airplane throughout the sparsely populated and isolated peninsula.

The DC-3 airplane originally acquired by AOG was used for occasional studies by other Scripps staff members until it was sold in 1974.

The amount and time of release of solar energy from the ocean and its transport to other parts of the globe are factors that determine climate and weather. Besides AOG, other groups at Scripps were looking into that transfer of energy.

Scripps Institution of Oceanography: Probing the Oceans

Several of them participated in the Barbados Oceanographic and Meteorological Experiment (BOMEX) in 1969, along with researchers from a number of other institutions. "Probably the most extensive air-sea investigations ever made of a large-scale ocean area," said one of the participants, BOMEX was set up to document the meteorological situation in a five-hundred-mile square in the Atlantic Ocean. The AOG researchers participated, using the DC-3, and "were rewarded by the making of some scientific history — the first airborne measurements of the total heat flow from the sea surface" — which was acknowledged by the Environmental Science Services Administration with a plaque of appreciation. *Flip* was also a participant in BOMEX, in the craft's first venture into Atlantic waters. Scientists from a number of institutions used the quiet platform, which was outfitted with special booms, for observations and measurements just above the surface of the sea. Among these researchers were Carl H. Gibson, Russ E. Davis, and Charles Van Atta, all of Scripps, who carried out measurements of air turbulence, temperature fluctuations, and wave parameters.

In the Pacific Ocean a major investigation of the causes of weather began in the 1960s, as an extension of the Marine Life Research program, under the guidance of John D. Isaacs. In the hope of determining the causes of fluctuations in fish populations in the California Current region, the MLR program expanded its scope to investigate large-scale air-sea interaction in the entire North Pacific Ocean.

The key to understanding local variations is recording weather parameters over a large area by continuous observation — a method that is tedious at best and expensive in ships and people. Isaacs's proposed plan — the North Pacific Buoy program — was to use unmanned, fixed instrument packages with continuous recorders, moored in deep water far from the coast. Isaacs and Willard Bascom had developed

Physical Oceanography

the first deep-moored recorders in 1952 for recording surface and near-surface explosion waves and, later, radioactive fallout from the nuclear-bomb tests.

In the 1960s MLR engineers George Schick and Meredith Sessions pursued the design problems of moored buoys for instrument stations — not only the instrumental problems, but such puzzles as how to discourage inquisitive fishermen from disassembling buoys, sea lions from hauling out on them, and sharks from biting the lines. The result — twelve feet long, twin-hulled, oval in shape, and striped in bright orange and black — was called the bumblebee buoy. The first two units were moored in 1968, north of Hawaii, in water 12,000 and 18,000 feet deep; they held instruments for continuous recording of wind speed and direction, water temperature down to 1,500 feet, barometric pressure, and solar radiation. Along with the Scripps units were installed "monster buoys," 40 feet in diameter, which had been developed by the Convair Division of General Dynamics. These giants telemetered information on weather phenomena and wave heights back to a receiving station (a converted bus) at Scripps for computer analysis. In 1972 two buoys were installed within the California Current, and current meters were placed on the sea floor adjacent to them.

Jerome Namias, formerly head of the Extended Forecast Division of the U.S. Weather Bureau, joined the Scripps staff on a part-time basis in 1968, to pursue means of deriving long-range weather forecasts by use of long-term interactions between the atmosphere and the ocean. Using historical files of records from merchant ships as well as buoy-gathered data, he has found correlations of ocean-surface temperature patterns with temperature fluctuations across the continent. He determined "that a major change in winter wind, weather and ocean surface temperature patterns took place between the roughly decadal periods

1948-57 and 1958-70. Clear evidence of the change shows up in temperature anomalies over the United States where in the earlier decade it was unseasonably cold in the west and warm in the east, but the reverse in the latter decade. It was demonstrated that these changes were associated with equally remarkable changes in North Pacific sea-surface temperatures and upper air wind patterns."[25]

Joseph C. K. Huang developed a numerical dynamic model to simulate the North Pacific Ocean, based on the hydrodynamic equations for fluid in a basin and incorporating the configuration and bottom topography of the region. "This model was developed for the study of air-sea interacting mechanisms in order to understand the physical nature of large-scale normal characteristics and anomalous changes in the North Pacific Ocean in response to the various seasonal meteorological conditions."[26]

From the broad base of the North Pacific Buoy program an even vaster project has emerged: NORPAX, the North Pacific Experiment, designed to study the interaction between the upper waters of the North Pacific and the overlying atmosphere. This multi-institution project, organized in 1972 and headquartered at Scripps, has drawn in researchers from several units at the institution.

An active participant in NORPAX is Charles S. Cox, who has been on the Scripps staff since he received his Ph.D. at the institution in 1954. His researches have been aimed at measuring microstructures — fine-scale fluctuations in temperature, salinity, and magnetism — within the ocean waters to determine the processes of mixing. Cox developed free-fall instruments that have been used on a number of expeditions. With such an instrument temperatures are recorded continuously as a tube sinks slowly through the water; it returns to the surface after completing the measurements and dropping its ballast. Such measurements provide a fine-scale picture of temperature variations

Physical Oceanography

with depth and thus information on the flow of heat within the water. These microstructure measurements have shown that individual water layers are separated by sharp interfaces, and single layers can be traced for short distances. Cox, along with Jean H. Filloux, also has measured electric and magnetic field fluctuations at the surface of the ocean, to determine features of ocean motion and of sea-floor structure.

OTHER PROJECTS IN PHYSICAL OCEANOGRAPHY

Other studies of motion in the ocean and the impact of the ocean on the land have been carried out throughout the years at Scripps.

Dale Leipper, for example, in 1947 and 1948 conducted a long study for the Navy of the formation of fog in San Diego. He also analyzed the long record of surface temperatures accumulated by the institution, especially by George F. McEwen, to determine the water temperature fluctuations along the southern California coast.

William G. Van Dorn began in mechanical engineering before earning his Ph.D. at Scripps in 1953 and continuing on the staff. In 1948 he devised a magnesium-rod release timer for a deep-current meter on a project for the Office of Naval Research. During the International Geophysical Year, Van Dorn directed the Scripps program of wave recording in the island observatory project. He devised a long-period wave recorder for tsunami prediction, and he studied the pattern of the destructive tsunami at Hilo, Hawaii, caused by the earthquake of 23 May 1960 near Chile, and the tsunami caused by the earthquake of 28 March 1964, off Alaska. In the mid-1960s Van Dorn undertook a study of the circulation of water around various Pacific islands, and he has also made observations and measurements of the pattern of breaking of waves in deep water, by means of controlled studies in the Hydraulics Laboratory.

Two Scripps researchers, Walter H. Munk and Myrl C. Hendershott, advised the city fathers of Venice, Italy, in 1971 and 1972, on the periodic flooding of that city. Venice is built on mud flats, slowly sinking because of industrial removal of ground water; also, sea level is gradually rising throughout the world. Flooding at Venice occurs during storms with strong winds from the south. Hendershott, with an Italian scientist, developed a mathematical model of the circulation of the Adriatic Sea to predict circulation patterns and dispersion of pollution. Munk has proposed that Venice use caissons to close the three openings into Venice lagoon during storms.

Hendershott joined the Scripps staff in 1965, after receiving his Ph.D. from Harvard, on studies of internal waves carried out partly at Scripps. He worked out solutions to Laplace's tidal equations by using the method of finite differences. Satisfactory models derived for the global tides had to include the actual bottom relief and the deformation of the solid earth from the weight of the oceanic tidal column.

NOTES

1. W. E. Ritter, "The Scientific Work of the San Diego Marine Biological Station During the Year 1908," *Science,* Vol. 28, No. 715 (11 September 1908), 332.

2. Robert S. Arthur, Denis L. Fox, Carl L. Hubbs, and Russell W. Raitt, *In Memoriam,* University of California (July 1975), 92.

3. H. U. Sverdrup, "Research Within Physical Oceanography and Submarine Geology at the Scripps Institution of Oceanography During April 1939 to April 1940," *Transactions of the American Geophysical Union for 1940,* 344.

4. Sverdrup, "Research Within Physical Oceanography and Submarine Geology at the Scripps Institution of Oceanography During April 1938 to April 1939," *Transactions of the American Geophysical Union for 1939,* 423.

5. W. E. Allen, "The Growth of a Marine Observatory," *Internationale Revue der gesamten Hydrobiologie und Hydrographie,* Vol. 39 (1939), 467.

6. Letter from Allyn Vine to Richard H. Fleming, 20 August 1941.

7. *Los Angeles Times,* 8 February 1973.

8. "MIDPAC – The First Big Step," manuscript, 17 August 1975.

9. "The Mechanical Bathythermograph: An Historical Review," paper presented at Silver Jubilee of the Instrument Society of America, 26-29 October 1970, Philadelphia.

10. "New Techniques in Undersea Technology," *IEEE Transactions on Aerospace and Electronic Systems,* Vol. AES-2, No. 6 (November 1966), 626.

11. Letter of 29 December 1958.

12. "The Mechanical Bathythermograph," paper, 26-29 October 1970.

13. "The University and the Pacific," *California Monthly* (March 1937).

14. "Preliminary Report, Shellback Expedition," SIO Reference 52-47, 9.

15. "SHELLBACK Expedition," *Naval Research Reviews,* Vol. 6 (May 1953), 4.

16. "Warren S. Wooster," *American Oceanography* (August 1968), 3.

17. Joseph L. Reid, Jr., "Zetes Expedition," *Transactions of the American Geophysical Union,* Vol. 47, No. 4 (December 1966), 558.

18. *California Cooperative Oceanic Fisheries Investigations Reports,* Vol. 17 (October 1974), 14.

19. SIO Annual Report, 1967, 5.

20. Letter from T. Wayland Vaughan to J. C. Harper, 6 August 1931.

21. Letter from George F. McEwen to T. Wayland Vaughan, 3 August 1931.

22. Letter to University President R. G. Sproul, 3 November 1931.

23. Letter from Gifford C. Ewing, 30 November 1961.

24. G. C. Ewing, "Slicks, Surface Films and Internal Waves," *Journal of Marine Research,* Vol. 9, No. 3 (1950), 161.

25. *California Cooperative Oceanic Fisheries Investigations Reports,* Vol. 17 (October 1974), 12.

26. *Ibid.*

12. Sand, Silt, and Sea-floor Spreading: Studies in Marine Geology

"In 1940," wrote H. William Menard in 1969, "it appeared that the sea floor was a relatively quiet place with minimal relief and that all mountain building and other important geological processes occurred on continents or at their margins."[1] At that time the known relief of the Pacific basin was a few deep trenches and bits of the mid-ocean ridge system, discovered on the few prewar explorations and cable surveys.

Then geologists began going to sea, many of them because the Navy asked them to go. They went with enthusiasm and great expectations, carrying an encyclopedic knowledge of the native customs on each tropic isle. In general, those at Scripps subscribed to Menard's theme: "There is no virtue in going to an unpleasant atoll if a beautiful one has the same geology."[2]

During the 1950s, "shattered beliefs soon became a commonplace as it developed that almost everything supposed about ocean basins was wrong."[3] That "relatively quiet place" was found to be gashed by trenches, humped with undersea volcanoes and flat-topped guyots, and arched by rises and ridges. To marine geologists the

Scripps Institution of Oceanography: Probing the Oceans

Himalayas became molehills and the Grand Canyon a ditch.

By the 1960s it had become apparent that in the ocean basins are found the most active areas of the earth's crust. Now it is believed that new crustal material slowly rises from the mantle at oceanic ridges and spreads to the sides, leaving its record in the reversals of magnetization of the rocks. The surface of the earth is divided into about ten major rigid plates which are moved in respect to one another by deep, as-yet-unknown forces. The land areas are carried by these motions toward or away from each other. Where two plates meet, one is thrust beneath the edge of the other, usually in a deep oceanic trench.

To reach these conclusions, geologists have joined forces with geophysicists. Together they have devised a remarkable roster of hardware, much of it developed or put to shipboard use quite recently, and some of it apparently susceptible to mal de mer. A geologically oriented expedition setting out now carries echosounders, heat-flow probes, gravity meters, corers, magnetometers, deep-sea cameras, grab samplers, dredges, and assorted seismic equipment. Huge winches and towering A-frames manipulate the gear, along with straining muscles.

Scripps Institution has been, of course, only one of many research centers that have contributed to the discoveries in geology (some of which are also covered in this book in chapter 4). Equipment and theories have been invented all over the world, and borrowed freely. Scripps scientists have ridden ships of other institutions and have reciprocated with their own ships. Scientists have moved from one institution to another, also, bearing their favorite ideas. Scripps has contributed the results from a great many expeditions throughout the Pacific Ocean, some in the Indian Ocean, and a few in other seas. The institution has also contributed to the floor of the ocean a great many

Marine Geology

pieces of hardware, and to the maps a great many new names of sea-floor features in honor of wives and colleagues and ships.

Marine geology on the west coast began with Francis P. Shepard at Scripps Institution and almost simultaneously on the east coast with Henry Stetson at Woods Hole Oceanographic Institution. Shepard entered the geology profession in the 1920s only to be "assured time and again that the major problems of geology had been solved"[4] — but he didn't believe it. A summer of sea-floor sampling from his father's yacht off New England quickly showed him that offshore sediments were not evenly distributed outward from coarse to fine as the textbooks had said. In 1933 he spent part of a sabbatical from the University of Illinois in La Jolla; he was then identified in the local paper as "one of the most energetic students in the U.S. . . . of marine bottom deposits and the configuration of the sea bottom."[5] Four years later he moved to Scripps as a visiting investigator to find out about sediments, incidentally bringing with him the largest grant ever awarded up to that time by the Geological Society of America: $10,000.

Simply by wading and rowing out from shore, Shepard became a marine geologist. Essentially his approach has been to extend land geology onto the sea floor, and he has continued to work chiefly on nearshore phenomena. In those primordial days of marine geology, Sverdrup looked at it differently: "The work of Shepard is not oceanographic," he noted in 1937. "We shall continue work in sedimentation which is a problem in oceanography, but to us details of the submarine topography are of small interest and the geological character of the rocks forming the bottom is insignificant."[6] Shepard continued his work at Scripps under the Geological Society grant, alternating between the seashore laboratory and the University of Illinois. In 1942 he joined the University of California

271

Francis P. Shepard beaming as he enters the Diving Saucer for a look at his submarine canyon. Photo by Ron Church.

Marine Geology

Division of War Research. There he undertook studies of the nature of the sediments and the effect of them on underwater sound waves; he also compiled sediment charts of Asiatic continental shelves from what meager material was available. After World War II, by which time marine geology had become thoroughly respectable, he became a permanent member of the Scripps staff.

On his arrival at Scripps in 1937, Shepard quickly had his eye and sextant on, his rowboat over, and his sounding line in Scripps submarine canyon, while awaiting the conversion of the *E. W. Scripps* into a research ship. With sounding line, with the help of a helmeted diver, and later of Scuba divers, with echosounder, with marker buoys, with underwater camera, and from diving saucers, he has mapped and pored over "his" canyon for almost forty years, until it has become certainly the best known in the world. (Shepard's long-used shore points for map coordinates for the Scripps canyon have baffled more recent students, who can scarcely be expected to know that "NFG" is the north front gable of the house in which Robert Dietz lived long ago, or that the "monstrosity" is the Moorish-castle-style house on Torrey Pines Road.)

The helmet-and-airhose diver who explored the submarine canyon at first hand in 1947 for Shepard was Frank Haymaker, a zoology major from UCLA who learned geology in a novel way. As Shepard said: "Diving into a narrow submarine canyon which has vertical and even overhanging walls is a hazardous undertaking."[7] Haymaker on the sea floor communicated with Shepard and the surface boat by two-way telephones. When Haymaker found interesting rocks, "a skiff was rowed over to the diver and a sledge hammer and a chisel lowered so that he could break off the rocks and send them up in a bag. . . . Haymaker obtained samples of the fill on the canyon floor by driving in pipes. The personnel at the surface would pull out the pipe, and

Haymaker would tie a cover over the pipe end to hold in the sample."[8] He also photographed bottom features, dispensing with a tripod and holding the underwater camera by hand to stay ahead of the cloud of sediment he stirred up.

During the early 1950s, Scuba divers replaced the helmeted diver and continued the firsthand observations of Shepard's canyon.

A debate over the formation of submarine canyons raged for decades. For quite some time, Shepard and others advocated that such canyons had been cut by rivers and insisted that sea level had to have been sufficiently lower to allow for aerial erosion. But Shepard became persuaded that the depths were too great, and he now believes that the deep gashes in the continental slope were shaped by a combination of moderate lowering of sea level, turbidity currents, and other currents and mass movements. He bowed gracefully to the accumulating evidence and modified his opinion with the comment that "it is monotonous trying to support the same old ideas."[9]

Shepard's interest in nearshore features has not kept him close to home; indeed, it has turned him into a world traveler, almost always to a delightful isle or shore with sun-drenched beaches. He selected the north side of Oahu, Hawaii, as a comfortable place to write *Submarine Geology* in 1946, and so on 1 April of that year he became an expert on tsunamis, earthquake-generated sea waves. He brought to the attention of the world that the first wave is not necessarily the highest — a point emphasized by his having lost his book manuscript and notes to the ninth giant wave. Himself he saved by scrambling up an ironwood tree and hanging on for dear life.* Thereupon, he set out to measure

---

*Mrs. (Elizabeth) Shepard had stayed on the high ground to which they fled after the first wave wakened them, while he returned to their house to salvage what he could.

the heights that the tsunami waves had attained around the Hawaii coast and to determine the offshore features that damped or reinforced them.

While touring coastal regions over the years, Shepard gathered samples for dating sea-level fluctuations throughout the world. By means of radiocarbon dating, done chiefly by Hans Suess's laboratory (see chapter 13), he established that for the last 17,000 years the sea level has risen an average of 25 feet every thousand years, but more slowly toward the present. In recent years he has installed series of current meters to record the movement of material down — and up — a number of submarine canyons along the Pacific coast. While the net movement is downward, considerable back-and-forth motion of the sediments has also been recorded.

In 1966, the Society of Economic Paleontologists and Mineralogists established the Francis P. Shepard Award for Excellence in Marine Geology; they defined Shepard's areas of prominence as the distribution and characteristics of sediments, marine geomorphology, and the structure of the continental margins. Shepard, said science writer Bryant Evans, "gives you the impression of being engaged in a delightful hobby."[10] His walk is ever jaunty, his tone is always enthusiastic, and his only variation from an ever-present smile is a beaming smile. Emeritus since 1966, he continues to pursue his hobby — very professionally — on beaches from Moorea in the Society Islands to La Paz, Baja California.

From 1951 to 1957 Shepard directed API Project 51, sponsored at Scripps by the American Petroleum Institute. This was a geologic study of the coastal waters of the Gulf of Mexico for the purpose of remedying "the admitted ignorance of geologists about the conditions of deposition of sedimentary formations similar to those in which oil is found."[11] From 1954 API Project 51 was administered through the Institute of Marine Resources (see chapter 6).

Scripps Institution of Oceanography: Probing the Oceans

The first area selected for detailed studies by API Project 51 was the northwest Gulf of Mexico, which has had rapid, large-scale deposition of sediments for a long time and is slowly subsiding. The project was "a joint effort by sedimentationists, biologists (including students of macroorganisms, foraminifera, ostracods, microfossils, and bacteria), clay mineralogists, chemists, and — to a limited extent — petrographers and physical oceanographers."[12] The field workers maneuvered in small vessels, a marsh buggy, and an air boat along the Gulf coast and Mississippi delta to gather cores, mud, sand, and rock samples. These were analyzed in field and laboratory, in the hand and under the microscope, with mass spectrometer and X-ray diffraction, for microfossils and macrofossils, for organic content and remanent magnetism, for grain size and roundness, for areal extent and local variation.

Said Shepard: "Perhaps more than anything else the project has shown how important it is to use a multiple approach in diagnosing environment characteristics."[13]

Fred B Phleger, who had been instrumental in bringing API Project 51 to the campus, called it "a liberal education in sedimentology for all of the numerous participants,"[14] and he summarized the accomplishments as: providing a detailed description of the sedimentary patterns of the Gulf coast region, analyzing the distribution of the organisms within them, summarizing the history of the Holocene rise of sea level on the continental shelf, and defining further problems in sedimentology.

As might be expected, the project continued. Tjeerd H. ("Jerry") van Andel became project director in 1957, as API 51 was turning to studies in the Gulf of California, the site of Scripps Institution's first venture into geologic oceanography in 1939. Other geologists and geophysicists at Scripps were also interested in the great trough that splits Baja California from the mainland, so the Marine Physical

Marine Geology

Laboratory, API Project 51, the Institute of Geophysics at UCLA, and other staff members all joined in a combined study of the Gulf of California. The major expedition there was Vermilion* Sea Expedition in the spring of 1959, using the *Spencer F. Baird* and the *Horizon;* the participants took many soundings, collected core and dredge samples, gathered biological specimens, recorded gravity measurements, shot seismic-reflection profiles and refraction lines, and with Scuba gear and echosounder explored the submarine canyons off Cabo San Lucas.

Visiting Danish biologist Henning Lemche was especially pleased with the capture of live *Neopilina,* a problematical mollusk-like creature that had been thought long extinct until a few live ones were trawled from a depth of 3,570 meters well off Costa Rica by scientists on the *Galathea* in 1952 and described by Lemche in 1957. He had visions of dredging other archaic forms alive from the depths of the Gulf of California with the aid of the newly developed Isaacs-Kidd deep-diving dredge that slid on underwater slopes as if on skis. Asides were made by some about the trilobite-canning expedition, but no other kinds of living fossils were hauled from the deep.

The seismologists were surprised to find that the crustal layer at the northern end of the gulf seemed to be continental while that at the southern end appeared to be oceanic. Might it be that the Baja California peninsula was being split away from the mainland by sea-floor motion?

While Vermilion Sea Expedition researchers explored the ocean depths, an overland trip along the marshy coastal plains of the eastern side of the Gulf of California was led by Joseph R. Curray, another of the early participants in API 51, one whom Shepard called a "remarkably good field

---

*The word is correctly spelled with either one "l" or two, and the participants in the expedition did not strive for unanimity on this point in their reports.

observer." Curray's field crew used trucks on land and dugout canoes by sea, one of which — owned by the richest man in that small town — even boasted an outboard motor.

API Project 51 ended in 1962, at the choice of Scripps geologists, who turned elsewhere for the funding that was becoming too much for industry. The American Petroleum Institute had contributed more than one million dollars to the eleven-year study.

Curray has continued his interest in coastal sediments and the geology of the sea's margins. Often in collaboration with David G. Moore (long with the Navy Electronics Laboratory and its successor, the Naval Undersea Center), he has described coastal profiles and the evolution of continental margins. Curray and colleagues have graduated from dugout canoes to the largest Scripps vessels, from which they employ seismic techniques to determine the thickness and nature of sediments of the continental shelves and slopes. Using seismic profilers in 1966, Curray surveyed the continental shelf and slope along eastern North and South America, and concluded that continental slopes are chiefly depositional features.

In 1967 Curray and Shepard coordinated Carmarsel Expedition to resolve questions on sea-level fluctuations. As Curray commented just before the trip: "In most areas, we cannot determine whether the land has been going down or the sea has been going up." The chosen area was the Caroline Islands and the Marshall Islands in the southwest Pacific, where uniformly submerged terraces had been reported. Geologists from Cornell University, the American Museum of Natural History, the U.S. Geological Survey, and Yale University participated, along with those from Scripps, including biologist William A. Newman. On Carmarsel, the group surveyed the topography of the islands, sampled marine terraces, collected shells for dating, and drilled the coral islands. The results were not simple; the group concluded that both subsidence and Pleistocene fluctuations in sea level had shaped and eroded the reef structures.

Douglas L. Inman, a physics major at San Diego State University, after World War II became a student of Shepard's and joined the Scripps staff upon completing his Ph.D. in 1953. Combining physics and fluid mechanics with geology, he has concerned himself with the processes in the wave zone, where man comes into conflict with the ocean. Where the sea meets the land, storm waves may undermine foundations, rip currents may drag swimmers seaward, favorite beaches gradually disappear, and harbors slowly fill with silt. These problems have become Inman's problems. He and his co-workers constitute the Shore Processes Study Group, which uses both instruments and Scuba gear to study the mechanics of the beaches. This program at Scripps is almost unique among oceanographic laboratories, and has become of increasing interest as man encroaches upon the sea.

Rip currents have been a subject of concern, and curiosity, to Scripps scientists for many years. Shepard, K. O. Emery, and Eugene LaFond measured and pondered them during the 1930s, and Inman and colleagues have continued studies in more recent years, finally to conclude that rip currents are caused by the interaction of incident surface waves with longshore edge waves. Following the suggestion by Inman that edge waves might have an important influence on nearshore circulation, Anthony J. Bowen presented his evidence for the formation of rip currents in 1967 at the only defense-of-doctorate ever held in the Hydraulics Laboratory, where he produced rip currents in the wave basin as part of his presentation. Edge-wave interaction has been found by Inman's group to form nearshore circulation cells and various common beach features such as beach cusps and crescentic offshore bars.

The transport of sand by waves was an early study of Inman's. He and Edward D. Goldberg devised a method of irradiating sand grains with phosphorus-32 and used it to trace sand movement. Inman and Theodore H. Chamberlain then found that the sand was moved by wave action much faster than had previously been suspected. Scuba divers determined that as waves swash back and forth they create ripples, churn up the sand, and also sort and distribute it according to the size of the grains. Most beach sand comes from coastal streams, which in recent years have been dammed for water supply and flood control. Sand that is moved along many of the world's coasts by wave action is steadily lost into submarine canyons, where it cannot easily be recovered. Thus, beaches slowly disappear. So far, Inman's recommendation that dams be placed along submarine canyons to hold the sand for retrieval has fallen on deaf ears. His group has gone on, however, to design and build a sand-transfer system that could be substituted for the dredging now used to keep harbors from silting. This less expensive system consists of a jet pump and suction mouth, which when located in a depression in the sea floor can collect sand and retrieve it for the beaches.

The Shore Processes Study Group is continuing to determine what happens in the churning wave zone.

> Not many years ago [wrote Birchard M. Brush and Inman], it was a logistic triumph for an oceanographer to dip one instrument in the ocean in order to acquire a pitifully limited amount of information on one or two parameters of the many he wished to know.
>
> The present program embraces the continuous and simultaneous acquisition of current velocity in two orthogonal directions, wind speed and direction, a directional array of accurate pressure sensors for determining wave directional spectra, a modern digital wave staff to measure surface displacement, and thermistor arrays for measurement of the thermocline.[15]

Anthony J. Bowen demonstrating the formation of rip currents in the wave tank of the Hydraulics Laboratory in 1967, for his doctoral committee and audience.

Scripps Institution of Oceanography: Probing the Oceans

Through these simultaneous measurements, the Shore Processes Study Group has concluded that "the important processes that operate in the nearshore waters of oceans, bays, and lakes are similar. . . . Moreover, it is becoming increasingly clear that processes in nearshore waters are driven by basic, interrelated forces that are systematic and essentially regular in form. These systematic driving forces lead in turn to the development of coherent processes such as the nearshore circulation cells and the longshore transport of sand that are basically similar the world over."[16]

The concept of uniformity of the coastal processes, which vary only in relative magnitude from place to place, has made it possible to predict the impact of a particular change — such as a structure — on a particular beach area.

The sediments of the deep sea have drawn the attention of a number of Scripps geologists through the years. These thick layers are derived from several sources and are rearranged by deep currents, by turbidity currents, and by burrowing creatures. Sand, mud, and rocks wash into the sea from rivers (and fall from melting icebergs); undersea volcanoes spew material into the ocean and above-sea volcanoes pour lava and fine ash into it; the winds carry minute dust particles far from land; and always a constant rain of dead organisms drifts downward to the sea floor. Tiny plants and animals of the surface waters occur in such vast numbers in some areas that their hard shells through the millenia pile up into a thick slurry of soft ooze. There are radiolarians, diatoms, and silicoflagellates — with shells of silica — and there are foraminifera, coccolithophores, and pteropods — with shells of calcite. Sponges dwelling on the sea floor at all depths leave siliceous spicules when they die, and from the fishes are left scales and teeth and bits of bone. The sediments in shallow water are rich in calcium carbonate, but many of the calcite shells are dissolved before

Marine Geology

they reach the deep-sea floor, where silica remains are more common.

One of the first puzzles about the deep-sea sediments was why they formed such a thin layer. In 1946 the Dutch geologist Ph. H. Kuenen estimated that the average thickness of ocean sediment should be three kilometers, if the earth were, as then estimated, two billion years old. On Midpac and Capricorn expeditions seismologist Russell W. Raitt found the sediment column only about 300 meters thick (east-coast seismologists at about that time found the sediments in the Atlantic about 450 meters thick). Cores of older sediments from the sea floor were rarely older than Upper Cretaceous; none were apparently older than Lower Cretaceous (135 million years before present). Where was the rest of the geologic column that should, in the undisturbed floor of the ocean, leave a record to the beginning of time?

It took many years of work by geologists and geophysicists to resolve the puzzle. They concluded that the floor of the ocean was certainly disturbed. The older sediments are now thought to have been systematically destroyed as new crustal material has been forced upward and outward along the sea floor.

The sedimentologists have had plenty of material to work with, however. The early cores, only five meters long, could represent many years of geologic time, and longer cores were soon available.

One of the earliest Scripps workers in this field was Milton N. ("Bram") Bramlette, who transferred to Scripps from UCLA in 1951. He continued to demonstrate "his ability to throw light upon very large problems by his study of very small objects, the remains of microscopic and submicroscopic life of ancient seas"[17] — which led to his receiving an LL.D. from the University of California in 1965, the first awarded in ceremonies at UCSD.

Scripps Institution of Oceanography: Probing the Oceans

Texas-born Bramlette had spent eighteen years with the U.S. Geological Survey before he joined the staff of UCLA in 1940. The Department of the Interior acknowledged his studies with the Distinguished Service Medal in 1963, when they cited "his study of the origin of the siliceous rocks of the Monterey formation, which won for him international recognition; his study of trans-Atlantic deep-sea cores, which resulted in the first trans-oceanic correlation of glacial deposits; and his understanding of the origin of the Arkansas bauxite deposits, which permitted effective exploration that led to the discovery of major sources of aluminum ore." By then, Bramlette was identifying coccolithophores, discoasters, and calcareous nannoplankton in Pacific cores also, from Capricorn Expedition, in which he had participated, and from later Scripps expeditions.

Fred B Phleger also joined Scripps in 1951, when he made permanent what had been a visiting position while he was on the staff of Amherst College and Woods Hole Oceanographic Institution. Phleger set up and headed the Marine Foraminiferal Laboratory, which has integrated geology and biology in its analysis of fossil foraminiferal shells in marine sediments and in the taxonomy, distribution, and ecology of living foraminifera that eventually add to those sediments. Phleger and Frances L. Parker, when they first arrived at Scripps, began studies of the successions of foraminiferal assemblages in cores from the North Atlantic Ocean and Mediterranean Sea, taken on the Swedish Deep-Sea Expedition in 1947 and 1948. Scripps expeditions were soon providing cores from throughout the Pacific Ocean. Frances L. Parker enlarged the studies of foraminifera to analyze the factors that influenced the assemblages on the sea floor and in the geologic record. She also worked on shallow-water foraminifera, stressing the ecological aspects of their distribution. A great deal of taxonomic revision was required in this complex group of simple animals.

Marine Geology

In more recent years Wolfgang Berger has undertaken studies on the solution of foraminifera shells and coccoliths in sea water, and the distribution of carbonate in the deep ocean. He determined that most of the dissolution takes place on the sea floor, not during settling through the water. Berger and colleagues have mapped the distribution of the present-day calcite-compensation depth for comparisons with other periods.

William R. Riedel began at Scripps by joining Northern Holiday Expedition in 1951 and becoming a staff member in 1956. His specialty has been the intricately shelled radiolarians, a complex group of organisms that he has found defies easy classification. They can, however, be used as indicators of geologic age, and by the mid-1960s Riedel was detailing the movement of fossil radiolarians away from the East Pacific Rise, a movement that became one more verification of sea-floor spreading. He has analyzed radiolarians from all the oceans and from marine sediments on land, and has, for example, been able to establish a radiolarian zonation that divides the Cretaceous period into seven parts. Since the mid-1950s Riedel also has been curator of the Scripps collection of sea-floor cores and dredged rocks.

Gustaf O. S. Arrhenius began his studies at Scripps on deep-sea cores and has gone into several other fields as well: chemistry, physics, and space science in particular. Born in Stockholm, Arrhenius was a staff geologist on the Swedish Deep-Sea Expedition of 1947 and 1948, and shortly after receiving his Sc.D. from the University of Stockholm in 1952, he accompanied Capricorn Expedition as a visitor. He joined the Scripps staff the following year.

Arrhenius has delved into theories on the origin of life from organic material that existed before the formation of the earth; on the nature of comets; on asteroids; and on the earth's moon and its composition. In 1966 he helped to establish the Institute for the Study of Matter, and in 1970 he became an associate director of the Institute for Pure and Applied Physical Sciences, also based at UCSD. In his ocean-oriented studies, he has looked into the formation of manganese nodules and the structure, properties, and composition of natural minerals in the sea. With postdoctoral J. S. Hanor in 1966, Arrhenius determined that barite in sediments was continuous from the East Pacific Rise into Baja California, and thus helped to demonstrate that the spreading center was a continuous feature. Arrhenius and his group set out to map the distribution of sediments in the oceans and set up a computer storage bank containing descriptions of sediment samples. They also analyzed chemical interaction of materials from the oceanic crust and upper mantle with the ocean floor. For this study they developed in the late 1960s an automatic X-ray energy dispersion spectrometer system for rapid and accurate chemical analysis of sediment samples.

Edward L. ("Jerry") Winterer transferred to Scripps from UCLA in 1962 and has also delved into the ages of sediments. "The ocean basins have a lucid but very short geologic memory – about 200 million years," Winterer once said, and he has gone also to continental rocks – which he once called "dusty and badly kept archives" – to help extend the geologic record. Winterer and Riedel in 1969 announced the recovery near New Guinea of an unusually complete core by the Deep Sea Drilling Program's *Glomar Challenger,* a core that contains "an almost complete record" of radiolaria, foraminifera, and nannofossils through 30 million years.

The structure and the topography of the Pacific basin have interested Scripps geologists since 1950, when Midpac Expedition set out. H. W. Menard was then in the Sea-Floor Studies group at Navy Electronics Laboratory, a cosponsor of the expedition. That group – Robert S. Dietz, Edwin L. Hamilton, Edwin C. Buffington, Robert F. Dill,

Marine Geology

David G. Moore, Carl J. Shipek, and Menard — was a lively one, all Scuba divers who often sought their geology with the aid of swim-fins. They have long had a close association with Scripps and have contributed a great deal to the theories of sea-floor geology.

In 1955 Menard was appointed to the Institute of Marine Resources at Scripps (see chapter 6). His early interest had been in turbidites, coarse sediments presumably transported from their land source by turbidity currents. He developed a reputation for being able to read echograms faster than anyone else and, in fact, he looked at more of them than anyone else. He soon was describing the topography of the sea floor and the origin of its features — from ripple marks and fan deposits to fracture zones, the Mendocino Escarpment, the East Pacific Rise, and, later, the geometry of tectonic plates.

In 1957, through IMR, Menard set up a bathymetric survey of the northeastern Pacific Ocean, to which the Bureau of Commercial Fisheries also contributed. Their interest was in aiding tuna fishermen, who had long known that tuna, and other fish, congregate near seamounts and submarine banks. Fishermen who came across such a feature usually kept the location secret and returned to it repeatedly. BCF persuaded the fishermen to yield their secrets, for the advantage of all. Gerald V. Howard, then director of the Biological Laboratory of BCF, noted in 1960 that the accidental discovery of Shimada Bank,* off the west coast of Mexico, led to large catches of yellowfin tuna; this set in motion the cooperative project of analyzing and plotting the large amount of sounding data gathered in the previous decade by Scripps vessels and by those of BCF,

---

*Named for Bell Shimada of the Inter-American Tropical Tuna Commission, who was killed in a plane crash in Mexico en route to join the *Spencer F. Baird* in June 1958. Also killed were Townsend Cromwell of Scripps and three women family members of ship's personnel.

the Coast and Geodetic Survey, and the Navy. Scripps personnel and BCF personnel assisted in the data compilation.

The project enlarged, so that by 1961 an equal-area bathymetric chart of the Pacific basin and a physiographic diagram of the northeastern Pacific Ocean had been completed. By the mid-1960s Stuart M. Smith had devised a computer program to handle the formidable task of coordinating the routine ship records of bathymetry, magnetics, and seismic data. These are indexed, correlated with navigation records, and filed in digital form on magnetic tape or on charts of the Pacific Ocean. Under the direction of Thomas E. Chase from 1970 to 1976, the Geological Data Center provides its data and charts to anyone interested — and fills many such requests from fishermen, oceanographers, oil companies, and Sunday sailors. "Eventually," the compilers say, "all oceans are to be mapped in a painstaking and patient accumulation of bits and pieces of data gathered by the world's oceanographic community."[18]

As a separate mapping project, Jacqueline Mammerickx (Winterer) has been compiling bathymetric charts of the South Pacific, on which she has identified three fracture zones and a series of magnetic anomalies in that area.

One of the chief problems in compiling a map of the ocean floor is simply knowing where the ship was that recorded the depths. Echosounders may provide precise depths to the sea floor, but a surface reference point is needed. Star and sun sights can be marred for days by overcast, and currents skew the dead reckoning figures calculated from course and speed. On Scripps expeditions, the "discussions" have often been quite sharp between scientists and officers on the bridge as to where the ship has been for a number of hours. "Part of the problem for many years," said one expedition leader, "was ever getting a captain to admit that any of his sights were less than perfect." On one short trip the captain of a Navy-operated ship said tartly, "If I had known you wanted navigation, I would have brought my sextant."

Two-ship operations enlarge the scope of the discussions. Menard gave an example of an attempt by *Argo* and *Horizon* in 1967 to find each other 22 hours after they had separated:

> Waving to each other we parted [at 0800 hours] and since each could clearly see the other, we absent-mindedly assumed that we were at the same point. . . . I think it a rude awakening to climb out of bed at 0600 hours, pour down half a cup of aging coffee, walk to the bridge to find that the dead-reckoning position is in doubt by ten miles, walk to the laboratory and learn that the depth is almost a mile shallower than expected, and radio *Horizon* to learn that according to their dead reckoning the ships should be almost side by side. Near us they were not. . . . I radioed *Horizon* to ask where it had been when we were last side by side. (Navigators get nervous if you ask them where they *thought* they had been.) A long pause followed. . . . The radioed position showed that, as we waved at each other, the navigators on the two ships had differed by about twenty miles in their opinions of our position. . . . The ships reversed course and somewhere in the mist and rain squall that frustrated our Radar, we hoped that we were headed toward each other for a rendezvous.[19]

Over the years, the development of Loran helped, in those areas within the network, but the greatest improvement has come through satellite navigation, which was released by the Navy for use by civilian ships in 1967. Even this has its limitations, and the sextant is not yet obsolete, nor yet absolute.

Scripps Institution of Oceanography: Probing the Oceans

Among the sea-floor oddities that attracted the attention of the roving geologists were the manganese nodules, soft rocks that look much like potatoes blackened in a campfire. Menard, under IMR auspices, in the mid-1950s began a survey of the abundance and distribution of manganese nodules throughout the Pacific, using dredges, corers, and deep-sea cameras. Chemical analyses by Edward D. Goldberg and Gustaf O. S. Arrhenius indicated significant amounts of manganese, iron, cobalt, nickel, and copper in the nodules. Goldberg analyzed by the ionium-thorium method the unusually large nodule that had been uniquely acquired on Northern Holiday Expedition (see chapter 15) in 1951; he found that its rate of growth was somewhat less than 0.01 mm per thousand years — "probably one of the slowest reactions occurring in nature in which a measure of the rate of the reaction can be ascertained."[20] Southeast of Tahiti on Downwind Expedition in 1957, Menard found nodules in sufficient quantity to represent a potentially minable resource, and he compiled photographs from Scripps and Soviet expeditions of the International Geophysical Year to determine the regions of greatest concentrations of these odd rocks. From sea-floor photographs Menard and Shipek in 1958 estimated that twenty to fifty percent of the deep-sea floor in the southwestern Pacific is covered with nodules. John L. Mero joined IMR (on the Berkeley campus) for several years to carry out feasibility studies and soon concluded that mining the nodules for valuable minor minerals was economically sound. He then went off to set up a company to try sea-floor mining.

Exploration of the deep trenches of the Pacific by Scripps began on Capricorn Expedition in 1952, when geologists on the *Horizon* surveyed the "challenging oceanic complex" known as the Tonga Trench for nearly a thousand miles, with "all of the best instruments and scientific ideas of the expedition."[21] Robert L. Fisher, whom Revelle called "trenchant," showed his enthusiasm then by crying "Olé" at each trench crossing until he was finally "blasé about being 30,000 to 35,000 feet from the nearest solid ground." To their disappointment the surveyors of the Tonga Trench "didn't find the deepest spot on earth but certainly came pretty close to it."[22]

For several years the declared deepest spot on earth was shifted among four trenches, depending upon which one was last visited by a research ship: the Kurile Trench, northeast of Japan; the Mindanao Trench, off the Philippines; the Tonga Trench, east of the Tonga Islands; and the Mariana Trench, southeast of Guam. Fisher finally resolved the question, from echo-sounding records that he took with precise sounders over the Tonga, Mindanao, and Mariana trenches and from soundings by Soviet oceanographers over the Kurile Trench. Challenger Deep in the Mariana Trench won the contest, with a record depth of $35,810 \pm 30$ feet; Tonga was second, with a maximum depth of $35,435 \pm 6$ feet, and Mindanao (or Philippine) third, at $32,910 \pm 20$ feet.* So the Mariana Trench was selected for the world's deepest dive, by the U.S. Navy-owned submersible *Trieste* in 1960, when Jacques Piccard and Lt. Don Walsh reached the bottom of the sea.

Fisher has returned often to the areas of the deepest trenches and to not-quite-so-deep ones as well, such as the Middle America Trench and the Peru-Chile Trench. He probably holds the record for time at sea among Scripps senior scientists. With the advent of the International Indian Ocean Expedition, he turned his attention to the Indian

---

*Cook Deep in the Mindanao Trench briefly held the deepest record, which was cited in some authoritative articles, but a few months after the record depth was announced the echo sounder was found by the Royal Navy Commander of the surveying ship to have been "producing" soundings about one-sixth deeper than it should have.

Scripps Institution of Oceanography: Probing the Oceans

Ocean, whither he has led several expeditions. Fisher's shipboard messages have always inclined toward subtlety, with puns – a tricky business by radio. For example, he relished confusing several people at home base as to whether he had extensively revised the schedule on Downwind Expedition with his remark in a late January message: "ETA Easter Sunday" – and he then arrived on Easter Island the following Sunday.

After many days at sea, any piece of land looks attractive, and the stunning islands of the Pacific look like paradise indeed. Scripps geologists have never had to struggle for an excuse to get ashore: on Hawaii, to visit Kilauea volcano; on Consag Rock in the upper Gulf of California, because it was rumored to be granitic (it isn't; the light color proved to be guano); on Bayonnaise Rocks, because a submarine eruption had occurred there a year before; on Easter Island, to see the giant carvings; on St. Paul in the southern Indian Ocean, to see the penguins; on the Galápagos, to pay homage to Darwin's isles; on Rapa; on Pitcairn; and more. Capricorn Expedition in 1952 led geologists to Ocean Island, an uplifted atoll with deposits of phosphate:

> We were somewhat interested in dating the uplift and inquired about possible fossils [wrote Fisher to Shepard]. The result was most gratifying. While we sat in the shade sipping Melbourne Bitter Ale the inhabitants brought us beautifully preserved fossil echinoids and clams they had been using for doorstops and paperweights.[23]

The explorers paused too at Rotuma Island, in the hope of determining on which side of the andesite line* the island

---

*For some years the basic rocks of the Pacific Ocean basin and the acidic rocks of its continental borders were considered separated by the andesite (from Andes) line. Detailed geologic studies over the years have made this simple demarkation less significant.

Marine Geology

lay. Harris Stewart and Robert Fisher went exploring:

> A native named George, who had been to school in Suva, knew where there were some outcrops and led us off through some of the lushest jungle I ever saw [wrote Stewart to Shepard]. Cocoanut palms and banana palms, banyan, breadfruit, papaya, and mangoes grew in profusion, frangipanis and hibiscus were the only flowers I could recognise, big land crabs sidled into their holes in the trail as we passed and then came back out to peer after us, birds called. . . . George led us perhaps a mile back in the jungle where we happily thumped away at the volcanic rocks, much to the amusement of George and the brood of urchins that tagged along after us. . . . [24]

From such brief island stops, tons of rock have been transported back to home port for analyses (and rock gardens).

Indeed, when Albert E. J. Engel moved from Caltech to Scripps in 1958 he observed that the institution collections of island rocks were enormous but that samples from the floor of the ocean were skimpy. He doubted that island rocks were typical of the Pacific basin, and he set out to gather quantities of sea-floor rocks to test his guess. His wife, Celeste, of the U.S. Geological Survey, carried out the analyses of these samples with him. "To our surprise, and delight," said Engel, "when we dredged in the Atlantic and the Pacific and the Indian oceans, we found the rocks extraordinarily unique, kindred to achondritic meteorites, and unlike all previous conceptions of basaltic and other mafic rocks of the volcanic edifices built upon the ocean floor. This was, perhaps the most exciting petrological find of its time."[25] Engel has been particularly interested in the history of the earth's crust. His quest has sent him far afield as well as afloat, for example to South Africa to find what he considered "probably the oldest, little-altered sedimentary rocks on Earth."

Hauling rock samples from the floor of the ocean, which Scripps has been doing for a quarter of a century, is not easy. Such rocks from the walls of trenches or from the slopes of seamounts where they are not covered by sediments can provide clues to the geologic history of the ocean basins and to the reflecting layers that are recorded by seismologists elsewhere beneath the sediments. The collecting method is by dredge, but, as Menard said:

> The whole idea of dredging is preposterous. We stop a ship over a spot where we have every expectation that the bottom is irregular solid rock; we then lower a dredge on a steel cable a fraction of an inch thick and a mile or so long and attempt to break the rock without breaking the cable.[26]

Indeed, sometimes all goes awry, as on an early attempt at dredging on Capricorn Expedition:

> One incident nearly broke our spirits [wrote Fisher to Shepard]. Running eastward up from a depth of 4800 fathoms, we came upon a seamount 100-120 miles east of Vava'u [Tonga Islands], which reached to within 225 fathoms of the surface. We attempted to dredge. The ½-inch line is old and snapped near the bail when the dynamometer hardly registered a serious stress, though it was obvious we were somewhat hung up. Slightly daunted, we assembled and wove up the remaining dredge, put it over, and, after it was on bottom 15 minutes, lost it in the same manner. Considering the experienced personnel — Menard,

Marine Geology

[Robert F.] Dill, [Harris B.] Stewart, and Fisher — we prefer to think the fates were against us.[27]

Such near-spirit-breaking is not entirely a thing of the past. On the second leg of Eurydice Expedition in 1974, a dredge was left in the Canton Trough south of the Line Islands, and 8,000 meters of wire followed it to the floor of the trough when the cable snapped. Graduate student Bruce Rosendahl, otherwise always cheerful, didn't smile again until the next day, when another dredging attempt was successful. On the third leg of Eurydice, another scientific party "left one dredge and 367 m of wire on flanks of Combe Bank."[28]

Nevertheless, Scripps geologists have persisted in collecting, principally by dredging, the hard rock samples that provide "ground truth" for the sea-floor spreading hypotheses about the movements or collisions of crustal plates and the composition of the lower crust. Collections of abyssal rocks available for study at Scripps Institution now are second to none. Fisher in particular has become adept at retrieving well-located and fresh specimens of ultramafic and mafic igneous rocks from the deepest walls of the major trenches of the western Pacific. His work in collaboration with Celeste G. Engel has confirmed not only the chemical and petrographic uniformity of the low-alkali basalts characteristic of the shallow layers of the oceanic crust, but also has established the extraordinary range of igneous rocks that, perhaps as sills or crudely stratiform bodies, comprise the deeper layers of the oceanic crust and possibly upper mantle: lherzolites and other peridotites, gabbros of several varieties, anorthosites and titanium-rich gabbros similar to those that have been found on the moon, and even plutonic rocks rich in silica previously thought to be characteristic of continental areas.

George Hohnhaus (left) and Edward L. Winterer showing off a particularly successful dredge haul in the southern California borderland, in 1964.

Marine Geology

The recent trend in marine geological studies is interpreting the history of specific regions. The ships crisscross anomalous areas, bearing questions and drawing out answers.

Menard, for example, planned Nova Expedition in 1967 "to try to determine the development and geologic history of the peculiar Melanesian region in the southwestern Pacific where the sea floor seems to be part continent and part ocean basin."[29] He summarized the cruise's wonders and woes in *Anatomy of an Expedition.* Thanks to Harmon Craig, when he was scientific leader on the *Argo,* that trip put Dixon Seamount and Hohnhaus Seamount on the map, in honor of two Scrippsians who made many an expedition's accomplishments possible. George W. Hohnhaus, sometimes called Big George, because he is, began working for Scripps in 1953, and proved well-nigh indispensable at handling shipboard oceanic equipment. He has spliced wire, rigged equipment, and hoisted more load than one person should. A former underwater demolition expert, he has served as explosives shooter aboard ship, and he was a qualified scientific Scuba diver. Fred S. Dixon, who began at Scripps in 1955, also proved adept at rigging equipment, ingenious at developing and building shipboard gear, and useful as a general expediter. He wrote many sections of the Marine Technician's Handbook, a manual continuously updated by Scripps personnel for shipboard work. Dixon is not quite as big as Hohnhaus, but he was "given" the larger seamount, which was found and surveyed by researchers on the *Argo* while Dixon was aboard. The two peaks are about halfway between Midway Island and the Fiji Islands. Dixon Seamount measures 38 miles across the tip and is 13,100 feet high; Hohnhaus Seamount is about 32 miles across and 12,400 feet high. The two honorees will never get more than a glimpse by photograph, if that, of their mountains, because the tops are almost a mile beneath the surface. The crew members and scientists then on Nova

Scripps Institution of Oceanography: Probing the Oceans

Expedition heartily celebrated the christening of the underwater peaks at a *magiti* — a Fijian feast — in Suva when *Argo* and *Horizon* reached port.

The vast Indian Ocean was a new area to peruse and describe after the first Scripps trip there in 1960 (see chapter 15). Robert L. Fisher, who coordinated and led several trips across that ocean, outlined the bathymetry. The Mid-Indian Ocean Ridge, "part of a world-girdling system of earthquake-prone submarine mountain chains where new crust is being created and emplaced as volcanic rock," was surveyed on Lusiad Expedition (1962-63), Dodo Expedition (1964), Circe Expedition (1968), and again on Antipode Expedition (1971).

Joseph R. Curray and David G. Moore on the latter two expeditions explored one of the world's great natural dumps: The Bengal Deep-Sea Fan in the Bay of Bengal. This pile of sea-floor sediments has been poured into the ocean from the great Himalayan Mountains by way of the Ganges and the Brahmaputra rivers. By surveying and through seismic reflection and refraction profiling carried out by Russell W. Raitt, the fan was found to be 3,226 kilometers long, 1,613 kilometers wide, and 16,460 meters thick at the landward end. Turbidity currents have left meandering and braided channels, rimmed with natural levees, much like river-delta features on land.

In the latter 1960s the trend in geology turned to sea-floor spreading, and fairly quickly every Scripps expedition was bringing in evidence. Menard presaged the turn in a letter to Director Nierenberg on 12 November 1966:

> I just returned from a remarkable meeting in New York and I cannot let a month pass before I return from the South Pacific without letting you know my views on the consequences for Scripps. Sea floor creation and spreading centered on the mid-ocean ridge

Marine Geology

and rise system is now demonstrated although the mechanism is still in doubt. This means that the history of the ocean basins is capable of being unraveled through mapping and sampling of magnetic anomalies and the rocks of the second crustal layer.

. . . It is my impression that marine geology and geology as a whole are at a turning point comparable to physics when radioactivity was discovered. It will be a very exciting time for participants but a sad time for onlookers. . . .

The participants — a number of geologists and geophysicists at various institutions, sparked by a dynamic group at Cambridge University in England — were putting together the details of the theory of plate tectonics. In 1967 Dan McKenzie, then at Scripps Institution, devised the idea of using rigid-body rotations to describe plate motions, and with Robert L. Parker, then newly arrived at Scripps, he published a significant paper on tectonics on a sphere.

Graduate student Tanya Atwater was drawn into the theorizing at that time. She later wrote:

> Sea floor spreading was a wonderful concept because it could explain so much of what we knew, but plate tectonics really set us free and flying.
> . . . From the moment the plate concept was introduced, the geometry of the San Andreas system was an obviously interesting example. The night Dan McKenzie and Bob Parker told me the idea, a bunch of us were drinking beer at Little Bavaria [in Del Mar]. Dan sketched it on a napkin. "Aha!" said I, "but what about the Mendocino trend?" "Easy!" and he showed me three plates. As simple as that! The simplicity and power of the geometry of those three plates captured my mind that night and has never let go since.

Scripps Institution of Oceanography: Probing the Oceans

> . . . The best part of the plate business is that it has made us all start communicating. People who squeeze rocks and people who identify deep ocean nannofossils and people who map faults in Montana suddenly all care about each other's work. . . .[30]

Scripps geological expeditions contributed to the developing theories. Graduate student Dan Karig said that data gathered on Circe Expedition in 1968 indicated that arcs of the western Pacific had migrated away from Asia. Also from Circe Expedition John G. Sclater theorized that Ninetyeast Ridge in the Indian Ocean may have formed as a result of the breakup of the single continent Gondwanaland some 30 to 70 million years ago. Robert L. Fisher and Celeste G. Engel dredged very fresh basaltic rocks from the Mid-Indian Ocean Ridge on Circe Expedition and suggested that these indicated an active spreading center. Graduate student Tanya Atwater on Piquero Expedition in 1969, after surveying back and forth across the Peru-Chile Trench, concluded that the trench represented a collision between two crustal plates. Edward L. Winterer on Seven-Tow Expedition in 1970 determined by magnetic surveys that a long chain of seamounts between Samoa and Honolulu had been transported by crustal shifting as much as 1,500 miles. From that same expedition Sclater and James W. Hawkins determined that the Tonga Islands had been moved eastward from the Fiji and Lau islands. Hawkins returned to the area on Antipode Expedition in 1971 to investigate the spreading apart of the sea floor in the Lau island arc. And so the search continues. . . .

THE MOHOLE

A project that involved a number of geologists and geophysicists was dramatically announced at the First

Marine Geology

International Oceanographic Congress in 1959: drilling a hole to the mantle of the earth. The idea began inauspiciously, gained considerable prestigious support, and died in politics. Yet the defunct Mohole project indirectly provided a great deal of geologic information, from the experimental Mohole drilling, and as a foster parent of the Deep Sea Drilling Project. Scripps was a principal participant.

Some say that the Mohole idea began in Walter Munk's patio at a champagne breakfast. Well, almost.

According to Willard Bascom, it began in a meeting of the Earth Sciences review panel of the National Science Foundation in March 1957. Somewhat discouraged by the lack of scientific breakthrough in the proposals under review, the committee adjourned. But committee members Walter Munk from Scripps and Harry Hess from Princeton asked themselves, "How could the earth sciences take a great stride forward?" Munk suggested that they should consider what project, regardless of cost, would do the most to open up new avenues of thought and research. He thought that the taking of a sample of the earth's mantle would be most significant.

". . . The scope could not then be imagined but obviously such a project would be a heroic undertaking costing a large sum of money and requiring new techniques and monumental equipment. Their own grand ideas, so far from realization, made them a little self-conscious. Hess suggested that it be referred to the American Miscellaneous Society for action."[31]

That sometimes-maligned and oft-misunderstood society was christened by Gordon Lill and Carl Alexis of the Geophysics branch of the Office of Naval Research in the summer of 1952 over a pile of proposals to that office that could be categorized no more closely than one at a time, i.e., they were all miscellaneous. Bascom described the society:

Scripps Institution of Oceanography: Probing the Oceans

Any scientist who has business with ONR's Geophysics Branch is likely to claim membership in the American Miscellaneous Society since there are no official membership rolls. In fact, there are no bylaws, officers, publications or formal meetings. Nor are there any dues, for funds are a source of controversy. The membership is largely composed of university professors or scientific researchers but the rumor that only persons can be admitted whose research proposals to ONR have been turned down because they are too far-fetched is completely false — it is merely a coincidence.[32]

Such whimsy and the society's general antics have brought the comment that it is "a mildly loony, invisible college of otherwise mature academicians . . . exceedingly democratic, but harmlessly anarchic."[33] Among the rare activities of the group is the occasional awarding of its albatross to deserving oceanographers. The bird is a mounted adult specimen, "a bit scruffy about the tail feathers." The recipient is obliged to transport home the bulky award, give it house room until the next presentation, and deliver it to the succeeding honoree at a meeting chosen usually for its great distance. Scripps recipients of this acknowledged but awkward honor have been John Knauss, Walter Munk, Victor Vacquier, Roger Revelle, and Sir Edward Bullard.

AMSOC, as the abbreviation goes, met informally for breakfast — with white but not sparkling wine — at Munk's home in La Jolla in April 1957. Among other subjects, the group discussed drilling a hole to the Mohorovičić discontinuity, i.e., to the earth's mantle.* According to

---

*The boundary was named for Croatian seismologist Andrija Mohorovičić, who discovered it from earthquake records in 1909.

Bascom, "They were not certain about the minimum depths to the Moho or of the maximum depths that had been reached in the search for oil, so they could not even make a good guess whether or not such a hole was possible. What they could do was talk about past experiences and who should be consulted and what such a hole might find."[34] In short order, Gordon Lill was elected chairman, and a committee was formed consisting of Roger Revelle; Joshua Tracey and Harry Ladd of the U.S. Geological Survey; Walter Munk; and Harry Hess.

"The project sounded so simple and logical at a breakfast meeting on a sunny patio. The members lazily looked down a desert canyon at the sparkling Pacific below and felt pleased with the morning's work. . . . That afternoon a delegation called on Roger Revelle to inform him about the grand new idea that had blossomed on his campus."[35]

Later that month William Rubey of the U.S. Geological Survey, Maurice Ewing, director of Lamont Geological Observatory, and Arthur E. Maxwell, chief of oceanography for the Office of Naval Research, were added to the committee.

"It was decided to ask the National Science Foundation for funds to make a feasibility study [continued Bascom]. With genteel horror that august organization declined, politely suggesting that such a distinguished group of scientists might be able to attach themselves to a more reliable group than the American Miscellaneous Society."

AMSOC applied to the National Academy of Sciences for sponsorship and became a committee of the Academy and National Research Council. The American Geophysical Union voted their support of the drilling idea, and finally the National Science Foundation granted $15,000 to begin a feasibility study of drilling to the Mohorovičić discontinuity.

Willard Bascom, "an inventive, restless, cocky oceanographer and mining engineer with long experience and unconventional ideas about deep-water engineering," [36] became the executive secretary of the project. Bascom had been at Scripps as an instrument engineer from 1951 until he joined the staff of the National Academy of Sciences in 1954. He christened the new project Mohole, in an article in *Scientific American* in April 1959. For some time he was the nominal director of the whole project. One of his greatest contributions, to the Mohole and to the present Deep Sea Drilling Project, was a dynamic-positioning system for the drilling ship. On the first ship this consisted of "four 200-horsepower outboard motors located around the hull and operated by a central joystick which activated them to compensate for shifts caused by wind and current."[37] Some said it wouldn't work. It did.

The drilling ship CUSS I (named for the members of Global Marine Exploration Company that owned it: Continental, Union, Shell, and Superior oil companies) was selected for the test and was modified for one and one-half million dollars to drill in deep water. CUSS I, according to novelist John Steinbeck, had "the sleek race lines of an outhouse standing on a garbage scow."[38] It, too, worked.

The first hole was drilled 25 miles off La Jolla, in March 1961, and it broke records: the 3,111 feet of water over which the ship floated was ten times greater than water depths previously attempted. Five holes were drilled in two days of testing, the deepest to 1,035 feet in sediments of the San Diego Trough. Then the clumsy rig was towed slowly and laboriously by a tug to a site 40 miles off Guadalupe Island, Baja California, Mexico. The Guadalupe site had been selected on the advice of Russell W. Raitt and George G. Shor, Jr., as an interesting spot geologically in an accessible range for CUSS I (which was based in Los Angeles) but well off shipping lanes. It was hoped that the drilling could reach the "second layer," a hard reflecting horizon presumed to be volcanic.

Marine Geology

At the Guadalupe site, Scripps ships *Horizon* and *Spencer F. Baird* stood by to serve as handmaidens on the radar and sonar buoys, and *Orca* served as a shuttle for a succession of distinguished visitors. Revelle was there, and Walter Munk, Gordon Lill, Sir Edward Bullard, *Life* photographer Fritz Goro, and John Steinbeck, who wanted to be there because of "some small experience in matters of the sea."[39] The nervous man in charge of the scientific program was William R. Riedel.

In waves up to 14 feet high and wind gusts up to 20 knots, CUSS I held its position while scientists held their breath. The water depth was 11,672 feet. It took half a day for the drill pipe to reach the floor of the ocean before it could even begin drilling. Excitement went up as the drill bit went down 600 feet into the sediments, to the approximate depth of the "second layer." Steinbeck felt "a joy like a bright light at having been there to see" and recorded the day in a "casual log":

> Easter Sunday, April 2 [1961] — Delight on the CUSS. We brought up a great core of basalt, stark blue and very hard with extrusions of crystals exuding in lines — beautiful under a magnifying glass. The scientists are guarding this core like tigers. Everyone wants a fragment as a memento. We have broken drilling records every day but now we have broken through to the second layer which no one has ever seen before. We figure this core cost about $5,000 a pound. I asked for a piece and got a scowling refusal and so I stole a small piece. And then that damned chief scientist gave me a piece secretly. Made me feel terrible. I had to sneak in and replace the piece I had stolen.[40]

President John F. Kennedy, in a telegram to the National Academy of Sciences, called the experimental drilling "a remarkable achievement and an historic landmark in our scientific and engineering progress."

The feasibility of drilling in very deep water and the practicality of dynamic positioning had been successfully demonstrated. CUSS I was returned to Global Marine, and the Mohole project turned to phase two. While that fiasco dragged on, Riedel and dozens of other scientists from several institutions pored over the invaluable cores from the Guadalupe site, a historic first.

As *Science* commentator Daniel S. Greenberg said: "Clearly, the engineering problems of Mohole were formidable, but they were beginning to pale alongside the organizational and political problems" — which he recounted in a series of articles in January 1964. Briefly: the National Academy of Sciences stepped out of the picture, Willard Bascom went off to other ventures, and the National Science Foundation struggled to find its own role in Mohole. Congressmen questioned the politics behind the choice of Brown & Root as the contractor to carry out the drilling, and before long they even more seriously questioned the spiraling cost estimates for a drilling barge as tall as a thirty-story building. Differences of opinion swirled among geologists as to whether the project should drill one single spectacular hole to the Mohorovičić discontinuity with an obviously expensive drilling vessel, or whether it should first drill a number of shallower holes in a variety of interesting places with a not-quite-so-expensive intermediate vessel.

Hollis Hedberg of Princeton, then chairman of AMSOC, spoke effectively for the latter program:

> I might say that a major reason for the recommendation of the intermediate vessel has been to save the taxpayer's money while at the same time guaranteeing him a goodly return for a relatively modest

expenditure. Ocean-drilling exploration is inevitably going to be a long-continuing operation for many years into the future. It should be planned carefully with a modest, orderly, and progressive approach to the more difficult aspects, and it should go no faster nor at any greater rate of spending than experience and achievements justify. There is no need for a wastefully expensive crash program when at least equally valuable results can be attained sooner by a more systematic procedure involving relatively modest annual expenditures which need go on no longer than results justify.[41]

While arguments continued and plans were being formulated for a giant drilling vessel, geophysicists pondered a reasonable location for drilling to the mantle. Among the members of the National Academy's site-selection committee were Raitt and Shor at Scripps; the others were J. Brackett Hersey of Woods Hole Oceanographic Institution, John Nafe of Lamont Geological Observatory, and chairman Harry Hess of Princeton. The criteria for choosing a site, said Shor in 1962, were: "The weather must not be too bad, the site must be near an active port. Deep-water currents must not be strong. In addition, we need to find areas where the discontinuity is shallow and the seismic velocities beneath it suggest average conditions."[42] High temperatures in the ocean floor at the location would also be undesirable.

East-coast scientists explored several promising areas in the Atlantic and, in the spring of 1962, the Mohole project funded Hilo Expedition by Scripps geophysicists to survey possible locations between California and Hawaii. The group completed 33 seismic stations and heat probes in 29 days, at the end of which exercise they reported themselves "all well but tired." From that expedition, the most favorable

Mohole site appeared to be a spot about 120 miles north-northeast of Maui, Hawaii. Early in 1965 the site-selection committee endorsed that location for the Mohole. The next year Scripps participated in the multi-institution Show Expedition to carry out a detailed survey of the area.

In a flurry of political infighting and concern over costs in 1966, Congress decided that the Mohole project should not be continued. Its cost to that point was estimated at 20 million dollars. Some of that was salvaged in scientific accomplishment and engineering design; some of it was not. Mohole became no-hole.

Greenberg's epitaph called Mohole "the albatross of the scientific community"[43] — and AMSOC two years later varied its custom of honoring oceanographers with *its* albatross by bestowing the bird on a former member of the Atomic Energy Commission for his "study of the oceans and other liquids after 5 p.m." Harmlessly anarchic.

DEEP SEA DRILLING PROJECT

The philosophy of the intermediate-drilling program espoused by Hollis Hedberg was picked up by others, even while the Mohole project was still very much in the running. Geologists interested in ocean sediments began advocating a drilling program that would provide a number of samples of the sedimentary column from a variety of different sites, in order to trace the history of the ocean basins and at last to determine why the ocean sediments were so much thinner than predicted.

The success of CUSS I made such a program appear feasible. However, it "would still be too complex and costly for one individual or even one institution to initiate and manage independently," as Tjeerd H. van Andel pointed out.[44] He described the complex, inter-institutional founding of what has become the Deep Sea Drilling Project:

> Early in 1962, after a proposal of Cesare Emiliani, from the Institute of Marine Sciences, University of Miami, to charter a drilling vessel for work in the Caribbean and western Atlantic, a committee (LOCO) [*Lo*ng *Co*re] was formed consisting of two scientists each from Miami, Lamont Geological Observatory of Columbia University, Woods Hole Oceanographic Institution, Scripps Institution of Oceanography of the University of California, and Princeton University. The LOCO committee, realizing that a formal organization was needed, considered a nonprofit corporation of individuals or institutions, but failed to agree on its charter. Later that year, Maurice Ewing of Lamont, J. B. Hersey of Woods Hole, and R. R. Revelle of Scripps formed such a corporation (CORE), which, in February 1962, submitted a proposal to carry out a drilling program as visualized in the intermediate phase of Project Mohole. This proposal was not endorsed by LOCO and was not funded, and both groups faded away.
> . . . Finally, in the first half of 1964, scientists from Miami, Lamont, Woods Hole, and Scripps decided to attempt once more to form an organization to initiate and carry out large drilling projects in the ocean. In May of that year, the directors of these four institutions signed a formal agreement called JOIDES (Joint Oceanographic Institutions Deep Earth Sampling) to cooperate in deep-sea drilling. It was the intent that JOIDES should prepare and propose drilling programs based on the ideas of broad segments of the oceanographic community.[45]

Soon after that, Revelle left Scripps to become director of the Center of Population Studies at Harvard; in December 1964 he pointed out that Fred N. Spiess — as the director of Scripps — should become the representative of JOIDES in his place. "As you can imagine, I take this action with great regret," Revelle wrote to the other members of the committee, "because dealing with the deep sea floor is something I have dreamed about ever since I was a graduate student at Scripps."[46] When William A. Nierenberg became director in 1965, he also became the Scripps member of the JOIDES executive committee.

A planning committee, consisting of one representative from each of the JOIDES institutions,* during the first half of 1965 "carried out exhaustive consultations with marine scientists within and outside the four institutions." They concluded that "the most rapid, significant, and generally valuable advances in our understanding of the history of the oceans can be made by examining the sedimentary column, and to a lesser degree the shallow basement, of truly oceanic areas."[47] Tjeerd H. van Andel and William R. Riedel prepared a draft of a proposal to the National Science Foundation for a three-year program: "drilling of sediments and shallow basement rocks in the Pacific and Atlantic Oceans and adjacent seas."

The first drilling under JOIDES auspices was on the Blake Plateau off southeastern United States from the ship *Caldrill*. This project, under the direction of Lamont Geological Observatory, during April and May of 1965 drilled six holes in shallow-water areas and whetted the appetites of sedimentologists.

During mid-1965, with the active participation of Nierenberg in his new post as director, the original

---

*Charles Drake from Lamont, J. Brackett Hersey from Woods Hole, Fritz Koczy from Miami, and George G. Shor, Jr., from Scripps (Shor was replaced in May by William R. Riedel and Tjeerd H. van Andel).

Marine Geology

Memorandum of Agreement of the four institutions was modified to provide that the JOIDES executive committees should function in a scientific advisory capacity and that the designated operating institution should assume complete responsibility for grants or contracts.

In 1966 Scripps Institution was designated the operating institution for the Deep Sea Drilling Project. William W. Rand was appointed project manager; he was succeeded by Kenneth E. Brunot as acting manager in October 1967, then as manager from April 1968 to the end of 1970. The scientific advisers at first were van Andel and Riedel; in December 1967 Melvin N. A. Peterson was appointed chief scientist until 1971, when he became co-principal investigator, and N. Terence Edgar became chief scientist until 1975, succeeded by David G. Moore in 1976. To house the big project, the Deep Sea Drilling building was built in 1970, on the Scripps campus slightly up the hill and across La Jolla Shores Drive from the main center.

In January 1967, Scripps signed a contract with the National Science Foundation to manage a drilling program in the Atlantic and Pacific oceans for at least eighteen months, to begin in mid-1968. A subcontract was let to Global Marine Exploration Company to provide and to operate a drilling ship to be built for the new program. On 23 March 1968, at Levingston Shipbuilding Company in Orange, Texas, "with a bottle of seawater mixed from the Atlantic and Pacific Oceans and symbolic of the far-flung operations scheduled for the scientific program,"[48] Mrs. William A. (Edith) Nierenberg christened the ship *Glomar Challenger*, a name proposed by van Andel.

The big vessel, which has been in continuous use since 1968, is 400 feet long, and is dominated by a drilling derrick 142 feet high amidships. M. N. A. Peterson and A. J. Field said that the ship "is a low profile hull, quite heavy for its size and designed for maximum mass radius of gyration and minimum angular motion when drilling. It has a remarkable lack of angular motion, even in seas up to 5 and 6 feet. In fact, the ship is so laid out that the main deck can be awash in storm action without disrupting drilling activities in the elevated superstructures."[49] She is outfitted with a satellite navigation system and with four computer-controlled tunnel thrusters for dynamic positioning, which have made it possible to hold the vessel in position even in 45-knot head winds and one-and-one-half-knot cross currents. Although she is remarkably stable, the vessel does roll and pitch in heavy seas, so an automatic pipe-racking device was installed to avert hazard to personnel. The pipe racker stacks 23,000 feet of drill pipe in 90-foot lengths topside, and additional storage space is available below deck. The ship can carry 70 people: crew, technicians, and scientists. Deep Sea Drilling Project personnel say that the ship is fitted out with some of the best laboratories ever designed for the study of geological materials at sea.

As soon as the *Glomar Challenger* was launched, outfitted, and accepted, she went to work, under the direction of Maurice Ewing and J. Lamar Worzel of Lamont — and struck oil. At least it was determined that the Sigsbee Knolls in the Gulf of Mexico are salt domes like those that sometimes contain oil, and traces of oil were identified before the hole was closed.

The early work of the *Glomar Challenger* was chiefly in the Atlantic Ocean, but in 1969 the ship moved into the Pacific. Scripps expeditions had helped pave the way there, by making detailed surveys of a number of the sites selected for drilling. Scan Expedition in 1969 sailed counterclockwise around the Pacific, pausing at 33 locations for reflection profiling, magnetometer readings, piston-core samples, heat-flow measurements, and sea-floor photographs. These helped establish the precise location for each

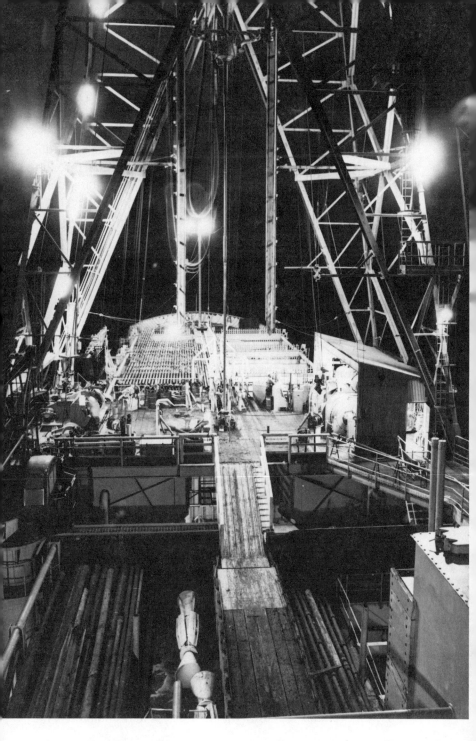

The *Glomar Challenger* of the Deep Sea Drilling Project, drilling at night.

Scripps Institution of Oceanography: Probing the Oceans

drilling spot. Later expeditions stopped for site surveys ahead of the *Glomar Challenger's* steady drilling schedule.

The vessel has gone on around the world and throughout the "seven seas" from the Arctic to the Antarctic and the tropics between. Each leg of its journeys is almost two months long, followed by a short stop in port, allowing a complete crew changeover. A high point in the drilling operation was reached on 25 December 1970, when a worn-out drill bit was replaced and re-entered in the same drill hole. This capability has made it possible to drill through the hard chert layers, which seem to be found so commonly throughout the world's oceans.

Preliminary analyses of the cores taken by *Glomar Challenger* are done aboard ship. "The first analysis is by paleontologists, who determine geologic age. Technicians X-ray cores, determine water content, and measure radioactivity. Then they study a core's composition, grain-size and general mineralogy. Cores are documented photographically, because some properties change in storage. Then the cores are carefully packaged and placed in cold storage." [50]

Preliminary reports on the cores are published soon after the completion of each leg of the cruise. A year after each leg, the samples become available to qualified researchers for further study. Lamont-Doherty Geological Observatory is the repository for cores from the Atlantic Ocean, the Antarctic area, and the Mediterranean; and Scripps Institution is the repository for cores from the Pacific and Indian oceans, under the curatorship of William R. Riedel. The quantities of cores required building an addition for core storage beside the Deep Sea Drilling building in 1974.

The original drilling contract for 18 months was extended by 30 months, and again by 36 months. Leg 44, "the culmination of over seven successful years of geological exploration," ended on 30 September 1975, and completed the third phase of the drilling project. The shopworn ship went into drydock in Norfolk, Virginia, for a complete overhaul. The fourth phase of the project commenced on 30 November 1975, as the International Phase of Ocean Drilling (IPOD), the purpose of which is to drill holes deep into the rocks beneath the sea-floor sediments in an attempt to determine their composition, structure, and geologic history, as well as to continue coring in the sediments. International it is, for through the years the member institutions of JOIDES have grown to include: the Bundesanstalt für Geowissenschaften und Rohstoffe, Federal Republic of Germany; the Centre National Pour L'Exploitation des Oceans (CNEXO) of France; the Department of Oceanography, University of Washington; Hawaii Institute of Geophysics of the University of Hawaii; the USSR Academy of Sciences; the Lamont-Doherty Geological Observatory of Columbia University; the National Environmental Research Council (NERC) of the United Kingdom; the Ocean Research Institute of the University of Tokyo; Oregon State University; the Rosenstiel School of Marine and Atmospheric Science of the University of Miami, Florida; Scripps Institution of Oceanography of the University of California at San Diego; Texas A & M University; the University of Rhode Island; and Woods Hole Oceanographic Institution.

To begin the new phase, during 1976 *Glomar Challenger* traveled more than 20,000 nautical miles, and 44 holes were drilled into the sea floor, to recover more than 1,000 cores. At a site west of Portugal a record penetration of 1,740 meters into the sea floor was attained.

In its eight years of effort, the Deep Sea Drilling Project has retrieved 145,728 feet of sediment (to May 1976). Those columns of sand, silt, volcanic ash, and debris, accumulated from 398 sites in the world's oceans, have "unequivocally established the geological youthfulness of [the oceanic] crust in comparison to most continental rocks and

Scripps Institution of Oceanography: Probing the Oceans

[have] helped confirm that sea-floor spreading and widespread crustal motions have occurred."[51] For example, it has been determined that the floor of the Pacific Ocean is moving northward beneath the equator, and that sediments in the eastern Indian Ocean have been moved northward. Vertical movements in the oceanic crust have also been found, the most dramatic of which is Ninetyeast Ridge in the Indian Ocean — once at sea level and now a mile beneath the surface. Cores from the Mediterranean Sea have shown that that basin was almost dry from approximately five to ten million years ago, during which thick layers of salt accumulated. Of future economic interest have been indications that oil and gas may have accumulated in deep areas of the Gulf of Mexico, and that metal-rich sediments occur just above the original igneous rock floor of the ocean. These contributions, along with the vast increase in detailed knowledge of the microfossils and the oceanic sediments, call forth the designation of the Deep Sea Drilling Project as "one of the most successful scientific expeditions of all time."

NOTES

1. *Anatomy of an Expedition* (New York: McGraw-Hill, 1969), 48. Used with permission of McGraw-Hill Book Co.

2. *Ibid.*, 4.

3. *Ibid.*, 48.

4. Reply to award of Wollaston Medal of the Geological Society of London, *Proceedings of the Geological Society of London,* No. 1634 (3 October 1966), 140.

5. Scripps Institution Notes, September 1933.

6. Private notes, 15-22 October 1937; SIO Archives.

7. "Terrestrial Topography of Submarine Canyons Revealed by Diving," *Bulletin of the Geological Society of America,* Vol. 60 (October 1949), 1599.

8. *Ibid.*

9. Reply to award of Wollaston Medal of the Geological Society of London, *loc. cit.*, 139.

10. *San Diego Union,* 23 April 1966.

11. F. P. Shepard, *The Earth Beneath the Sea* (Baltimore: Johns Hopkins Press, 1959), 210.

12. F. P. Shepard, "Stratigraphic Research That May Pay Off," *Oil and Gas Journal* (21 February 1955), 159.

13. *In:* Francis P. Shepard, Fred B Phleger, and Tjeerd H. van Andel, editors, *Recent Sediments, Northwest Gulf of Mexico* (Tulsa, Oklahoma: American Association of Petroleum Geologists, 1960), 3.

14. *In: Ibid.*, 365.

15. "Coastal Processes and Long Range Planning," 8th Annual Conference of Marine Technology Society, *Preprints*, 1972, 217.

16. Douglas L. Inman and Birchard M. Brush, "The Coastal Challenge," *Science*, Vol. 181 (6 July 1973), 21. Copyright 1973 by the American Association for the Advancement of Science.

17. Citation for LL.D. at UCSD, 1965.

18. IMR Annual Report, 1972.

19. *Anatomy of an Expedition*, 221-22.

20. E. D. Goldberg, "Chemistry in the Oceans," *Oceanography*, Washington, D.C., American Association for the Advancement of Science, 1961, 591.

21. Radio message, 1 January 1953.

22. Radio message, 6 January 1953.

23. Letter of 11 December 1952.

24. Letter of 11 December 1952.

25. Letter to the author, 27 November 1974.

26. *Anatomy of an Expedition*, 170.

27. Letter of 10 January 1953.

28. Radio message, 18 November 1974.

29. *Anatomy of an Expedition*, 4-5.

30. *In* Allan Cox, compiler, *Plate Tectonics and Geomagnetic Reversals* (San Francisco: W. H. Freeman & Co., 1973), 410 and 535-36.

31. Willard Bascom, *Hole in the Bottom of the Sea* (Garden City, New York: Doubleday, 1961), 47-48. Copyright © 1961 by Willard Bascom. Reprinted by permission of Doubleday & Company, Inc.

32. *Ibid.*, 30.

33. R. G., "Do Oceanographers Have More Fun?" *Science*, Vol. 181, No. 4103 (7 September 1973), 926.

34. Willard Bascom, *Hole in the Bottom of the Sea* (Garden City, New York: Doubleday, 1961), 48.

35. *Ibid.*, 49.

36. D. S. Greenberg, "Mohole: The Project That Went Awry," *Science*, Vol. 143, No. 3602 (10 January 1964), 116.

37. *Ibid.*, 118.

38. "High Drama of Bold Thrust Through Ocean Floor," *Life*, Vol. 50, No. 15 (14 April 1961), 111. Copyright © 1961 by John Steinbeck. Originally appeared in *Life* Magazine. By permission of McIntosh and Otis, Inc.

39. *Ibid.*

40. *Ibid.*, 122.

41. Letter in *Science*, Vol. 143, No. 3612 (20 March 1964), 1275. Copyright 1964 by the American Association for the Advancement of Science.

42. News release, 7 March 1962.

43. "Mohole: Aground on Capitol Hill," *Science*, Vol. 153, No. 3739 (26 August 1966), 963.

44. "Deep-Sea Drilling for Scientific Purposes: A Decade of Dreams," *Science,* Vol. 160, No. 3835 (28 June 1968), 1420. Copyright 1968 by the American Association for the Advancement of Science.

45. *Ibid.*

46. Letter of 14 December 1964.

47. Draft of proposal to National Science Foundation, 23 June 1965.

48. SIO Annual Report, 1968, 17.

49. "Deep Sea Drilling with D/V Glomar Challenger," MS of talk for Challenger Expedition Centenary, 13 September 1972, 4.

50. *Ibid.,* 10.

51. "Deep Sea Drilling: Research Lags Exploration," *Science,* Vol. 181, No. 4098 (3 August 1973), 428. Copyright 1973 by the American Association for the Advancement of Science.

13. Within the Waters and Muds: Studies in Marine Chemistry

"One of the most striking observations of marine biology is the fact that some parts of the ocean are very fertile while other parts are quite barren. There must be chemical factors which determine fertility, and an explanation of this was perhaps the first serious question which oceanographers asked the chemist. In the year 1930 there were probably no more than a dozen professional chemists in the world who were actively interested in the ocean, and practically every one of them was trying to answer this question."[1]

One of those in 1930 was Erik G. Moberg at Scripps Institution, a chemist whom Director Vaughan described as "careful, thorough, and reliable," and whom he credited with having placed Scripps among the leaders in research on the chemistry of sea water. Swedish-born Moberg had emigrated to the United States in his teens, and had received his B.A. at the University of North Dakota and his Ph.D. from the University of California, for work carried out mostly at Scripps, before joining the staff of the institution in 1925.

Scripps had no chemical laboratory at that time, and Moberg "with his own hands" proceeded to build one. By

Erik G. Moberg taking a water sample from a Nansen bottle, aboard the *Scripps,* probably in the 1930s.

Marine Chemistry

1934, according to Vaughan, Moberg had "published on the chemical composition of marine plankton in the southern California region, the hydrogen ion concentration in sea water, the phosphate, silica, and fixed nitrogen content of sea water, the interrelation between diatoms and chemical environments and upwelling in the sea off the coast of southern California, and the distribution of oxygen in the Pacific. . . . One of the most important investigations that has been conducted at the Scripps Institution [continued Vaughan] is the study of the buffer mechanism of sea water with reference to the solubility of calcium carbonate. Dr. Moberg took the lead in this investigation and he was able to associate with him such men of ability as Dr. D. M. Greenberg and Dr. Paul Kirk. Although their investigations reached an advanced stage, Dr. Moberg was not satisfied, because he was convinced that other weak acids in sea water exerted an influence in determining the conditions. Therefore he held up the publication of a rather large manuscript in the hope that they would be able to ascertain the boron content of sea water. This was ultimately done. . . . After the amount of boron in sea water had been determined the paper on the buffer mechanism of sea water was revised . . . [and] has been accepted for publication. . . . This study is, I think, by far the best of its kind that has as yet been made."[2]

By the late 1930s, when John Lyman was Moberg's graduate student, the chemistry department at Scripps had what Lyman called "a lot of service functions. We did Kjeldahls for Roger Revelle, maintained and calibrated the reversing thermometers, and ran all the chlorinities. At sea, we did phosphates and oxygens. Moberg knew all this forwards and backwards. . . ." Through bitter experience, Moberg "learned that the only way to bring back good salinity samples is to put them into citrate of magnesia bottles. . . . After he severed his connection with SIO there were people who attempted to use lighter containers (such as polyethylene) or cheaper ones (such as beer bottles), with the result that there are some Scripps cruises in this period, alas, for which there are no salinities."[3]

Moberg spent considerable time at sea on various research ships of other laboratories, as well as on the boat *Scripps,* of which he was essentially in charge. He also was extensively involved in the conversion of the *E. W. Scripps* into a research vessel, and he then used it for collections at a regularly scheduled series of stations out to the offshore islands. Moberg left Scripps in 1945.

In 1946 Norris W. Rakestraw was invited to Scripps from Brown University and Woods Hole Oceanographic Institution, to plan a research program in chemistry. Already noted as one of the nation's leading chemistry educators, Rakestraw continued his expertise at Scripps, where for some years he was dean of students, and in 1960 he became the first dean of the graduate division of UCSD. He also ably edited the *Journal of Chemical Education* for fifteen years. Rakestraw became professor emeritus in 1965.

During the decade following Rakestraw's arrival at Scripps, marine chemistry at the institution expanded greatly and turned to highly sophisticated techniques. In 1949 Edward D. Goldberg joined the staff, followed by George Bien, who began with API Project 51 (see chapter 12) in 1951, and by Theodore R. Folsom in 1952. In the mid-1950s Roger Revelle was envisioning an expanded university in the San Diego area, and — before that concept had been fully approved by university officials — he invited Hans E. Suess from the U.S. Geological Survey, Harmon Craig from the University of Chicago, and physicist Walter M. Elsasser from the University of Utah* to become the first appointments of the new campus. Craig has continued

---

*Elsasser transferred to Princeton in 1962.

Marine Chemistry

at Scripps, while Suess became a member of the UCSD chemistry department after holding a joint appointment there and with Scripps for several years. Charles D. Keeling was added to the Scripps chemistry group in 1956, and Tsaihwa J. Chow in 1960.

The work of these scientists has gone well beyond "service functions." As Goldberg said in 1960: "The paths of marine chemistry have also been fashioned by the advances of chemistry itself. . . . Whereas in the nineteenth century but twenty elements in sea water had been adequately assayed, today a knowledge of the concentration of fifty or sixty has been established and attention is now centering upon the distribution of isotopes of the elements."[4] Analyses of sea water have led marine chemists into studies of man's abuse of the sea as the world's largest waste basket. Dating by means of isotopes has drawn them into determining the ages of sediments, the reactions where sea water meets mud or shell, and the ages and distribution of water masses. They have worked with biologists, geologists, and physical oceanographers in solving the complexities of the sea, the air, and the land.

Edward D. Goldberg, for example, who joined the Scripps staff just after receiving his Ph.D. at the University of Chicago in 1949, began his researches along with Scripps geologists on the composition of manganese nodules and on sedimentation in the ocean. In the mid-1950s he used ionium-thorium ratios to date cores from various expeditions in the Pacific. Goldberg and colleagues found that radioactivity did not decrease steadily with depth, but increased in the lower layers. Barium was found in some Pacific sediments in greater concentration than in igneous rocks or in land sediments, and could be correlated with areas where biological activity was highest, so that concentration by siliceous organisms such as diatoms and radiolarians seemed the most likely explanation. In 1959, with

Scripps Institution of Oceanography: Probing the Oceans

Devendra Lal (then in Bombay, later also at Scripps) and with Minoru Koide, Goldberg announced the discovery of silicon-32 in siliceous sponges. By means of an omegatron, a highly sensitive mass spectrometer, Goldberg and his group in the 1960s determined the concentrations of some of the noble gases in sea water.

With graduate student Robert W. Rex, Goldberg analyzed quartz grains in ocean sediments, and they concluded that Darwin's suggestion that wind-borne dust furnished material for extensive ocean sediments was correct. They found that more than half the deep-sea deposits in the North Pacific are derived from wind transport from arid areas of Europe and Asia.

The wind was carrying other forms of dust, also, found Goldberg: talc, for example, which was surprisingly widespread in air samples even over remote parts of the ocean. The source proved to be land-sprayed insecticides, in which talc was the carrier. High concentrations of insecticides had already been found in tunas and in penguins and shearwaters far from sources of pesticides.

Such studies of widespread man-introduced material in isolated as well as populated regions led Goldberg on to studies of pollution, studies into which he was drawn by Roger Revelle's concern over man's alteration of the marine environment. "Man," said Goldberg in 1968, "has become a geological agent, competing with natural processes, in the alteration of seawater composition. . . . Excess carbon dioxide from the burning of such fossil fuels as oil, coal and gas, synthetic detergents from municipal discharges, pesticides, lead, radioactivity from the detonation of fission and fusion devices are found in measurable quantities in the oceans."[5] Mercury, he also pointed out, was introduced into the ocean at a greater annual rate by man's activities than by natural transfer. From samples taken from the Greenland ice sheet, Goldberg, Koide, and Herbert V. Weiss (of the Naval Undersea Center) found that the mercury content has doubled throughout the years.

In 1970 Goldberg and William A. Newman hosted a two-day symposium in La Jolla on "Man's Chemical Invasion of the Ocean: An Inquiry," which was attended by more than five hundred persons. Goldberg also has called for extensive monitoring of pollutants, and in 1973 helped to establish jointly with Soviet scientists an environmental monitoring network for measuring the effects of pollution on marine organisms. Along other lines, Goldberg took an early interest in the establishment of UCSD and served as the first provost of Revelle College from February 1965 to July 1966, at which time he returned to his chemical researches at Scripps.

As nuclear-bomb testing began in the early 1950s, various people, including Roger Revelle, became concerned over the effects of the introduction of radioactivity into the ocean. Theodore R. Folsom was drawn into these studies. A native San Diegan, he earned his B.S. and M.S. at Caltech in physics and in 1952 received his Ph.D. at Scripps, when he also joined the institution's staff; he retired in 1975. John Isaacs recalled one of Folsom's contributions to Project Wigwam in 1955:

> On the first and only deep underwater nuclear explosion, much of the upwelled radioactivity was distributed in extremely thin, submerged laminae; the first observation of such layering, I believe. We were frustrated — for while we could detect the radioactivity with the lowered probe, we could not obtain a sample of the radioactive water because the sampler was a meter above the detector and it was impossible to sample these layers from a rolling ship. Overnight and single-handedly, Ted took over the ship's shop and

GEOSECS engineers at console controlling computerized underwater telemetry systems aboard the *Melville*, 1972. Left to right: Arnold Mantyla, Arnold Bainbridge (at console), Jack Spiegelberg (in background), Rick Ackermann, and Bob Fong.

created a self-contained sampler, with its own internal logic that could snap a sample of a layer the instant it encountered one. It was only then that we were able to obtain direct evidence and samples of these astonishingly thin layers.[6]

By means of a closed sampling system, free of contamination, Folsom was able to sample deep-water layers and so determine that the radiocarbon ages were dependable and that fission products were confined to the upper layers of the ocean, not penetrating into deep water as rapidly as had been supposed.

During 1960, said Folsom, "we rushed about collecting specimens suitable for establishing the radioactive background in the sea during the weapons moratorium. We believed this to be a unique opportunity; perhaps the last time the fallout input would be simple to study."[7] In 1961 nuclear testing resumed and "this pleasant period ended" with the advent of "specimens noticeably contaminated with nuclides characteristic of young fallout." Traces of the 1961 and 1962 nuclear explosions were followed for several years through sea-water sampling in their slow movement around the North Pacific.

In a study for the city of Los Angeles, for its Hyperion sewage treatment plant, Folsom, with G. K. J. Mohanrao of Caltech, in the late 1950s set up a monitoring system by means of several isotopes, with special attention to the long-lived cesium-137. They were able to correlate variations in the cesium in the sewage treatment system with nuclear weapon testing.

To determine the effect of discharging radiocesium into the ocean by way of sewage, Folsom set out to establish the normal level of this isotope in the Pacific. This required developing a technique that readily distinguished cesium from other alkali metals. Sea-water samples from the Scripps pier and from sites far out in the Pacific were thus analyzed. On one occasion, a ton of sea-water samples from one of the expeditions was shipped by air from Honolulu to San Diego — as excess baggage, accompanying a passenger — all packaged in boxes just within the airline limit on suitcase size. There was grumbling at the airport, but the boxes went through. Folsom later devised a sampling procedure for cesium by the use of potassium cobalt ferrocyanide that allowed continuous measurements along the ship's track. In addition to providing more information, it reduced the headaches (and backaches) of airlines personnel.

Folsom and his co-workers expanded the Hyperion project into a comparison of radioactivity in sewage at twelve cities, including Los Angeles, which led into defining the problems inherent in monitoring sewage plants. They also undertook a study of the effects of discharge of radioactivity on marine organisms living near the Hyperion outfall and a comparison with similar organisms elsewhere along the Pacific coast. The study quickly showed considerable variation of radioactive isotopes among organisms within small areas. Albacore tuna, for instance, concentrated cesium in their tissues to more than a hundred times the level of their surroundings. Marine fishes, in fact, were found to concentrate much higher levels of natural radioactivity than any terrestrial animals.

Folsom's group established a reference of the concentration of several metallic radionuclides in various organisms along the coast and throughout the Pacific Ocean in order to be able to monitor changes over the years. They also analyzed the quantities of several other metallic trace elements in the tissues of fish and found the radioactive concentrations high for some years after their introduction into ocean waters.

In the mid-1950s some geochemists were looking into another aspect of man's impact on the environment: what they called the greenhouse effect. They were becoming concerned that the increasing use of fossil fuels was creating a blanket of carbon dioxide that prevented long heat rays radiated by the earth from escaping into space while allowing light rays and infrared rays to penetrate to the earth. There was concern that the surface of the earth would become warmer, and a rise of only two or three degrees would make a considerable increase in the melting of polar ice, with a consequent rise in sea level. Revelle warned:

> Estimates by the United Nations indicate that within the next 50 years we will have produced 1,700 billion tons of new carbon dioxide from combustion of industrial fuel. This astronomical sum is 70 per cent of the carbon dioxide now in the atmosphere. In this way we are returning to the air and the sea the carbon stored in sedimentary rocks over hundreds of millions of years. From the standpoint of meteorologists and oceanographers, we are carrying out a tremendous geophysical experiment of a kind that could not have happened in the past or be repeated in the future. If all this carbon dioxide stays in the atmosphere it will certainly affect the climate of the earth and this may be a very large effect.[8]

Revelle and Hans E. Suess, Harmon Craig, and James R. Arnold* and Ernest C. Anderson, from their background work in several laboratories, in 1957 published their independently derived figures on the rate of transfer of carbon dioxide from the atmosphere into the oceans: a residence time of approximately ten years for carbon dioxide in the atmosphere before it is dissolved in the oceans.

*Who in 1958 became a staff member of what was to become UCSD.

Scripps Institution of Oceanography: Probing the Oceans

Revelle called the study of carbon dioxide in the atmosphere and ocean one of the major Scripps projects for the International Geophysical Year of 1957-58, and announced the establishment of a radiocarbon laboratory at Scripps for processing samples of sea water for radiocarbon and tritium. Radiocarbon analysis had a twofold interest: determining isotope exchange and transfer of carbon dioxide between the atmosphere and the ocean, and using radiocarbon to trace the movement of deep ocean water and the mixing of water masses.

On the expeditions of the IGY, Norris Rakestraw and George Bien set up the shipboard sampling program, in which they collected surface and deep-water samples from 40° south latitude to 15° north latitude in the Pacific Ocean. The samplers, devised by the Scripps Special Developments Shop, were stainless steel barrels of approximately fifty-gallon capacity, with spring-loaded doors that could be tripped by a messenger traveling down the wire. Various models were tested — and some were left on the floor of the ocean. On the expeditions, once the water samples arrived on board, Rakestraw or Bien settled down to long continuous hours of shipboard analysis. ("They were real heroes," said Harmon Craig, who was aware of his colleagues' shipboard hours because he was putting in similar hours of time on gas analysis!) "Radiocarbon measurements [of the IGY samples, said Bien, Rakestraw, and Suess] have shown for the first time that it is possible to determine true age differences of water masses by the determination of radiocarbon in the dissolved bicarbonate."[9]

From the IGY expeditions and from later expeditions, which sampled both the Pacific and the Indian oceans, Bien, Rakestraw, and Suess concluded:

> In the Pacific Ocean the ^{14}C [carbon-14] content decreases constantly from south to north well into the

northern hemisphere; thereafter the gradient seems to flatten out. This is consistent with [John A.] Knauss' circulation pattern. . . . Extensive mixing probably also takes place.

In the Indian Ocean, the situation is approximately similar but not so clear. ^{14}C decreases from south to north, though not so regularly. . . . As in the Pacific Ocean, the results here are consistent with the assumption of slowly rising deep water in the northern parts of the ocean.[10]

From continued radiocarbon investigations in surface sea water, Bien determined that latitudinal variations in surface-water concentrations are due both to the location of atmospheric nuclear tests and to ocean mixing processes. The tests have been conducted primarily in the northern hemisphere, where maximum radiocarbon concentrations are also found.

Revelle also enticed Charles D. Keeling to Scripps in 1956 to pursue the study of carbon dioxide in the atmosphere and ocean, and Keeling has continued analyzing and monitoring carbon dioxide very precisely. He determined that the increase of carbon dioxide in the atmosphere was only about one-half the amount that would occur if all man-caused carbon dioxide stayed in the atmosphere. Much of the remainder, he has concluded, is absorbed by the surface waters of the oceans. To monitor carbon dioxide levels, Keeling and his colleagues in the early 1960s established three isolated recording stations: in Alaska, at the South Pole, and atop Mauna Loa in Hawaii. In 1969, Arnold E. Bainbridge developed a more sensitive analyzer for continuous measurement of carbon dioxide, which was installed at Mauna Loa to replace the equipment there. In cooperation with New Zealand scientists, Keeling established several sampling stations in the early 1970s in the southern hemisphere to monitor carbon dioxide. In 1972 Keeling installed a station on the Scripps pier, which very quickly showed a correlation of higher carbon dioxide levels with smog (mostly drifting down from the Los Angeles area).

Small variations in ocean conditions, such as slight changes in temperature or in barometric pressure, can affect the amount of carbon dioxide absorbed by sea water. Keeling and his colleagues have been correlating such variations with changes in atmospheric carbon dioxide. In 1968 Keeling prepared a chart of the distribution of the partial pressure of carbon dioxide in surface waters for the Atlantic, Pacific, and Indian oceans, based on two years of continuous measurements from Scripps expeditions. He found pronounced belts of high pressure near the equator in the Pacific and Atlantic, but not (at least for the summer measurements) in the Indian Ocean. Through very precise data from expeditions and several island stations, Keeling and colleagues have been able to find seasonal and short-term periodicities in the distribution of carbon dioxide. These have been correlated with the Southern Oscillation, a cyclical fluctuation in zonal wind circulation.

The radiocarbon laboratory that was announced by Revelle before the IGY was the laboratory of Hans E. Suess, who had established a similar laboratory for the U.S. Geological Survey earlier for determining ages of geological and archaeological events. Suess's major improvement on the technique of radiocarbon dating — which was based on the discovery in 1947 of the isotope carbon-14 in nature by Willard F. Libby — was the use of acetylene as the counting gas. That, combined with meticulous laboratory procedures, enabled him to reduce the experimental error of samples dated to less than one percent.

The La Jolla Radiocarbon Laboratory, as the new facility

was called, was established under funds from the Atomic Energy Commission and was ready to handle samples with its newly constructed acetylene counter in time for the IGY in mid-1957. Bien became a member of the group in 1958 and continued there until retirement in 1971 (he died in 1975). He proved to be an especially meticulous and precise analyst.

Suess's first interest had been in the dating of glacial advances in the United States, work that he has called "exceedingly enjoyable" and "most rewarding." Early measurements on logs buried by advancing ice sheets showed the last glaciation on the American continent to have been only 19,000 years ago, much more recent than had previously been estimated by geologists. Suess went on to compare recent radiocarbon dates with tree-ring-dated wood in order to determine the variations in atmospheric carbon-14 throughout the past 9,000 years. His determination that the radiocarbon had been decreasing since about the turn of the century, because of the burning of fossil fuels, has been termed the "Suess effect."

For other researchers, at Scripps and elsewhere, the Radiocarbon Laboratory for some years handled samples to provide dates on prehistoric man, on fluctuations in sea level, the time of formation of lagoon sediments, and of deep-sea oozes and muds. Charcoal, shell material, coral, foraminifera — even green-blue mud and llama dung — were submitted to the laboratory for dating events of significance to biologists and geologists.

Some funds for the Radiocarbon Laboratory were provided by the university's statewide Water Resources Center, established in 1957, for studies on isotopic variations in natural waters, carried out by Harmon Craig, and for studies of climatic cycles in California, carried out by Suess and Carl L. Hubbs. Samples dated under the latter project provided dates of southern California aboriginal populations as far back as 7,000 years, and gave a first certain date (14,500

Scripps Institution of Oceanography: Probing the Oceans

years before present) for the La Brea tar deposits. They also offered evidence that the rainfall in the southern California area had been considerably greater during the period from about 3,480 to 860 years ago, so that the region had then been able to sustain a fairly large number of inhabitants. Dating of shells from coastal locations provided supporting evidence that sea level had undergone a slow rise during the past 6,000 years.

After the establishment of UCSD and its medical school, the researchers at Scripps who were studying minute traces of natural radioactivities found the background level rising, from increased use of radioactive materials in various research projects. Folsom and Suess began seeking a "non-radioactive retreat," and, when the bunkers on Mt. Soledad that had served as an Army intelligence and radar station during World War II became available, Folsom especially urged that the almost-underground, isolated buildings be acquired. The property was awarded to UCSD in 1965 from the assets of the Templeton Foundation. Folsom moved his laboratory into the bunkers after they were renovated, and Suess's Radiocarbon Laboratory moved into an adjacent, separately built structure later.

Also drawn to La Jolla by Revelle in 1955 was Harmon Craig, who had received his Ph.D. in geology-geochemistry at the University of Chicago, where he had been in charge of the mass-spectrometer studies under Harold C. Urey for analyses of hydrogen, carbon, and oxygen isotopes.* At Scripps, Craig's laboratory eventually grew to contain five mass spectrometers: Samson, Delilah (of course), Micah, Gad, and an unchristened portable unit. For several years in the latter 1950s one of Craig's projects was measuring the variations in the concentrations of hydrogen and oxygen isotopes in natural waters.

*Urey was attracted to the new campus in 1958, as a professor at large, through conversation with Craig.

Marine Chemistry

Isotopes in ocean waters turned up some surprises: an unexpectedly high proportion of the helium-3 isotope, for instance, which Craig found on Nova Expedition in 1967. He concluded that this isotope was derived from within the mantle and was presumably leaking into sea water from the seafloor-spreading centers.

In the 1970s Craig has been developing a means of using concentrations of radon and helium as earthquake precursors. He and his colleagues have set up a monitoring network in thermal springs and wells along four major faults in southernmost California.

In a land-based project in 1972, Craig, assisted by his wife Valerie, developed a mass-spectrometer technique for identifying specific quarries of Greek marble through the proportions of carbon-13 to carbon-12 and oxygen-18 to oxygen-16, which is enabling archaeologists to identify the source of individual marble statues. The husband-and-wife team made detailed collections in the four major quarrying areas used by the Greeks from the archaic period to Roman times. Marbles from each quarry could be distinguished by isotope analysis.

Craig and his colleagues – Yu-chia Chung, Ray F. Weiss, and Manuel Fiadeiro – have also been measuring the radium distributions in the world oceans, studying the mixing processes in the bottom waters, and determining the distributions of dissolved argon, nitrogen, and total carbon dioxide in sea water.

Some of these studies have been within the Geochemical Ocean Sections Study (Geosecs) program. "Geosecs began," wrote Craig, "with the recognition by Henry Stommel [of MIT] that the full potential of geochemical tracers for the study of deep-ocean circulation could only be realized by a maximum collaborative effort in which simultaneous studies of as many significant properties as possible could be made over a large section of the oceans."[11]

Scripps Institution of Oceanography: Probing the Oceans

The multi-institutional program of mapping the geochemistry of the world oceans was established 1 March 1971 as the first major program of the International Decade of Ocean Exploration, sponsored by the National Science Foundation. Arnold Bainbridge of Scripps became project manager of the Geosecs Operations Group in 1972. The group is located in rented buildings in Sorrento Valley, about six miles from the campus, and has incorporated the Data Collection and Processing Group (DCPG) begun years before under the Marine Life Research Program (see chapter 3). An elaborate computer program was set up to handle the great volume of measurements of a great many chemical parameters throughout the column of water.

For Geosecs, Fred Dixon helped design a new balanced conducting wire for deep STD (salinity, temperature, depth) measurements, and Craig's group and Dixon built a new hydrographic winch and several items for carrying out shipboard chemical measurements. Thus equipped, Craig and shipmates on Antipode Expedition in 1971 gathered closely spaced hydrographic casts and made detailed geochemical studies. These resulted in the discovery of a major density discontinuity — which they named the benthic front — in the South Pacific Ocean, beginning east of New Zealand and sloping downward to the north and east, a discontinuity that separates deep water from bottom water.

The first major Geosecs expedition was in 1972-73 in the Atlantic Ocean on the Woods Hole Oceanographic Institution ship, the *Knorr,* and included various Scripps chemists and physical oceanographers. The Pacific's turn for study came in 1973-74, with a 35,000-mile expedition on the *Melville* from the Bering Sea to the Antarctic — "one of the most technologically advanced oceanographic expeditions ever carried out," according to participant Weiss.

Also participating in Geosecs projects has been Devendra Lal, whose varied studies have dovetailed with the work of

Marine Chemistry

several other geochemists and geologists at Scripps. Physics-trained Lal visited Scripps in 1958-59 and has had a part-time appointment at the institution since 1967; he was on the staff of Tata Institute of Fundamental Research in Bombay, India, from 1952 to 1972, and, besides his Scripps appointment, has been director of the Physical Research Laboratory at Ahmedabad, India, since 1972. A specialist in cosmic-ray-induced nuclear effects, Lal has analyzed lunar samples, working with James R. Arnold of UCSD and with Gustaf Arrhenius. He has used the fossil-track method to date sea-floor basalts and volcanic-ash layers, in cooperation with James W. Hawkins.

With Craig and others, Lal has participated in analyses of the composition and radioactivity of particulate matter in the oceans. He devised a sampling device that uses cotton filters and pressure-activated switches to turn on the filtration at specific depths. On Geosecs expeditions the device has filtered 5,000 to 15,000 liters of sea water at depths from 50 to 2,000 meters in the South Pacific.

For some years chemists at various universities had been carrying out studies on lead and its isotopes. In 1960, one of these chemists came to Scripps from Caltech, Tsaihwa J. ("Jimmie") Chow. Born in Shanghai, he had received his B.S. from National Chiao-tung University and his Ph.D. from the University of Washington. His early studies were on elements in sea water and the composition of marine sediments and of meteorites. Among those elements, Chow found lead occurring in greater quantity than expected. He analyzed samples from polar snowfields and later, with Goldberg, from layered sediments cored in several offshore basins, and so derived a history of the increasing accumulation of lead with industrialization. Chow perfected methods for distinguishing various sources of lead from their isotopic composition. In 1970 he announced that leaded gasoline was the major source of airborne lead, and he has been advocating the abolition of lead from gasoline ever since. In analyzing fish for lead content, Chow found that those caught near the California coast had almost twice as much lead as fish caught 200 miles offshore. In 1967 he set up four monitoring stations throughout San Diego County — from the Scripps pier to atop Mt. Laguna — to collect air, dust, and water samples for lead analysis. In 1971 he helped establish the baseline concentration of lead for a program, sponsored by the International Decade of Ocean Exploration, of establishing the background concentration level of various man-made pollutants in the marine environment.

In the latter 1960s new additions to the chemistry staff at Scripps included Joris M. Gieskes, D. John Faulkner, and Jeffrey L. Bada. Netherlands-born Gieskes came to Scripps in 1967 and turned to studies of the chemistry in interstitial waters of marine sediments and of the activity coefficients of mixed electrolytes. Faulkner, who started at Scripps in 1968, set up a program to isolate and identify previously unknown chemicals from marine sources, especially for useful pharmaceuticals. The researches by Faulkner and his associates have resulted in the identification of antibiotics in sponges, the first known instance of antibiotic activity in a tunicate, a complex mixture of toxins in starfishes, and the storage of toxic chemicals by the sea hare from its algal diet. Faulkner's group has also experimented with insect growth regulators on barnacles, one of which (ZR-512) caused young barnacles to metamorphose into adults before they attached to a surface, thus causing their death by starvation. Bada, a San Diego native and UCSD graduate who joined the Scripps staff in 1969, had devised a new method of dating sediments and fossil bone by means of the racemization reaction of amino acids. He has been using this technique for dating calcareous sediments from

Marine Chemistry

the North Atlantic and the Caribbean, and for dating bones from various sites throughout the world, including early man in America and hominoid remains from Olduvai Gorge.

NOTES

1. Norris W. Rakestraw, "The Chemist in Oceanography," *ICSU Review,* Vol. 3, No. 4 (October 1961), 166.

2. Letter to University President Robert G. Sproul, 16 January 1934.

3. Letter to E. N. Shor, 21 December 1975.

4. "Chemists and the Oceans," *Chymia,* Vol. 6 (1960), 162.

5. "New Chemistries in the Ocean," manuscript for talk.

6. Talk at Staff Council meeting, 13 May 1975.

7. Memorandum to Joseph Hutchinson, 25 October 1961.

8. Testimony to House of Representatives Appropriations Committee considering appropriations for the International Geophysical Year, 1956.

9. "Radiocarbon Dating of Deep Water of the Pacific and Indian Oceans," *Radioactive Dating,* International Atomic Energy Agency, 1963, 159.

10. "Radiocarbon in the Pacific and Indian Oceans and Its Relation to Deep Water Movements," *Limnology and Oceanography,* Vol. 10, Supplement (November 1965), R29-R30.

11. "The Geosecs Program: 1970-71," *Earth and Planetary Science Letters,* Vol. 16 (1972), 47.

Part IV
Out to Sea and Back Again

14. Marine Facilities and the Fleet

"The long arm of the oceanographer is his ships and his groping fingers, the cable. Without ships to test and to explore, the hypotheses and laboratory discoveries of the marine researcher become dry and insubstantial and the researcher blind and isolated."[1]

Since 1937 Scripps Institution has had a long arm, and since 1948 it has maintained the largest oceanographic fleet of any research institution.

The first ship of that fleet was the graceful schooner *E. W. Scripps,* given to the institution by Robert P. Scripps in 1937. Then came the hardworking tug *Horizon* in 1948, the first of many former military ships. From 1949 through 1968 she made 267 scientific cruises, spent 4,207 days at sea, and logged 610,522 miles.

From 1948 "new ships were regularly added from Navy or other surplus sources, up to the *Alexander Agassiz* in 1961 and the tug *Oconostota* in 1962. Soon after that Scripps began acquiring ships that were specifically designed for oceanographic research: the *Ellen B. Scripps,* modified from offshore oil-supply boats; the *Alpha Helix,* a floating laboratory; and two of the ships designed by the Navy as

Scripps Institution of Oceanography: Probing the Oceans

Auxiliary General Oceanography Research vessels, the *Thomas Washington* and the *Melville*. There are some long-time Scrippsians who have ridden both the *E. W. Scripps* and the *Melville;* even in a haze of nostalgia, they prefer the latter.

The service to operate the fleet, Marine Facilities (Mar-Fac), came into being in 1948. Finn Outler was then technical superintendent for the Marine Physical Laboratory, and among other duties was responsible for the conversion of the *PCE(R)-855* and *PCE(R)-857* into research vessels for the Navy Electronics Laboratory. When the onset of the Marine Life Research program enabled Scripps to acquire three ships, all of which required conversion for research use, Outler was asked to take charge. He found himself yearning for the help of a certain fellow submariner acquaintance from prewar Navy days before each had been separately assigned to China patrol. Quite by accident he came across that man's name in the San Diego telephone book: James L. Faughn, who, Outler found, was then out of the Navy and attending law school in San Diego. The persuasion of Outler and of just-departing Director Sverdrup brought Faughn to Scripps on a long useful career: as ship's captain and superintendent, technical administrator, coordinator for Naga Expedition, supervisor of the construction of the *Alpha Helix,* and staff officer.

As technical superintendent in engineering at Marine Facilities, Faughn took charge of finding available Navy ships for the Marine Life Research program in 1948 and of directing their conversion to research. On his recommendation the *Horizon* and the *Crest* were acquired, as well as the fishing boat *Paolina-T.* At the same time, yachtsman Clemens W. Stose, who had been the prewar captain of the *E. W. Scripps,* continued at Scripps after the war. In 1950, when it appeared that Stose might be retiring, Acting Director and shipmate Roger Revelle commented:

Launching of the *Melville*, sideways, at Defoe Shipbuilding Company, Bay City, Michigan, on 10 July 1968. Photo by James Pollock.

Scripps Institution of Oceanography: Probing the Oceans

Since 1937, when he supervised the transformation of the E. W. SCRIPPS from a movie star's yacht into a sturdy research vessel and became her first skipper under the University flag, Clem has retired several times to my knowledge. . . . I well remember, in his first days at Scripps, Clem occasionally wondered aloud what those scientists were up to, and whether it was necessary to mar a beautiful vessel with all that queer gear. But he never complained and always did his best to make sure that the maximum possible amount of scientific work at sea was done. During the years he has become accustomed to the peculiar behavior of oceanographers and now he almost acts like one himself.[2]

Instead of retiring then, Stose continued at Scripps as marine superintendent (hull) until 1953.

Clifford W. Colbeth was hired in 1951 as a ship's officer to replace Gus Brandel, who was also at times skipper of the *E. W. Scripps*. Colbeth was advanced to ship captain the following year, and in February 1954 he became marine superintendent. When Peter G. Trapani replaced him as marine superintendent in July 1956, Colbeth returned to serving as ship captain until 1958.

Trapani's familiarity with the sea was a long one: he enlisted in the Navy at the age of seventeen in 1926 as an apprentice seaman, rose through the ranks in the electrician series, was commissioned an ensign in wartime 1942, and advanced to commander before his retirement in the spring of 1956. In July of that year he became marine superintendent at Scripps and continued to direct the expanding Marine Facilities until his retirement in 1972. Peter S. Branson, who had retired as captain in the U.S. Coast Guard in 1968 after a number of years of service afloat and ashore, took charge of Marine Facilities at that time.

Marine Facilities and the Fleet

In its early years the fleet had no home. The *E. W. Scripps*, for example, was berthed at the yacht harbor in San Diego Bay before World War II. After the war, Scripps ships tied up at the Navy Electronics Laboratory dock at Point Loma, by courtesy of the Navy. When space was not available there, one or more of the ships, when in home port, had to be berthed at San Diego city docks or at National Steel and Shipbuilding Company.

A permanent home was finally acquired in 1965, when six acres of land were leased from the Navy (which gave the land to Scripps in 1975) alongside the Navy Electronics Laboratory on Point Loma. Jeffery D. Frautschy was a major participant in the design of the utilitarian buildings, jointly called the Chester W. Nimitz Marine Facilities, which were dedicated in March 1966. The Office of Naval Research and the National Science Foundation provided the funds for the million-dollar facility, which includes an administration building, an electronics shop, a general carpenter, welding, and machine shop, a warehouse and stores building, a storage yard, and — at first — a floating pier made of two large barges acquired through state surplus in 1962. After completion, it was just barely possible to tie up, at the pier and marginal wharf, almost all the ten vessels that Scripps then owned, the sole exception being *Flip*. In 1974, through funds provided by the National Science Foundation, the floating-barge pier was replaced by a larger concrete pier and extension of the marginal wharf; the new pier has sewer and bilge-water connections for the ships, and connections for telephones, compressed air, and fresh water. *Flip*, which had been berthed at the B Street Pier at the Embarcadero on San Diego Bay for many years, was finally able to join the rest of the fleet, as was ORB, the ocean research buoy of the Marine Physical Laboratory. The National Marine Fisheries Service ship *David Starr Jordan* regularly berths at the MarFac pier, and space can be provided, when needed, to visiting research ships.

The Scripps ship facility was named for Fleet Admiral Chester W. Nimitz, retired commander in chief of the Pacific Fleet, just a month after his death. In addition to his many years of distinguished service in the United States Navy, Nimitz had served as a university regent from 1948 to 1956 and long before then had organized NROTC on the Berkeley campus, where he had served as professor of naval science and tactics from 1926 to 1929.

To provide for the needs of Scripps research at sea, on expeditions long and short, the number of people at Marine Facilities over the years has fluctuated between 100 and 200, operating six to ten vessels. The group includes the ship captains, other ships' officers, crew members, and a shore-support staff.

Providing ship services to scientists is not an easy role for a sailor. Only under research circumstances is a ship *not* under the sole command of the ship's captain, for, in service to science, the captain shares with the chief scientist the decisions as to when and where the ship will sail. All responsibility for the safety of the ship lies with the captain, who occasionally has to remind scientists of that point. Scripps ships have run aground (one captain did seem to have a predilection for that), but not often. James L. Faughn, captain on Midpac Expedition and on many later ones, and Noel Ferris — whose skill in ship maneuvering drew awed admiration among crew, scientists, and jaded dock-workers — stand out as ship captains who have worked especially well with the scientific party.

In the years just after World War II, many of the Scripps scientists and technicians had served in the Navy and were already familiar with ships and the language of them. Not so many of today's oceanographers have that background, which sometimes leads to a breakdown in communication.

After all, as writer Fred Hoctor noted during Scan Expedition in 1969: "It is particularly difficult for a seaman to understand why anyone who has the brains to earn a college degree would leave a coffee cup unanchored on a table when the ship is rolling 30 degrees, or why anyone would hold a discussion in a narrow passageway, or leave a port open when green water cascades over the quarterdeck."[3]

The ship's cook has problems too when trying to serve science and scientists. "I remember our frustrated cook's exasperation [on Midpac Expedition, wrote Edward S. Barr] while he stood in the galley doorway with a full meal served in the empty messhall, while all the scientists were aft on the stern examining an unscheduled newly-caught sea creature!"[4] That was but the first of many chow-time interruptions.

When the *E. W. Scripps* made her first long research expedition to the Gulf of California in 1939, oceanography was an all-male profession. (Women graduate students at the institution then were in biology and were expected to stay ashore.) When Mrs. Roger Revelle and Mrs. Francis P. Shepard met the *E. W. Scripps* in Guaymas in 1940, they were forbidden by the captain to set foot aboard ship. The philosophy lingered, although after World War II women were sometimes tolerated on one-day trips. In 1949 Revelle expressed the philosophy of the era:

> . . . Obviously, certain inconveniences arise when women passengers are aboard for overnight trips on a small vessel. These are usually greatly exaggerated however.
>
> . . . On the other hand, our ships are not yachts; they are essentially fishing vessels. Hence the men aboard are liable to be dirty, smelly, profane and possibly even offensive to any women who think Queen Victoria's Albert was the ideal husband. In attempting

to be clean, odorless, refined and inoffensive, our crew members, marine technicians, and scientists might possibly lose efficiency and the scientific work might suffer. Thus it is necessary to consider each situation on its merits and to think up reasons whenever possible to discourage women from participating in the work at sea.

I am against the establishment of any formal policy with respect to women visitors. You know how perverse women are; if we made a policy prohibiting them from going out on the ships, we would find two or three in the chain locker every time we cleared the whistling buoy. An unwritten policy which does not prohibit but subtly discourages their presence will best achieve our rather dubious ends.[5]

Revelle did consider each situation on its merits. In 1952 he invited Rachel Carson, whose book *The Sea Around Us* had come out the previous year, to join Capricorn Expedition. She accepted and later regretfully declined. During Capricorn Expedition Revelle invited Helen Raitt to return home aboard the *Spencer F. Baird* when she met the ship in Tonga. By then, other women had spent shorter periods of time aboard Scripps ships. Although there have been complications at times, women are very regularly members of scientific parties aboard ship now, and in 1973 the first of several women crew members was hired (none of whom, however, stayed very long).

A jack-of-all-trades who performed heroic service for the Scripps fleet for many years, in an advisory capacity, was Maxwell Silverman. He once called himself "a seagoing pipefitter," but he was really an oceanographic engineer, before the term was invented. Silverman began at Scripps in 1951 for a year, then worked at Navy Electronics Laboratory for three years, and returned to Scripps as an assistant

Marine Facilities and the Fleet

engineer. Along with others on Capricorn Expedition in 1952, he became an expert in handling sea-going equipment. In those early days of "on-the-job training," everyone aboard ship contributed to improvements in design. Silverman was the deviser of the distinctive wishbone shape of the stern-mounted A-frame, copied on many research vessels at various institutions. On a number of the early expeditions he was the explosives handler, and a very safety-conscious one.

Silverman became more and more involved with the design of oceanographic ships, beginning with the conversion of the *Argo* in 1959. He also became concerned with the regulations imposed on those ships by the U.S. Coast Guard, and he helped form a workable set of regulations for this unusual — and rather contrary — category of ships. When the Navy began constructing research ships for the oceanographic institutions, Silverman became adviser, critic, and watchdog. His approach, said one observer, combined "courage and courtesy." While the first ships of the Navy's AGOR series were being designed, Silverman once carried to Washington a list of 104 items that he considered deficient in design, location, or utility. Backed by other Scripps officials, he persuaded the Navy to change more than 90 of those items. Silverman, on loan from Scripps, supervised the construction of the *Atlantis II* for Woods Hole Oceanographic Institution, and then supervised the *Thomas Washington* for his own institution. He participated in the design of the *Melville* and her sister ship, the *Knorr*, of Woods Hole. Over the years he spent a great deal of time in Washington as liaison officer on ship design, and he finally left Scripps in 1973 to work for the oceanographer of the Navy, until his death by heart attack four months later. Said Jeffery Frautschy: "Clearly, modern oceanographic ships are memorials to Max's dedication, competence, and effectiveness."

Scripps Institution of Oceanography: Probing the Oceans

THE FLEET

The following alphabetical list includes all ships that are owned or have been owned or operated by Scripps Institution since its beginning (but not rented or borrowed vessels). The list includes a number of Navy-owned vessels, as the Navy has provided research ships to the institution often since 1948, because of its keen interest in oceanographic studies. The *Glomar Challenger* is owned and operated by Global Marine, Inc., under contract to Scripps (chapter 12).

L is length, B is beam, D is draft, CS is cruising speed, R is cruising range in nautical miles, E is endurance (days at sea without resupplying).

Alexander Agassiz (the first) (1907-1917)
L 85 feet; B 26 feet; D 6.3 feet. Crew 5; scientists 4. MBA-owned.

This ketch or yawl was built by San Diego boat builder Lawrence Jensen for research work* for the Marine Biological Association, which founded Scripps Institution. The association acquired the ship on 21 August 1907, with funds provided by Ellen B. Scripps. From five names suggested by Scripps staff members, the donor chose to honor the Harvard geologist-oceanographer who had visited the young institution in 1905.

The ship, as described by her master, Captain W. C. Crandall, in 1912: "is schooner-rigged, and as originally built was a 'ketch'; that is, a boat with deck area forward of the mainmast large and unencumbered, the wheel being placed behind the rear mast. Her foremast was at first 65 feet high, carrying a boom and large mainsail, and her mizzen-mast 39 feet, rigged with a boom. She has a spoon bow and a 15-foot overhang. . . . The 'Agassiz' began work in

---

*This ship is perhaps the first one designed and built specifically for ocean research by an American nongovernmental institution.

Nimitz Marine Facilities (buildings in right foreground) and the fleet. The Point Loma buildings of the Marine Physical Laboratory are in the left center, beyond NOSC (formerly Navy Electronics Laboratory).

Scripps Institution of Oceanography: Probing the Oceans

June, 1908, and the first season made it clear that her rigging was too heavy; that the wheel should be forward; that the scientific work should have better accommodations on the after deck; and that the galley was too small."[6] A number of changes were made and the vessel was considered much more comfortable and practical.

In January 1917, Scripps sold the ship to Pacific Coast Trading and Shipping Company, which used her as a coastal trader along Mexico. In 1918 she was seized briefly by the American *Yorktown* as a suspected German raider, but she was soon released. In 1920 she ended her days by running aground at the entrance to San Francisco Bay.

Alexander Agassiz (the second) (1962-1976)

L 180 feet; B 32 feet; D 10 feet; CS 11 knots; R 3,600 miles; E 17 days. Crew 18; scientists 13. UC-owned.

This former Army freight and supply vessel (Army FS-208) was acquired by Scripps on 18 April 1962 from the State Educational Agency for Surplus Property.* She had been built in 1944 by Higgins in New Orleans, Louisiana. The name was selected through a campus contest, won by Elizabeth N. Shor, who was given the honor of signaling the unveiling of the new name plate at the dedication ceremony.

This vessel was used primarily by the Marine Life Research program (see chapter 3). On 8 November 1976 she was sold to Marine Power and Equipment Company, Seattle, Washington, to be used as a crab-fishing vessel.

Alpha Helix (1966-)

L 133 feet; B 31 feet; D 10.5 feet, CS 11 knots; R 6,200

---

*The first ship acquired at this time was the *FS-227*, and plans were begun on conversion. The *FS-208* became available and required less expensive conversion. The *FS-227* then was acquired by Texas A & M and converted to research as R/V *Alaminos*.

miles; E 30 days. Crew 12; scientists 12. UC-owned.

This biological research vessel was built by J. M. Martinac Shipbuilding Company, Tacoma, Washington, for the Physiological Research Laboratory (see chapter 8). After launching herself, she was christened by Susan (Mrs. Per) Scholander. She was acquired by the institution on 26 February 1966, with funds provided by the National Science Foundation. The ship is operated as a national facility for biological, especially physiological, research.

After considering several other names, Per F. Scholander named the ship for the helical configuration of proteins and genetic material.

Argo (1959-1970)

L 213 feet; B 39.5 feet; D 15 feet; CS 13 knots; R 8,000 miles; E 60 days. Crew 32; scientists 24. Navy-owned.

This former Navy rescue and salvage tug (U.S.S. *Snatch*, ARS-27) was built in 1944 by Basalt Rock Company, Napa, California, and was provided to Scripps by the U.S. Navy on 20 July 1959. Scripps returned the ship to the Navy on 11 June 1970.

The institution refused to accept the Navy's name, so it cut the red tape and named her *Argo*. The name given by Scripps was "after Jason's Argo that some 3,000 years ago sailed eastward from Greece in search of the golden fleece of Colchis – one of man's first explorations of the sea," and also as a reminder of California's early explorers, the Argonauts.

Argo's maiden voyage for Scripps was marred by mechanical problems (see chapter 15, International Indian Ocean Expedition).

After the Navy took the ship back, she sat in mothballs in Vallejo, California, for some time; she was finally towed to Taiwan to be scrapped.

Scripps Institution of Oceanography: Probing the Oceans

Buoy Boats

Scripps had several buoy boats and one or more picket boats at various times from the late 1940s to the 1960s, for nearshore work. Some were provided to the institution by the Navy for contracts. One buoy boat ended its career on 20 September 1953, while being towed by the *Horizon* with recording equipment in use; the log of the *Horizon* reads:

> This weather is too rough to be towing an open boat with no automatic bilge pump or cockpits watertight. . . . Sea & swell are such that turning around is impossible. . . . Bouy [sic] boat in tow on 340 meters of wire, taking sounding every 15 min. . . . lights out on bouy [sic] boat but still in tow, headed for the beach. 0227 [hrs] tow line parted, in 550 fms water, approx position 29°-31'-45" N 115°-37'30" W. . . . Apparently one leg of Bridle let go causing buoy boat to broach to, thereby filling with water & sinking. Cable not of sufficient strength to tow buoy boat to shallow enough water for salvage operations.

Crest (1947-1956)

L 136 feet; B 24.5 feet; D 6 feet; CS 10 knots; R 4,000 miles; E 30 days. Crew 13; scientists 8. UC-owned.

This former harbor minesweeper was built by Associated Shipbuilders, Seattle, Washington, in 1944 for the U.S. Navy, which transferred her to Scripps Institution on 4 December 1947. The ship proved to need extensive overhaul, which was done in a shipyard in San Pedro. Her first work for Scripps began in January 1949. In May 1956 Scripps returned the ship to the Navy.

The Scripps name was chosen from a contest on campus, simultaneously with the *Horizon*. The winner was Alfreda Jo Nixon, who received ten dollars and a large photograph of the ship.

Marine Facilities and the Fleet

Dolphin (1973-)

L 96 feet; B 22 feet; D 7 feet; CS 12 knots; R 1,700 miles; E 6 days. Crew 5; scientists 7. UC-owned.

This twin-screw diesel yacht was built in 1968 by Breaux Baycraft, Inc., in Loreauville, Louisiana, and was given to Scripps Institution by Robert O. Peterson on 21 December 1973. Several Scripps staff members had used the vessel for research cruises as guests of Peterson before he donated her to Scripps. The name was given by Peterson.

Ellen Browning (1918)

The minutes of the board of directors of Scripps Institution for 17 September 1918 read: "Mr. Crandall stated that the Navy had commandeered the boat 'Ellen Browning' and had fixed a value of $3000 therefor. . . ." A bill of sale to the Navy, which accompanied the report, said: "This boat was built by W. C. Crandall, Business Manager of the Institution, and paid for out of funds provided by Edward W. Scripps for the joint use of said Institution and Edward W. Scripps." A letter in 1931 by George F. McEwen referred to the "speedboat" *Ellen Browning,* "used [some years earlier] for making collections and observations at sea for the Scripps Institution."

Ellen B. Scripps (1965-)

L 95 feet; B 24 feet; D 6 feet; CS 9 knots; R 6,480 miles; E 30 days. Crew 5; scientists 8. UC-owned.

This research ship, built to plans modified (especially with advice from George G. Shor, Jr., and Maxwell Silverman) from those for offshore oil-supply boats, was built by Halter Marine Services, New Orleans, Louisiana, in 1965. Scripps acquired her in August 1965 from Dantzler Boat and Barge Company (Pascagoula, Mississippi), on a one-year lease with option to buy, and then bought her. The vessel's distinctive feature is a large afterdeck on which

359

portable laboratory vans and equipment can be placed. Living quarters are rather crowded.

The name was given by George G. Shor, Jr., for the institution's early benefactress. The ship was dedicated on 1 October 1965, by Ellen Clark (Mrs. Roger) Revelle, grandniece of Ellen Browning Scripps. The dedication was on the same day as the dedication of Revelle College of UCSD.

E. W. Scripps (1937-1955)

L 93.7 to 104 feet (depending on method of measurement); B 21 feet; D 12 feet; CS 9 knots, or — under full sail and a stiff Santa Ana wind — 12 knots; R 2,000 miles. Crew 4; scientists 6. UC-owned.

This auxiliary schooner — the largest constructed on San Francisco Bay to that time — was built for racing and pleasure cruising by J. H. Madden and Son, Sausalito, in 1924 for Russell Clifford Durant, son of W. C. Durant (who founded General Motors). She was christened *Black Swan;* she was black in color and "palatially fitted" in teakwood and mahogany. The scantlings were of Oregon pine, apitong, Port Orford cedar, and teak.

The next owner was Irving T. Bush of New York. Then actor Lewis Stone, a star in Andy Hardy films, bought the schooner. She went through name changes to *Aurora* and *Serena.*

In 1937 Robert P. Scripps purchased *Serena* to give to Scripps Institution. According to John Lyman, the grateful staff decided to rename the ship *E. W. Scripps,* but the donor commented: "There's too much Scripps around La Jolla already." So the name "Matthew F. Maury" was selected, but Robert P. Scripps then objected, as a loyal son, and the original choice was retained.

For research work the 100-foot masts were cut to 88 feet, diesel fuel tanks were added, two winches were

Marine Facilities and the Fleet

installed, a deckhouse laboratory was built, and some staterooms below were converted to laboratories.*[7] From 1938 to 1941, the ship was used a great deal for research cruises, including two expeditions to the Gulf of California (see chapter 2).

During World War II the ship was borrowed by the U.S. Navy for use by the University of California Division of War Research. At that time the deckhouse laboratory was doubled in size and a new diesel engine was installed. "The additional speed and weight of the new engine greatly reduced her rolling and pitching, and she became decidedly more comfortable to work in, thus increasing the efficiency of the personnel by decreasing the effects of mal de mer."[8]

After the war the ship required a major overhaul and was limited to fairly nearshore work. Her last cruise for Scripps was in April 1955, to the San Benito Islands.

Michael Todd bought the *E. W. Scripps* to use in the movie "Around the World in Eighty Days" (as the paddle-wheel steamer *Henrietta),* and later she was bought by Walter S. Johnson, Jr., for Pacific island trade from Tahiti to Raratonga. Her name then was *Tiare Maori.* On 4 January 1961 in the slipway in Papeete — where she was up for sale — the ship caught fire. A fireboat from Metro-Goldwyn-Mayer (then filming "Mutiny on the Bounty") pumped water on the ship until she foundered and sank in six fathoms of water.[9] Scripps personnel on Monsoon Expedition in Papeete in March 1961 recognized the masts of their old ship above the water line.

Flip (1962-)

L 355 feet; B 12.5 - 20 feet; D 13.7 feet horizontal; D 300 feet vertical; CS up to 11 knots when being towed horizontally; E 14 days. Crew 6; scientists 10. ONR-owned.

*During the conversion, according to Lyman, "about two dozen pairs of ladies' shoes were recovered from her bilges."

Scripps Institution of Oceanography: Probing the Oceans

This manned ocean buoy, which can be raised to a vertical position by flooding the ballast tanks, is not a self-powered vessel. See chapter 4 for information on *Flip* and on some of the other unusual craft of the Marine Physical Laboratory. *Flip* was built by Gunderson Brothers Engineering Corporation, Portland, Oregon, in 1962, was christened by Sarah W. (Mrs. Fred N.) Spiess, and was acquired by Scripps Institution on 6 August of that year. The Office of Naval Research has contracted her to the Marine Physical Laboratory.

The name was given by MPL as an acronym for *fl*oating *i*nstrument *p*latform.

Gianna (1973-)

L 55.5 feet; B 14.3 feet; D 3.5 feet; CS 16 knots; R 480 miles; E 2.5 days. Crew 2; scientists 4. UC-owned.

This diesel pleasure cruiser was built by Cantiere Navale di Chiavari in Italy in 1969 for Howard B. Lawson of Newport Beach, California, who gave her to Scripps Institution on 27 December 1973. She was named by Lawson for the Italian version of his wife's name, Jane.

The Golden One (1962-1963)

L 38 feet; B 11 feet; D 3 feet. UC-owned.

This boat, built in 1930 by Matthews Boat Company, Port Clinton, Ohio, was given to Scripps by Elmer Bernstein of Los Angeles in December 1962; she was not put into use by the institution and was sold to Charles Watson of San Diego in August 1963.

Horizon (1948-1969)

L 143 feet; B 33 feet; D 13.5 feet; CS 11.5 knots; R 6,800 miles; E 48 days. Crew 19; scientists 16. UC-owned.

This formerly Navy-owned ocean tug (ATA 180) was built in 1944 by Levingston Shipbuilding Company, Orange,

Marine Facilities and the Fleet

Texas. Scripps Institution took custody of her on 30 March 1948, especially for the Marine Life Research program, and in 1950 she opened the era of exploration on Midpac Expedition.

The name was given following a contest on campus, simultaneously with the *Crest*. The winner was J. F. T. ("Ted") Saur, who recalls that he won ten dollars and a drawing of the ship.

The *Horizon*, said H. William Menard, "was small and not above reproach. . . . People liked *Horizon* and she was always, as far as I know, a happy ship. Perhaps for that reason the charts now record Horizon Guyot in the central Pacific, from which were dredged some of the oldest rocks yet found in the ocean basins; Horizon Depth, which is the second deepest place in the ocean; Horizon Channel, in the floor of a flat plain in the Gulf of Alaska, and Horizon Bank, a drowned atoll east of the New Hebrides Islands in the southeastern Pacific. Few other ships have been so honored."[10]

Scripps sold *Horizon* on 5 September 1969 to Pacific Towboat and Salvage Company in Long Beach. A later owner offered some of the ship's equipment to the institution, which re-purchased the A-frame and other items at scrap prices. In 1975 she was still in use in San Diego, owned by California Molasses Company.

Hugh M. Smith (1959-1963)
L 128 feet; B 29 feet; D 14 feet; CS 9 knots; R 10,000 miles; E 45 days. Crew 14; scientists 8. USFWS-owned.

This yacht (YP-635), built in 1945, was borrowed from the U.S. Fish and Wildlife Service by Scripps Institution from 23 June 1959 until 4 November 1963, chiefly to replace the *Stranger* during the two-year Naga Expedition. The name honored Hugh McCormick Smith (1865-1941), who served as U.S. Commissioner of Fisheries from 1913 to 1922.

Scripps Institution of Oceanography: Probing the Oceans

When Scripps gave up the ship, the governor of American Samoa requested her, and former Scripps captain Marvin Hopkins delivered the ship to Pago Pago. She operated from there for several years.

Macrocystis (1958-1964)
L 22.7 feet; B 8 feet; D 1.8 feet; CS 20 knots; R 75 miles. Crew and scientists 2. UC-owned.

This boat was built by Jeffries Boat Company, Venice, California, in 1958 for the kelp project of the Institute of Marine Resources (see chapter 6), and was named for the giant kelp. In November 1964 she was sold to LaRoy B. Wickline of San Diego.

Melville (1969-)
L 245 feet; B 46 feet; D 15 feet; CS 12 knots; R 9,840 miles; E 41 days. Crew 25; scientists 25. Navy-owned.

This research vessel, designated AGOR-14 (Auxiliary General Oceanography Research) by the Navy, was built at the Defoe Shipbuilding Company in Bay City, Michigan, in 1969, for research use by the Scripps Institution, which received her on 2 September 1969. She is distinguished by two vertically mounted, multi-bladed, cycloidal propellers, which enable her to move forward, backward, sideways, and around her own axis.

The name was given for Rear Admiral George Wallace Melville (1841-1912), engineer in chief of the U.S. Navy, who participated in three Arctic voyages, and was a contributor to improved equipment on Navy ships.

Oconostota (1962-1974)
L 100 feet; B 25 feet; D 10.5 feet; CS 11 knots; R 4,500 miles; E 16 days. Crew 8, scientists 6. Navy-owned.

This former harbor tug (YTB-375, later YTM-375) was built by Gulfport Boiler and Welding Works, Port Arthur,

Marine Facilities and the Fleet

Texas, in 1944, and was provided to Scripps Institution in October 1962. She was acquired chiefly to tow *Flip* but proved not satisfactory for that purpose, so she became a regular ship of the line until 1974, when she was returned to the Navy and was transferred to Moss Landing Marine Laboratory in April 1975.

The name was given by the Navy for Cherokee Indian chief Oconostota, who in colonial days was first friendly with the British but later allied with the French.

Orca (1956-1962)

L 100 feet; B 23 feet; D 7.5 feet; CS 8.5 knots; R 3,000 miles; E 18 days. Crew 7, scientists 9. UC-owned.

This one-time Coast Guard patrol vessel (YP42), built in 1926 by Defoe Boat Building Company in Bay City, Michigan, was bought by Scripps Institution in June 1956 from the J. W. Sefton Foundation. Scripps sold her to Murphy Marine Service, San Diego, in October 1962. Texas A & M later used her for research work. J. W. Sefton had named her for the killer whale.

Paolina-T (1948-1965)

L 80 feet; B 22 feet; D 9.7 feet; CS 8.5 knots; R 2,450 miles; E 30 days. Crew 9, scientists 5. Navy-owned.

This purse seiner was built in 1944 at the Colberg Boat Works, Stockton, California, and was purchased through Navy funds by Scripps Institution from fishermen brothers Michele and Guiseppe Torrente on 15 June 1948. Scripps returned her to the Navy on 17 March 1965. In 1975 she was under private ownership in the Monterey Bay area, with the name *New San Joseph.*

The original name was given by the Torrentes. In 1952 a campus contest was held to rename the *Paolina-T* and the newly acquired *Spencer F. Baird,* in order to release the latter name to the U.S. Fish and Wildlife Service. Three "learned and impartial judges" (Joel W. Hedgpeth, Carl Eckart, and Carl L. Hubbs) selected "Petrel" for the *Paolina-T* and "Seamount" for the *Baird,* from sixty-odd entries. Three individuals and one group shared the ten-dollar prize for the name "Petrel," and seven individuals and one group shared similarly for the name "Seamount." Among the names rejected, but specifically remarked upon by Hedgpeth's committee, were "Dramamine" and "Jolly Roger." The names of the two ships were never changed, however, because the Navy was the actual owner of those ships and prevented the change.

Red Lion (1959-1963)

This 22-foot gasoline-powered launch was given to Scripps Institution by Dr. C. C. Curtis in January 1959, but was never put into use by the institution. In February 1963 she was given to Escuela Superior de Ciencias Marinas in Ensenada and was delivered to them aboard the *Alexander Agassiz* while en route to Cabo San Lucas.

ST-908 (1961-1973)

L 45 feet; B 12.5 feet; D 5 feet; CS 9 knots; R 655 miles; E 4 days. Crew 2; scientists 3. UC-owned.

This former Army harbor tug was built in 1945 by Burger Boat Company in Manitowoc, Wisconsin, and was acquired by Scripps in 1961 through the State Educational Agency for Surplus Property, at the same time as the *Alexander Agassiz*. From 1973 until 1976 she was loaned to Moss Landing Marine Laboratory, which called her *Artemia.* In August 1976, she was sold to William H. Richter of San Diego.

Scripps (1925-1936)

L 64 feet; B 15 feet; D 6.7 feet. Crew and scientists 10. UC-owned.

Marine Facilities and the Fleet

This purse seiner, then named *Thaddeus,* was bought by Scripps Institution in September 1925, and was renamed *Scripps.* The following year she was painted University of California colors: a blue hull with gold stripes.

On 13 November 1936 an explosion and fire in the boat sank her at the dock. Living aboard were the captain, Murdock G. Ross, and the cook, Henry Ball. Ball died of his injuries a week later, and Ross never fully recovered. The salvaged hull was later put into use as a garbage scow, and Director Sverdrup asked the owner to change the name, which he did, to *Abraham Lincoln.* (For additional information, see Raitt and Moulton.[11])

Spencer F. Baird (1951-1965)
L 143 feet; B 33 feet; D 13.5 feet; CS 11.5 knots; R 6,800 miles; E 48 days. Crew 20, scientists 15. Navy-owned.

This former Army tug (LT-581) was acquired by Scripps from the U.S. Maritime Commission in August 1951. The *Baird,* as she was usually called, and the *Horizon* often sailed together on two-ship expeditions, and a certain amount of rivalry developed at times. The ship was returned to the Maritime Commission in October 1965. She was later in use by the Vietnamese government, under the name *Tien Sa.*

Her name had been given by the Navy for the outstanding naturalist, Spencer Fullerton Baird (1823-1887), who was the first director of the U.S. Fish Commission (see *Paolina-T).* The ship had been used as a research vessel by the U.S. Fish and Wildlife Service prior to being acquired by Scripps.

Stranger (1955-1965)
L 134 feet; B 24 feet; D 14.5 feet; CS 12 knots; R 6,000 miles; E 40 days. Crew 14, scientists 10. UC-owned.

This yacht was built in 1938 by the Lake Union Dry

Scripps Institution of Oceanography: Probing the Oceans

Dock and Machine Works, Seattle, Washington, for Fred Lewis, who supplied the ship plans. Lewis was a rancher from Wyoming who also owned a small island off the coast of British Columbia. His wife had been born and raised in Hawaii and was interested in mollusks (for a sea-shell collection) and other ocean creatures. Said L. H. Hughes, Plant Superintendent of Lake Union Drydock, who had helped in construction of the *Stranger:* "There was sixty ton of lead ingots installed on the keel for stability, but after using the vessel a short time it was found to be very tender, and rolled very easily. It was brought back to the plant, drydocked and the hull was sponsoned out about eighteen or twenty inches on each side. This gave the vessel a lot more stability."[12]

Those who rode the *Stranger* on trips to the Gulf of Alaska in 1956 and 1961 were not favorably impressed with her stability. Alan C. Jones commented after three weeks of Chinook Expedition in 1956:

> The crew finally managed to untie the knots that held us in San Diego, and away we rolled in the great yacht. Twasn't long before BTs, oscillators, and roast pork were flying all over, especially in the living room.
>
> Two days later a few of the Stranger Rangers emerged from the basement to become lounge lizards on the sun porch. By the fourth day all were recovered from the strange ailment that hit almost everyone off Point Concepcion.
>
> . . . Ah yes, it's hard to beat this gracious living out aboard a yacht. The gentle motion (rolling up to and including 52 degrees!*), the attractive, spacious living room (if one is really determined he can crawl through the maze of wires and tubes that completely fill the

---

*Once during this expedition a roll of 58 degrees was registered.

Marine Facilities and the Fleet

labs!), and the delicious meals expertly cooked to please the slightest whim of the happy tourist. (I've been trying for three weeks to get a soft boiled egg.) . . . One fine, moonlightless night the *Stranger* inmates had to scramble up the walls (or was it down?) to avoid being battered by flying objects when the yacht decided to lie on its side for a while. We were glad the yacht wasn't in the trough of the waves because the sun porch might have gotten damper than it did. [13]

In all fairness to the *Stranger*, her performance on Naga Expedition in the Gulf of Thailand and South China Sea was very good (see chapter 15).

From 1941 to 1947 the ship was used by the University of California Division of War Research, under the name U.S.S. *Jasper* (PYC-13). Scripps Institution acquired the ship as a gift from N. A. Kessler on 22 April 1955, and on 9 March 1965 sold her to Charles H. Briley of Newport Beach, California. She was later in use by Teledyne Corporation and for them revisited Thailand. In 1976 she was operating as a cruise ship under the name *Explorer*.

T-441 (1955-1969)

L 65.6 feet; B 18 feet; D 6 feet; CS 10.5 knots; R 1,830 miles; E 5 days. Crew 5, scientists 4. Navy-owned.

This former Army cargo and passenger T-boat was built in 1953 by National Steel and Shipbuilding Corporation, San Diego, and was provided by the Army to Scripps Institution on 23 March 1955. Scripps gave her up on 20 August 1969, and the Navy provided her to the University of Connecticut.

Thomas Washington (1965-)

L 209 feet; B 39 feet; D 13.7 feet; CS 12 knots; R 10,000 miles; E 29 days. Crew 25, scientists 17. Navy-owned.

Scripps Institution of Oceanography: Probing the Oceans

This AGOR-10 (Auxiliary General Oceanography Research) was built in 1965 by Marinette Marine Corporation, Marinette, Wisconsin, and was provided by the Navy to Scripps Institution on 29 September 1965. The *Washington* and her sister ship *Thomas Thompson* (of the University of Washington) had considerable engine problems during the first few years and had to operate at very reduced speeds until new engines were installed.

The name was given by the Navy for Admiral Thomas Washington, Navy Hydrographer from 1914 to 1916, "who served with distinction in the Navy for 40 years before his retirement in 1929."

Utility Boat (1959-)

L 32 feet; B 13.5 feet; D 3.5 feet; R 200 miles; E 1 day. Crew and scientists 2. Navy-owned.

This small vessel, built by Jeffries Boat Company, Venice, California, was requested by H. William Menard as a workboat on the *Argo,* when that ship was being converted for oceanographic work in 1959. The boat was used to some extent as a "second ship" for seismic work with the *Argo* but proved of limited use in open ocean work. She was kept for nearshore work, chiefly by the diving program, after the *Argo* was returned to the Navy.

Some tried to call this boat the *Argonaut,* but the name has never caught on.

Most of the fleet home for Christmas in 1974, at the then-new dock of Nimitz Marine Facilities. Counterclockwise from left: *Flip, Alpha Helix, David Starr Jordan* of the National Marine Fisheries Service, the barge ORB (at end of dock), *Alexander Agassiz, Melville.* Angled behind *Melville* are *Oconostota, Ellen B. Scripps, Dolphin.* Not present are *Thomas Washington* and *Gianna.*

NOTES

1. "Proposed Development of Marine Biology at the Scripps Institution of Oceanography," proposal to Rockefeller Foundation, 15 August 1953, Appendix III.

2. Memorandum to staff, 22 December 1950.

3. "Odyssey of the Argo," *Oceans,* Vol. II, Nos. 3 and 4 (September-October 1969), 28.

4. Manuscript, "MIDPAC – The First Big Step," 17 August 1975, 7.

5. Memorandum to Carl Eckart, 20 April 1949.

6. *In* William E. Ritter, "The Marine Biological Station of San Diego; its History, Present Conditions, Achievements, and Aims," *University of California Publications in Zoology,* 9 (9 March 1912), 176.

7. E. G. Moberg and J. Lyman, "The 'E. W. Scripps,'" *Records of Observations,* SIO, Vol. I, No. 1 (July 1942).

8. Completion Report of UCDWR, 1946, 163.

9. *Pacific Islands Monthly,* Vol. 29, No. 6 (January 1959), 108.

10. "The Research Ship *Horizon,*" SIO Reference 74-3 (1974); 1.

11. Helen Raitt and Beatrice Moulton, *Scripps Institution of Oceanography: First Fifty Years* (Los Angeles: Ward Ritchie Press, 1967), 121-22.

12. Letter to Elizabeth N. Shor, 30 January 1974.

13. Letter of July 1956.

15. Oceanography is Fun: A Glimpse of the Expeditions

Scripps ships and trips depart so frequently and so far afield nowadays that it is next to impossible to keep track of them. That hasn't always been so.

In the early 1950s each expedition departure (except a few classified ones) drew a crowd of Scripps people and several reporters to the dock, and the arrival was equally hailed. Each trip had its "first": a new region, a new kind of equipment, a new discovery.* "All these cruises," said Revelle, "had essentially the same purpose: they were voyages of discovery in which new instruments for oceanographic exploration were pitted against the vast unknown of the Pacific Ocean."[1]

Exploration was only an excuse. After all, as Revelle also said:

> Oceanographers are not such a serious-minded lot that they keep asking themselves why they are doing their job. The spiritual ancestor of most of them was

---

*Some of the expeditions created personal complications, which I feel obliged to omit — even though some of them were very much part of Scripps history.

Ulysses. He was called the Wanderer, because he was the first to venture into the River Ocean, out of the salt and fishy Sea-between-the-Land, the wine-dark Mediterranean. Perhaps he disliked administration, hated farming, and was bored by Penelope. In any case, Ulysses managed to spend a great deal of time away from home. He never stated his reasons very clearly, but he still lives in the hearts of oceanographers.[2]

The spiritual descendants of Ulysses who found their sea legs just after World War II were a special breed. Many had seen Navy or Marine Corps service. Some had taken wartime courses in meteorology or radar or electronics. They were a bit suspicious of sophisticated "black-box" instruments, certain that they knew a better way to do the job, and eager to try. Their motto became: "Why couldn't we do it this way?"

At sea they gathered records and samples and specimens for science; ashore they gathered kava bowls, tapa cloths, cowry necklaces, vicuña rugs, carvings, "antiquities" — and memories.

Among similar institutions, Scripps has the advantage of having selected as its bailiwick the "south seas," the ocean that holds the tropical islands "which lie like carelessly tossed necklaces on the velvet sea."[3]

A few days of snorkeling on the reef of a deserted atoll in the Tuamotus [wrote H. William Menard]; a boat ride through basking sea turtles in the Revilla Gigedos; a stop at Robinson Crusoe's cave on Más a Tierra (now Isla Róbinson Crusoe), a climb on a giant stone image on Easter Island — these are not everyday pleasures, and so cherished more.

. . . I have stepped, waded, and swum ashore from a *Horizon* smallboat to many a beautiful, tropical

A cordial meeting on Ocean Island during Capricorn Expedition in 1952-53.

island — named but virtually unknown to the outside world. At Ocean Island we joined in a happy Micronesian wedding. At Lifuka we saw Polynesians riding horses through the surf as in a Gauguin painting. At Nuku Hiva we walked through the silent valley of Typee, made famous by Melville when it teemed with cannibals. At Vanua Mbalavu we saw the grass walks being swept clean of leaves each dawn.[4]

Harris B. Stewart, then a graduate student, waxed lyrical over Tonga:

Christmas dinner we had on the beach with two Tongan families. . . . A lei about my neck, a drinking cocoanut before me, a slice of breadfruit in one hand and a slice of pineapple in the other (the only way to eat breadfruit, a bite of one and a bite of the other), and a young native girl using a cocoanut palm frond to wave away the flies, warm sun filtering through the acacia trees above us, and the incessant booming of the surf on the barrier reef some 200 yards offshore, this was really living — oceanography at its best.[5]

Remote ports became collectors' items. Honolulu, for example, is only considered romantic by wives and girl friends who have never been there, but Hilo is somewhat novel. Before *Mutiny on the Bounty* was filmed in Tahiti (during which time, and since, swarms of visitors introduced alien customs into that charming land), Papeete was *the* port. Expedition planners pointed out, with a grin, that it was the only port in the south-central Pacific where fuel was available. Rapa Island was considered a high point, although the party on the *Horizon*, which wasn't even supposed to get there, didn't see much of the island, as the ship caught on an uncharted coral head on the way up the

A Glimpse of the Expeditions

channel and took hours to winch herself off. A few enthusiasts are fond of Adak, "the emerald island washed by the cool blue waters of the Bering Sea"* ("cool" being barely above freezing); but not everyone yearns to visit Adak "National Forest" in the invariable fog — and the pine trees, planted by military personnel, are only five feet high. The few who have been to Pitcairn Island can always draw a respectful pause in maritime conversation by merely mentioning it. The Galápagos Islands are a favorite stop for readers of Charles Darwin. Easter Island is held worthy of a detour. Scripps ships found Port Victoria in the Seychelles before the tourists did.

To a certain extent Scripps people feel that the entire Pacific basin — one-half the surface area of the world — is their domain. *Sotto voce* they sometimes mutter, "What are they doing in *our* ocean?" when they hear of an east-coast oceanographic vessel heading into the Pacific. When long-time Scrippsian Warren Wooster was departing to become director of the University of Miami's Rosenstiel School of Marine and Atmospheric Sciences in 1973, he was presented with an ornate scroll that ended with the hope that he would find happiness in Miami, "in spite of being three thousand miles away from The Ocean."

Scripps certainly has no monopoly on sunsets at sea, but Scrippsians enjoy them as much as do other sea rovers. Would-be poet Harris B. Stewart on Capricorn Expedition wrote of "another glorious sunset as the plug was pulled along the western horizon and all the light and color drained from the sky." Fifteen years later Baron Thomas recorded in his log a sunset that was "enough to make a sailor out of any confirmed landsman as it was magnificently enhanced by the dark blue shades of the Sea as the day's last shimmering light was cast through each swell." One crew member

---
*The motto of the local radio station.

specialized in taking artistic photographs of sunsets, and one scientific leader persists in praising sunrises over sunsets, to the annoyance of those on a different watch schedule.*

Sunset hour is the relaxed time of day aboard ship. Shortly after supper, people start arriving on the fantail, one or two at a time, for a smoke or a cup of coffee, visiting quietly and gazing at the kaleidoscope of sky and sea. No photograph or string of words can capture those sunsets, because neither can encompass the entire sky or the gamut of nostalgia that makes a sunset at sea so special to each person.

Some at Scripps call them sea trips, others say cruises. Officially they are expeditions. Their purpose is to solve the mysteries of the deep. According to Willard Bascom, Roger Revelle liked to say, "You must go to the sea with a question."[6] Robert L. Fisher said, "At sea you are betting that you and your people can get an answer. There's no better satisfaction."

So expeditions have sailed from Scripps with questions and have come back with answers — and with more questions to be resolved another day. All have had their successes, their frustrations, and their sea stories. Memories of them are as kaleidoscopic as sunsets, and are enhanced by the distance of time.

When was the first Scripps expedition? In a way, it was three-quarters of a century ago, when some of the staff members took short trips on the yacht *Loma* that was owned by E. W. Scripps. A trip to the offshore islands then was an expedition, and an experience. In the 1930s T. Wayland Vaughan hoped to draw Scripps Institution into exploring the entire Pacific Ocean. His successor, Harald U. Sverdrup, was finally able to begin that exploration,

---

*This veteran radioed from a summer trip through the Kamchatka Basin that "magenta sunrises at 0200 are a tourist must."

Completing a net haul at sunset on Tasaday Expedition in 1974. Photo by Elizabeth Venrick.

when he acquired the schooner *E. W. Scripps* in 1937. The first long expedition was to the Gulf of California for two months in 1939. But to most Scrippsians today the expeditions began after World War II, with Crossroads and Midpac.

Operation Crossroads was a military project to test the effects of atomic bombs against naval vessels and to determine the environmental impact of the explosion on Bikini Island in Micronesia. The aim of the team of one thousand biologists, geologists, oceanographers, and technicians was "to carry out an integrated investigation of all aspects of the natural environment within and around the atoll: the currents and other properties of the ocean and lagoon waters, the surface geology, the identity, distribution, and abundance of living creatures, and the equilibrium relationships among all these."[7]

Roger Revelle was the officer in charge of oceanographic studies, called at the time "the most complicated laboratory experiment ever undertaken." Much later Revelle wrote: "The results, combined with those of later investigations in 1947 and following years, gave the clearest, most detailed pictures of an atoll and its flora and fauna that we possess, even today."[8] He went on:

> In order to make the tests as comprehensive as possible, Admiral [W. H. P.] Blandy and his technical assistant, Rear Admiral William S. Parsons, agreed that we should try to learn as much as we could about the possible effects of the bombs on the ecology and geology of the atoll. We also wanted to learn about the waves that would be produced by the air and underwater explosions and about the dispersion of radioactive materials in the lagoon and ocean waters.
> Scientists from the U.S. Geological Survey, the Fish and Wildlife Service, government laboratories,

A Glimpse of the Expeditions

Scripps, Woods Hole, and the Universities of Michigan, Southern California, Washington and California joined the effort. An ancient hydrographic survey ship, USS BOWDITCH, was assigned to the task of making a pretest geological, biological, and oceanographic survey.

[After the *Bowditch* had sailed, the historian of the project noted: "The entire country, because of the demands of this group and the radiological group, is now largely drained of oceanographic equipment and personnel since oceanography as a small prewar specialty has not had time to catch up with the demands put upon it by wartime developments."[9]]

Three high photographic towers were constructed on Bikini Island [continued Revelle] to take time-sequence photographs of the waves, and heavily cased, wave-pressure recording devices built to withstand very high shock pressures, which we called 'turtles,' were planted on the lagoon bottom throughout the array of target vessels.

The theory of explosion-generated waves in shallow water was unsatisfactory, and in order to learn what to expect, we carried out a series of model studies in Chesapeake Bay, using various amounts of TNT. We also made seismic refraction and reflection measurements of the depth of the coral and the nature of the material beneath the atoll.

. . . The waves from the underwater explosion were about what we had predicted from our model studies, producing 12-foot high surf on Bikini Island. But completely unexpected was the behavior of the giant jet of foaming seawater that shot a mile into the air, and then fell back in a huge, doughnut-shaped mass of spray nearly 1,000 feet high which rapidly spread out and enveloped the entire array of target ships. The initial outward velocity of this 'base surge' was so great

Scripps Institution of Oceanography: Probing the Oceans

that for a few moments we thought it was the water wave produced by the explosion, and that it would break right over the rim of the atoll.

Scripps scientists were well represented in this cooperative study, and some close associations were formed with other institutions. In the spring of 1947 Revelle directed a resurvey of Bikini Atoll for the Joint Operation Crossroads Committee of the Department of Defense. This was intended primarily to determine delayed or long-term effects of the underwater nuclear explosion. As a separate project at that time, several deep holes were drilled on Bikini Island, under the direction of the U.S. Geological Survey, to try to determine the geologic history of a coral atoll.

Revelle returned to the institution from his Navy service in the summer of 1948, and was soon gazing beyond the California Current that was under scrutiny by the Marine Life Research program. So little was known, so much was possible at Scripps with a fleet of ships, research money, and new equipment.

MIDPAC

So, in the fall of 1949 Revelle proposed the University of California Mid-Pacific Expedition:

> By contrast to the Atlantic, the central and eastern Pacific is almost unexplored from a modern scientific point of view, but sufficient is known to indicate the existence of problems, the solution of which would be of the greatest importance in understanding the history of the earth.[10]

Revelle may have been thinking of the expedition by the *Atlantis* of Woods Hole Oceanographic Institution in

A Glimpse of the Expeditions

the summer of 1947, to explore the Mid-Atlantic Ridge. It was time for Scripps to go to sea.

The plans expanded to include the Navy Electronics Laboratory and its Navy-operated ship, *PCE(R)-857*, thereby allowing for seismic-refraction studies which require two ships, and other programs. The Institute of Geophysics at UCLA also provided support. Letters and memos proliferated.

Midpac, Scripps Institution's first major postwar expedition (with a nod to the steadily continuing CalCOFI monthly cruises of that time), probably led to the custom of naming expeditions. In a typically brief memo for typically efficient reasons, Finn W. Outler, technical superintendent of the Marine Physical Laboratory, wrote to Revelle: "In order to save several reams of paper, a few typewriter ribbons and many hours of clerical time, I suggest we adopt 'OPERATION MIDPEX' in lieu of 'JOINT UNIVERSITY OF CALIFORNIA–U.S. NAVY ELECTRONICS LABORATORY MID-PACIFIC EXPEDITION.'" Revelle answered: "How about MIDPAC to distinguish from other things already in existence?"

The Scripps custom of naming expeditions is unique among oceanographic institutions. Nautical phrases (Shellback, Downwind) and regional winds (Chubasco) — or the lack of them (Doldrums) — have dominated the list, but some prefer Greek mythology (Tethys, Amphitrite) or plain whimsy (Northern Holiday, Swan Song). The custom leads to confusion among the uninitiated, who are baffled to hear, "When I was in Lahaina on Show . . ." or "When I was out on Limbo. . . ."

With James L. Faughn as captain, the *Horizon* sailed away on Midpac Expedition on 27 July 1950, after a delay for engine repairs. She carried nineteen in the scientific party, and the *857*, under the command of Lt. Comdr. D. J. McMillan, carried thirteen, of whom three were from

Scripps Institution of Oceanography: Probing the Oceans

UCLA and three from the University of Southern California. The ships also carried a quantity of new and improved equipment for probing the ocean's mysteries.

"Nick" Carter manned the radio shack on *Horizon,* and Revelle helped keep him busy relaying the shipboard news back home. The first report was on 14 August:

> ... We have found that exploring the ocean floor by any method in water three miles deep with current velocities up to two knots and winds up to thirty-five miles an hour is no picnic. But we are getting pretty well shaken down into an efficient working system. ... We have a daily newspaper complete with cartoons. Movies in the evening and good radio communications, so we are not completely out of touch with the great world. Chow is plentiful and good, supplemented by an occasional albacore or dolphin.* ...

In about a month the explorers reached Honolulu, where, said Revelle in the next shipboard message, "the whole atmosphere in the islands was so delightful and relaxing that we found it difficult to remember the urgency of getting back to sea again. Almost everyone blossomed out in a bright-colored Hawaiian shirt, and nooks and crannies all over the ship are stuffed with Hawaiian things to be taken back to the families at home." (Thus began another Scripps custom, that has been followed on many a voyage to many an island; Scripps homes abound in "native crafts.")

The *857,* in fact, did *not* remember the urgency of getting back to sea, and was detained in Honolulu for repairs to the clutch. (Many of the families of the Navy crew members of the *857* met the sailors in Honolulu, and

---

*This would be the dolphinfish *(Coryphaena hippurus),* called mahimahi by Hawaiians, not the marine mammal, the dolphin.

A Glimpse of the Expeditions

Scripps participants on *Horizon* suspected that the ship might find another excuse to return to the port, so the rendezvous point for the two ships was moved farther and farther along after the *857* left Honolulu, to get her beyond the point of practical return.)

Without pausing to await their companion, the scientific party on the *Horizon* set out to explore the 50,000-square-mile area from 700 to 1,050 miles west of Hawaii, and from 18 to 20 degrees north, throughout which they ran sounding lines at right angles to the regular shipping tracks. On 9 September they proudly announced the discovery of a new mountain range, the Mid-Pacific Mountains:

> . . . a great elongated mountain range extending in a northeasterly direction towards Necker Island in the Hawaiian chain and some sixty miles wide. These Mid-Pac mountains are dominated by a scarp on the southern side averaging a mile high and sloping downward toward the deep Pacific floor south of the mountain range at an average angle of about six degrees. . . . In one basin Jeff Frautschy's corer showed alternating layers of globigerina ooze and red clay overlying a well-sorted sandy gravel with rounded pebbles of volcanic rock. Seismic reflection measurements by Russ Raitt indicate a sediment thickness in these basins of about 1300 to 2200 feet. [James M.] Snodgrass and [Arthur E.] Maxwell find that the temperature appears to increase with depth in the sediment about one degree Fahrenheit per twenty-five feet. We are now surveying a flat topped sea mount 920 fathoms deep near the southwestern end of the range. Our first dredge haul on this sea mount, skillfully supervised by Bob Dietz, brought up fragments of shallow water limestone to all appearances like the beach rock at high tide level near Kaneohe or Bikini. . . . For the moment

385

our big winch, appropriately nicknamed Massey's* Tinker Toy, is behaving reasonably well. . . . We are all gratified that we have such consistently excellent radio communications through WWD, thanks to Frank Berberich and Nick Carter. It helps to bring a small feeling of home to this otherwise lonely ocean. All on board are well and as happy as can be expected this far from home. . . .

The *857*, also working en route after its delayed departure from Honolulu, caught up with the *Horizon*, and both ships continued on to Bikini for a seismic-refraction profile of that atoll and of nearby Sylvania Seamount, as a follow-up of Operation Crossroads. They then sailed on to Kwajalein, from where Revelle and eleven others returned home by airplane, and Raitt took over to continue seismic-refraction profiles. Although much disappointed at not finding the explosives he had been promised at Kwajalein, Raitt was able to complete almost all of the planned survey of the atoll with explosives supplied by the Navy base there. While the seismic work was being done near Bikini and Kwajalein, field parties camped ashore for chemical and geological studies. Raitt and Dietz flew home from Kwajalein on the second port call there, and the ships turned homeward, sailing by sun and star. For the Navy's air-sea rescue system, they regularly set off four-pound Sofar bombs, one of which discharged by the *Horizon* provided a distance record for propagation of underwater sound of 3,500 miles and was called "the shot heard round the Pacific."

Captain Faughn kept home base up to date on their progress. On 16 October he sent word to Revelle:

Crossed 180th meridian at noon today. . . . Everything going well and everybody happy to be on God's

*John Massey was the ship's chief engineer.

A Glimpse of the Expeditions

side of the time meridian X Do you want to join the anchor pool* and if so, what is your best E.T.A.?** X Passed our first ship in the night, quite a friendly sight.

Revelle guessed 26 October at 2359 hours for the anchor pool, but he was overly optimistic, for the *Horizon* pulled in on 28 October. The *857* arrived six days later. Science writer Andrew Hamilton found the "pair of salt-encrusted vessels" at the dock:

> Not substantially larger than Magellan's *Trinidada* and *Vittorio,* the little ships were loaded to the scuppers with statistic-filled notebooks, newly-drawn charts and graphs, slim, plastic tubes of mud, odd pieces of rock, coral and fossilized shells, hundreds of bottles of seawater and pickled fish, and an assortment of bent and battered scientific gear that would delight any junk dealer.[11]

The scientists had already been counting their accomplishments. The Mid-Pacific Mountains were the most dramatic discovery, although not completely unexpected. Before the expedition H. William Menard and Robert S. Dietz (then both with the Navy Electronics Laboratory) had plotted the many known guyots and shoal spots within the area to be surveyed, but not until the *Horizon* zigzagged across the area with a continuous echo-sounder did the vast extent and continuity of the geological feature become known.

Menard was pleased to have confirmed his guess that the Mendocino Escarpment — a submerged cliff half a mile high

---

*A wager as to when the ship will drop anchor at the dock.

**An abbreviation that an oceanographer's wife learns early: estimated time of arrival.

Scripps Institution of Oceanography: Probing the Oceans

that lies seaward from northern California — did indeed extend a full thousand miles from shore, as, perhaps, an extension of California's San Andreas fault. Manganese nodules dredged from seamounts interested the geologists, who were impressed by the great area of ocean floor covered by them. Could the nodules be mined? Probably not economically, they concluded, if it took as many hours of dredging time as Midpac scientists had spent laboring to retrieve their nodules.

With their improved Kullenberg corer, the geologists had acquired ten cores up to 24 feet long from as deep as 2,800 fathoms, and they retrieved a number of shorter cores with the gravity corer, but problems with the big winch on the *Horizon* prevented the taking of really long cores. "We used to bet a bottle of beer* on what we would find in any particular core," wrote Revelle to Edward Bullard; "the chances always seemed about even that we would get globigerina ooze, red clay, volcanic ash or various modifications and combinations of these."[12]

Raitt's seismic-refraction profiles near the atolls provided strong evidence that such features are indeed coral reefs around volcanoes. These profiles, in combination with similar work done earlier on Project Crossroads, and with drilling done by the Atomic Energy Commission and Los Alamos Scientific Laboratory in 1950-52, verified the existence of lava beneath the coral, and so provided the final proof of Darwin's theory of the origin of coral atolls as volcanic peaks that had slowly submerged. Raitt also gained a thousand miles of seismic-refraction profiles in the deep ocean. Previous calculations had indicated that at least two miles of sediments should have accumulated over the millenia. But the seismic work indicated that only 1,000

---

*Scripps ships carried beer at that time, were "dry" for many years, and since 1971 have carried beer and wine.

A Glimpse of the Expeditions

feet of sediments lay on the ocean floor, leaving another puzzle to be resolved.

An improved underwater camera, designed and manned by Carl Shipek of the Navy Electronics Laboratory, provided photos of the remote sea floor 20,000 feet down that showed ripple marks presumed to be from previously unsuspected deep currents. A borrowed geomagnetic electrokinetograph — nicknamed the "jog log" because it required right-angle measurements, hence jogging, by the ship — provided surface-current measurements that showed eddies instead of the presumed straight-line currents toward the equatorial region. Meteorologists sent radio-equipped balloons 100,000 feet into the air and found the weather there as variable as that which surrounded them.

The biologists found the equatorial waters rich in plankton, and noted the great numbers of fish and squid that darted into the light of lanterns hung over the side at night. They collected hatchetfishes, with rows of night-sparkling photophores along their sides, and other midwater creatures.

Richard Morita found surprising life elsewhere: bacteria, in a state of suspended animation, from cores taken from the bottom 9,000 feet down, which, when brought to "normal" atmospheric pressures, began to grow rapidly. He kept these alive aboard ship, and Claude ZoBell continued to culture them in a specially developed pressure chamber at Scripps after the expedition.*

In moments of whimsy the oceanographers taped eggs to the hydrographic wire and lowered them as deep as a mile and a half; the eggs returned to the surface unbroken — but scrambled inside. Oranges similarly treated, and

---

*Biologists haven't quite forgiven Revelle for the "beer incident" on Midpac. He removed some of Morita's bacteria samples from the refrigerator in order to cool the beer. Morita naturally protested, to which Revelle replied, "Young man, you must realize that there are times when beer is more important than biology."

appearing normal, were passed off to unsuspecting shipmates, who found them thoroughly salted.

A historic first on Midpac was measuring the flow of heat through the ocean floor.* It was the climax of a long-term aspiration of Edward ("Teddy") Bullard, then professor of physics at the University of Toronto, who spent the summer of 1949 as a visitor at Scripps. He had been seeking the use of a research ship with a suitable winch, and this he found at Scripps. Engineer James M. Snodgrass and graduate student Arthur E. Maxwell joined the project, and, using the facilities of the Scripps workshop and their own hands, the three fashioned a device for measuring temperature gradient, following Bullard's design — in spite of much contrary advice from kibitzers.

The heat-flow instrument devised that summer was a "very clumsy apparatus . . . thoroughly old-fashioned," according to Bullard, who said that he preferred to depend upon known equipment: galvanometers from the British Admiralty, "which were very robust," photographic recording, thermocouples, standard reversing switches, and *"no electronics."* The probe consisted of a thick, hollow steel tube about fifteen feet long, plus a recording chamber with a galvanometer to measure temperature differences between the thermocouples. As an innovation, the unit was made watertight with 0-rings, which proved very effective.

The first tests were made from a platform — "a sort of gallows," said Bullard, built in the marsh at Mission Bay. These demonstrated that the long probe would indeed penetrate mud to its full length. But the allotted time was

*Scientists on the Swedish Deep Sea Expedition in 1947 and 1948 had tried for such measurements, but succeeded in getting only two they considered reliable. Bullard believes that they probably measured only the warming of the probe by friction, especially as they jiggled the instrument up and down to prevent it from becoming stuck.

James M. Snodgrass (left) and Arthur E. Maxwell straightening the heat probe after a lowering on Midpac Expedition, 1950.

running out by the time the equipment was built, calibrated, and tested, and ship time had become scarce. Two attempts at measurements at sea were made in 1949, and both were unsuccessful, although everything appeared to be working properly. Bullard departed for England to become Director of the National Physical Laboratory — and soon to be knighted. There he set about making a second probe, with the aid of the facilities and the personnel of the laboratory's workshop.

Maxwell modified and rebuilt the original instrument and joined Midpac Expedition in the summer of 1950. Time after time the clumsy device was lowered laboriously to the sea floor, where its own weight drove the long probe into the sediments. It was then left in place for half an hour to dissipate the heat of friction from its plunge into the mud. Invariably, the probe returned to the surface with a sharp curve from the pull of the cable while the instrument was in place, but the probe could be straightened and used repeatedly. At each station a sediment core was also taken so that the thermal conductivity of the material could be determined in the laboratory.

> We managed to get at least seven good measurements of temperature gradients in the bottom muds [wrote Revelle to Bullard]. All but one of these indicate a gradient around $0.12°$ C per meter. If your conductivity estimates can be used, this means that the heat flow from the sea floor is the same as that from the continents. To me, at least, this is a very surprising result, but apparently you slyly suspected it all along. . . . In any case, it is obvious that measurements of this type are of great interest and should be made in many different parts of the oceans. They are not at all easy to do, however, and much credit is due Jim Snodgrass, John Isaacs and Arthur Maxwell for successfully pushing them through this summer.[13]

A Glimpse of the Expeditions

News releases about Midpac Expedition brought prompt response from feature writers and from an inquisitive public. Revelle answered many of the letters, for he believed that "if the story of such expeditions is properly told it cannot help but quicken the imagination of the people of California and increase their understanding of the need for further work of this kind."[14] He admitted to being "most anxious to convince the people of California, and particularly the Regents of the University, that one of the proper functions of this Institution should be to explore the entire Pacific Ocean." He was a true disciple of Harald Sverdrup.

So Revelle and others who had sailed on Midpac Expedition spoke to scientific and civic groups far and wide, and provided information to many a magazine writer. They patiently replied, too, to the disciples of Colonel James Churchward, who believed in a lost continent of Mu. This vast land was supposed to have occupied much of the central Pacific Ocean and to have been inhabited by 64 million people in "fine large cities at the mouths of mighty rivers." The Lemurian Fellowship, located near San Diego, was established on a belief in Mu, which the members feel was destroyed by "one grand cataclysm" 26,000 years ago. Revelle refuted the Mu enthusiasts by pointing out that fossils dredged from the Mid-Pacific Mountains were of reef animals that had become extinct 30 to 40 million years before, not of the remains of an early civilization. He gave a similar reply to believers in Atlantis, who were even placing it in the wrong ocean.

Oceanography had reached a new high, Revelle felt, as he wrote to Louis Slichter, director of the Institute of Geophysics at UCLA, which had also provided funds for Midpac:

> In general I feel that the summer's work amply justifies our belief in the scientific interest and importance of exploring the deep Pacific with modern

instruments. It is quite evident that the Pacific sea floor has been a scene of great and continuing geological activity, and it would appear that we cannot go much further in understanding the history of the earth until we know a good deal more about what has happened under the ocean. Much can be learned with presently available instruments and techniques. I know of few fields in geology or geophysics where the potentialities are so great.[15]

Midpac Expedition was only a beginning. As Revelle said to feature writer Andrew Hamilton, "We've only scratched the Pacific's bottom."

Much of the scratching had been done from the end of a worn wire. The good piece of hydrographic wire was lost on the first lowering, but Frautschy had foresightedly put aboard all the junk wire he could find on the day of sailing. He spliced the pieces together into the wire that was used for the rest of the expedition. On examining it afterward, Faughn found that the wire had been spliced nine times, had seven kinks, and needed eight more splices.

NORTHERN HOLIDAY

The following year, 1951, brought Northern Holiday, the expedition name that introduced subtlety into the custom. A "holiday" in Navy parlance is an area in which work has been left undone; to oceanographers the Gulf of Alaska was *that* kind of holiday.

Departing on 28 July, with John Isaacs as expedition leader, the *Horizon* first surveyed the Mendocino Escarpment, which was found to extend 200 miles farther west than previously known, and then the ship continued with another survey of the Mid-Pacific Mountains. The participants looked at more than geology:

> Night before last [radioed Isaacs] we passed over great shoals of salps and comb jellies that slipped under our bow and boiled out of our wake feebly protesting with flashes from their cold blue lanterns, pelagic flints, sparked by the steel of the *Horizon's* hull. Today at noon a shout brought everyone from chow; not 100 yards off our beam glowed a great green brown Japanese glass float. Swiftly it was astern and steady on its solitary course. We would have liked to have stopped and picked it up but we are not very maneuverable underway with cables streaming astern to the jog log and thermitow. These are invaluable but unbeautiful instruments that give us a wealth of oceanographic data. They stop us from chasing glass bubbles.[16]

Flanked by hungry goony birds (Black-footed Albatrosses), the *Horizon* turned northward into the Gulf of Alaska, where Harris B. Stewart celebrated his birthday by "bringing in the biggest seamount of the trip . . . the one they have been saving the name Scripps for." And Scripps Seamount it officially became — 11,400 feet from seafloor to summit. In the "interminable Aleutian fog and drizzle" of Kodiak, Warren Wooster took over as expedition leader. "The grey wet skies seemed to merge with the grey wet ocean," he complained, "and we were never certain whether our casts were from the surface downward or from some intermediate position."[17]

The innovators devised a new method of collecting seafloor samples, described in a poetic message by Wooster:

> The sea is an unpredictable mistress. Yesterday she soothed us into relaxation, pouring over us warm blue

waters, breathing gently, deluding us with her beauty. Last night as we slept she cast off the veil of pretense, raged at us bitterly, flung salt in our eyes, even now is tossing restlessly beside us. While we were on good terms with the sea we sought the westward extension of the Mendocino Escarpment, were not successful until 70 miles south of our last crossings a month ago. Here we found the same depression, crest and then drop off from 2400 to 3000 fathoms. To the north and south cores of red clay were obtained. The northern core was accompanied by an incredible manganese nodule weighing more than 100 pounds. Neatly tied up in 10 meters of hydrographic wire.

Harris B. Stewart entered the event in the coring and dredging log of 10-11 September 1951:

On Station #47 — With more faith than the circumstances warrent [sic], we lowered rig over at 2116 [hr]. . . . we let out the full 6300 m. [of wire] available, & at 2150 started her back up. At 6159 [m.] the winch ground to a halt, groaned, & shivered, & the accumulator accumulated to beat the band. . . . At 0235 [hr] with 112 m still out one of the weirdest things I ever saw occurred. A great splash & dripping occurred at the wire, we were dipnet fishing & stared unbelieving as a great rock rose on the wire above the surface. We shreiked [sic] at George to stop the — — [sic] winch & there twisting lazily — securely wrapped with 4 or 5 strands of wire was what is undoubtedly the finest geological specimen ever brought up from the abyss. We had him raise it to the top of the bucket — my heart, I am sure stopped beating until we had it securely aboard — Reidel [William R. Riedel] kept saying 'don't let it fall, don't let it fall.' Hains [Robert

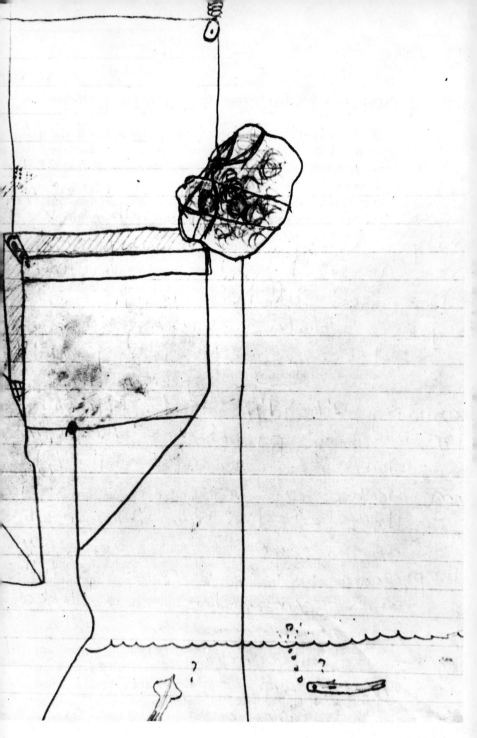

Drawing by Harris B. Stewart of the manganese nodule acquired by a "unique bottom sampling method" on Northern Holiday Expedition in 1951.

B. Haines] brought a wire comealong & took the strain of the 100 m & the coring tube off the bottom & we gently — even lovingly — undid the wire & gently lowered the thing to the floor of the bucket. Great manganese globules measuring as much as 5 inches in diameter were pyramided on the top of the specimen, a piece measuring 2'1" X 1'9" and 13 inches high, weighing probably 120-140 lbs. We are restraining ourselves (with difficulty) from breaking off a chunk to see if it is MnO_2 all the way through. It looks as though it might be, but we will preserve it as it came up till older & wiser oceanographers have looked & marveled. . . .

Wooster's radio message continued:

This unique bottom sampling method was tried on the next station, resulted only in snarls and jumbles. Yesterday's station featured a hydrographic cast to 4000 meters. The core liner contained enough water for additional samples at a depth of 5500 meters. Almost all wire was off the winch and deeper cores will surely require abnormally low tides. Surface waters are warm and very low in nutrients. A twenty minute net tow yields enough plankton to cover one's thumb. Dip netting is unproductive although squid and dolphinfish are frequently seen. If one puts the midwater trawl in the water, the wind begins to blow within the hour. . . . [18]

Among the acquisitions of Northern Holiday Expedition were some fine catches from the just-developed Isaacs-Kidd midwater trawl: from 2,200 fathoms came "small jeweled lantern fish, bright red shrimp, long-fanged viper fish, deep purple jellies, transparent squid and a host of other creatures,

A Glimpse of the Expeditions

scalded no doubt by the surface water and sunlight for the water where they live is almost freezing."[19]

The return leg of the expedition also added a new geologic feature: a straightline boundary between regions of different depths, similar to the Mendocino Escarpment but less extreme. Stewart directed the crossing of this suspect feature eight times — in spite of the crew becoming "a bit teed off" at what they felt might be a delay in their return home on September 26 — and thus outlined the Murray fracture zone.

SHELLBACK

The next expedition, in the opposite direction, was almost called Southern Holiday, but wiser heads (with an eye on funding sources) prevailed. As the cruise was to cross the equator half a dozen times — the first of those also a first for a Scripps ship — it became Shellback (and used the same abbreviation, SH). Another cooperative venture, under the heading of Operation Co-Op, this study of "one of the least known oceanic areas of the world" — west of Central America and northern South America — included the Scripps ship *Horizon*, NEL's *PCE(R)-857* again, and the *Gilbert*, the *Hugh M. Smith*, and the *Cavalieri* from the U.S. Fish and Wildlife Service. Personnel from the Inter-American Tropical Tuna Commission and the U.S. Weather Bureau also participated.

The *Horizon* sailed on 17 May 1952, with Warren Wooster as expedition leader. "The first SIO vessel to work in the southern hemisphere hesitated briefly Sunday as she crossed the line, then plunged bravely into the southern winter. The hesitation was caused by a visit from Neptunus Rex and his court, a villainous lot who quickly converted a large undisciplined mob of pollywogs into a sturdy band of shellbacks. The ceremonies made a deep and lasting impression on the pollywogs' minds and anatomies."[20] (Since that first one, Scripps equator-crossing ceremonies have been oft repeated. The traditional ceremonies end with the presentation of a certificate that confers the degree of Doctor of Shellbackosophy. The original design, enlivened with cheerful sea creatures and marine instruments, still in use years later, was drawn by Sam Hinton.)

In Lima, the *Horizon* hosted an open house for 200 Peruvian guests, and when the ship came into Guayaquil, Ecuador, some of the ship's party accepted an invitation to fly to Quito to greet President Galo Plaza Lasso, an alumnus of the University of California. Local papers gave wide and cordial coverage to Co-Op's ships in each port. Enroute home the shellbacks paused at the Galápagos Islands, where, among other explorations, they snared "20 assorted orange, red, white, green, and black and white fish."[21]

The ship reached home on 27 August, laden with fishes, hydrographic and plankton samples, records on current movements, live marine iguanas, and other souvenirs. One of the captured iguanas had leaped overboard far at sea and had been last seen trying vigorously to rejoin the ship.

CAPRICORN

Capricorn Expedition followed, in the fall of 1952. The workhorse *Horizon* and the *Spencer F. Baird* slipped away without fanfare, for the first part of the trip was devoted to Operation Ivy, the first thermonuclear tests. Several of the Scripps scientists who had participated in the atom bomb tests were also on that trip. John Isaacs recalls sitting in a small skiff with three others — the four of them alone in an endless ocean; as the first burst came and the mushroom cloud began to spread, Willard Bascom phrased it aloud:

A Glimpse of the Expeditions

"My God! They really *can* do it." Because of some poor advice from the project's meteorologists, the *Horizon* was directed straight into the fallout; she was the "hot" ship of Scripps for some years.

The participation of Scripps in this work brought a formal commendation to the institution and also one to Willard Bascom from the Chief of Naval Research, Rear Admiral C. M. Bolster, for "outstanding performance" in a "herculean project."

After the classified work had ended, the ships turned to more customary oceanography. Helen Raitt joined the expedition in Tonga, at the invitation of Revelle, after she had met the ship in the hope of spending the short stay in port with her seagoing husband. Her book, *Exploring the Deep Pacific,** cheerfully recounts the portion of Capricorn on which she sailed. For that cruise, the ships carried new deep-sea echo-sounders, and the *Spencer F. Baird* was outfitted with a newly built winch and enough wire to reach the greatest ocean depths then known.

Although the ocean mysteries appealed to the Capricorn adventurers, so obviously did the land. Harris B. Stewart wrote of a drive in Fiji:

> Up rugged mountain spurs lush with tropical vegetation on a narrow twisting gravel road, down into tangled green valleys, every turn with a view more beautiful than the last. Flame red flamboyant trees and wild bouganvillea, fern trees and cocoanut and banana palms, mango, papayas, and breadfruit. An occasional patch of dalo with its big arrowhead-shaped leaves and a native village just the way you feel it should look with its thatch-roofed huts, naked children, a stray

*W. W. Norton, 1956. Revelle's talk to civic and scientific groups about Midpac Expedition was often presented under that title, which Helen Raitt used for her book on Capricorn with Revelle's approval.

dog or two, palms, shore, and a lazy curl of smoke from a cook fire fighting its way up through the heat of noonday....[22]

Capricorn was the first Scripps expedition to carry scientific divers, who enthusiastically explored coral reefs and a submerged shallow peak, Falcon Shoal, "one of the volcanic up-again down-again islands of the world."

The captain who had carefully navigated us [wrote Willard Bascom] to the position of the island shown on the charts, was on the bridge scanning the horizon for land when the echo sounder showed the rock rising almost vertically toward the keel. He leaped to ring the engine room telegraph for full astern. Unmarked rocky pinnacles hundreds of miles from land are enough to scare anyone and the R. V. *Baird* churned to a stop, throwing a turbulent foam out over water less than twenty feet deep. The ship then retreated to an ultrasafe distance, where it sulked and the scientists who wanted to have a close look had to row a long way to see Falcon.

Using self-contained diving equipment for the first time in the Tongas, we dove down to have a look. The fresh dark basalt of the recent volcano was studded with little coral colonies just getting their start in life and already tiny angel fish swam through their branches. Walter Munk even jammed a thermometer into the soft volcanic rubble to see if there was any trace of volcanic warmth remaining.

Although no one had really doubted that corals do attach themselves to volcanoes and grow in this way, it is a sight that few men have actually seen. We were a little awed to be present at the birth of an atoll,

Roger Revelle (left) and Jeffery D. Frautschy gloating over the wire cable delivered to the *Spencer F. Baird* on the eve of departure on Capricorn Expedition, 1952.

the greatest structure ever built by any animal, including man....[23]

Interest was high at that time in the mysterious deep scattering layer, discovered by Russell Raitt, R. J. Christensen, and Carl F. Eyring in 1942, but ten years later still an enigma. Near the Tonga Trench one evening, as the layer's upward motion was being watched on the echo-sounder, eager divers begged to go overboard to gather sample denizens for identification. Permission was granted, and quite soon an exultant diver surfaced, holding aloft his collecting bag, and calling, "I've got it! I've got it!" He had carefully selected, by hand, the larger objects barely visible in the underwater murky gloom – but alas, quick scrutiny in better light on deck showed that someone had just flushed the head.

As the years went by, so did the Scripps expeditions. In 1953, on the centenary of Commodore Matthew C. Perry's visit to Japan, Transpac Expedition transited the Pacific to Japan. At an audience with Emperor (and biologist) Hirohito at the palace laboratory, members of the expedition presented him with a rare mollusk that they had dredged not far off his shores. Robert L. Fisher, expecting that the audience would be shoeless, selected his most brilliant crimson socks and, as he bowed, he was rewarded with an approving gasp from the distinguished host. At Bayonnaise Rocks – never scaled before or since – Warren Wooster, Fisher, and seaman Stanley J. O'Neil risked life and limb to collect fresh rock samples not far from where a Japanese research ship had disappeared during a violent eruption the previous year.

Transpac was followed by Cusp Expedition, along the west coast, and by Chubasco, which sailed southward, both in 1954. That year and the next took most of the institution's fleet – *Spencer F. Baird, Horizon, Paolina-T*, and

A Glimpse of the Expeditions

the wallowing *T-441* — to participate in another classified project, Operation Wigwam, an underwater test of a nuclear-fission device. As had been the case with Operation Crossroads, environmental studies were necessary beforehand. Scripps scientists helped make the selection of a site near 29° north latitude and 126° west longitude, "in a biological desert, some distance from any commercial fishing areas, where transport of contaminated water is away from fishing grounds."[24] Alfred B. Focke, of the Marine Physical Laboratory, was scientific director of Operation Wigwam, Gifford C. Ewing was deputy director, Paul L. Horrer was in charge of the physical oceanography studies, and Milner B. Schaefer was in charge of the biological program. Areal surveys were carried out in the spring of 1954, and after the test in mid-May of 1955 additional field and laboratory tests were made to monitor the effects. Various minute organisms, fish eggs and larvae, tunas and midwater fishes were gathered by net, trawl, and longline to determine the uptake of fission products and their concentration in and within the creatures.

After these trips came Norpac, Eastropic, and Equapac expeditions; then Chinook, to the Gulf of Alaska; Acapulco Trench, to Central America; Mukluk, into the Bering Sea.

Then came the year that was eighteen months long.

INTERNATIONAL GEOPHYSICAL YEAR

On 5 April 1950 a dozen scientists conversed sociably on the unsolved mysteries of the earth at the home of James A. Van Allen in Silver Spring, Maryland. Floyd McCoy of Pitcairn Island, descendant of a *Bounty* mutineer, knew naught of their conversation then, but because of them he was working to help solve the earth's mysteries several years later. Also because of them, the palm trees on remote Clipperton Island were rearranged.

Scripps Institution of Oceanography: Probing the Oceans

At the Van Allen home, Lloyd V. Berkner of the Carnegie Institution proposed an International Year for a frontal attack on the unknowns of the earth and its atmosphere, similar to the Polar Years of 1882 and 1932. The suggestion caught on, and so committees were formed, nations joined in, and on 1 July 1957 began the International Geophysical Year (IGY), which, with casual disregard for the calendar, was designated to continue until December 1958. The time was selected to coincide with the peak of the eleven-year sunspot cycle, with a concurrent eclipse, and as the silver anniversary of the second Polar Year. It was not, said Revelle, designated as "a year in which we should all be kind to geophysicists" – although, considering how much they had to do, perhaps we should have been.

The international committee that coordinated the IGY was the "Comité Special de l'Année Geophysique Internationale" (CSAGI). Most of the sixty-plus participating nations set up their own national committees; in the United States this was under the auspices of the National Academy of Sciences-National Research Council, and was chaired by Joseph Kaplan of UCLA. For the many scientific fields to be covered, separate panels were established, to consider proposals of special projects. In the United States these were managed through the National Science Foundation.

As the year opened, Kaplan defined its scope: "Scientists of the world are going to take a long and special look at our earth – at its wrinkled crust, its hot heart, its deep seas, its envelope of air, its mighty magnetism, its relationship to outer space."[25]

So, during the IGY, rockets and satellites went up, probes and dredges went down. Meteorologists, physicists, astronomers, glaciologists, oceanographers – all had observations to make. Projects were established from the equator to the poles. The "year" began auspiciously two days early,

A Glimpse of the Expeditions

with a spectacular sun flare that was promptly monitored by eager astronomers.

In and beneath the vast seas lay some of the answers to the riddles of the earth: How does water move throughout the ocean? How fast is the exchange of water from the Antarctic to the equator? Can the ocean contain radioactive wastes safely and harmlessly? How do the oceans affect weather? How great is the change in sea level throughout the world? What causes the long waves and storm surges that invade beaches from time to time? What is the shape of the sea floor? What lies beneath it? Hoping for answers to these questions, during the IGY 75 research ships plied the high seas; eight of them were American, and three of those, Scrippsian.

In addition, Scripps was in charge of an island observatory program, which established 16 stations throughout the Pacific Ocean to measure sea level fluctuations, ocean temperatures and salinities, and carbon dioxide in the atmosphere. William G. Van Dorn also set up equipment for measuring earthquake-caused tsunamis. Under coordinators Jeffery D. Frautschy and June G. Patullo, a call went out for personnel to man the island stations. For the station on Tahiti, "a veritable flood of applicants" came forth, said June Patullo, but with a smile she noted that "for some reason we are unable to understand, we have no volunteers for Jarvis." For that bit of land, two miles long and one mile wide, 1,500 miles south of Hawaii, a volunteer was finally found: Otto Horning, a man who preferred solitude, and who had the mechanical ingenuity to keep the equipment operating. He had previously lived alone on equally lonely Palmyra Island. Horning died on Jarvis Island while in the service of the IGY, and his native assistant (whom Horning had recruited from another island) notified British authorities. Audley A. ("Al") Allanson of Scripps and his wife completed the IGY observations on Jarvis.

Scripps Institution of Oceanography: Probing the Oceans

The most important man to the survey [concluded June Patullo] was the individual observer. On Arorae, a small speck of British land just south of the Equator in the western Pacific, Neemia Matiota and his Gilbertese neighbors kept a tide gage working for more than two years. Five hundred miles to the west Captain Steve Dexter, with the assistance of the British Phosphate Commissioners, ran what is probably the best complete station of the Scripps-operated stations: Ocean Island, where both sea level and temperature and salinity observations were observed for more than two years. A couple of thousand miles southeast of these two, a French citizen of Tahitian-Polish ancestry, Malinowski by name, found time between his duties as weather observer, radio operator, and postmaster on Rurutu to take a good series of both kinds of data at that hilly green island. A young man working on Guam as Observer in Charge of the U.S. Coast Survey's geomagnetic station received a box of equipment from us and then wrote us a puzzled letter asking what to do with it. We explained to our embarrassment that our supposed arrangements had somehow gone awry; nevertheless Dave Newman [the young man] agreed to take the boat observations we requested and did a first-class job.[26]

In the Tuamotu Islands one time, the observers' boat sank a mile from shore, to which two of the three men (one American and one Polynesian) swam; they returned with an outrigger canoe to rescue their older companion, who had remained clinging to an equipment box. They radioed for replacement supplies and soon resumed work.

Another figure in the island program was Martin Vitousek, who with his wife ran the supply boat from Fanning Island to the lonely outposts of Jarvis and Palmyra.

Drying plankton nets on Transpac Expedition, 1953.

Scripps Institution of Oceanography: Probing the Oceans

Park Richardson of Scripps had as much variety as anyone in the IGY island program: after spending eight months with the tide station on the equatorial Galápagos, he left for a stint at Point Barrow, Alaska. There he replaced M. Allan Beal, who during his tenure reported an "epidemic" of polar bears which had been attracted by walrus corpses left behind by Eskimo ivory hunters.

On Pitcairn Island, Floyd McCoy serviced a sea-level recorder, collected sea-water samples, and measured the vertical temperature structure of the water throughout the IGY and for a year beyond it. This fourth-generation descendant of one of the original *Bounty* mutineers was saving his earnings for a trip to the United States with his Australian-born wife. In 1960 the McCoys spent six months touring the United States, the first couple from Pitcairn to visit this country. Their final stop was at Scripps to visit with acquaintances made through the IGY.

Downwind Expedition was the first of three that Scripps sent out as part of the IGY. From October 1957 to February 1958, the *Horizon* and the *Spencer F. Baird* logged 40,000 miles throughout the southeastern Pacific on a track laid out to take advantage of following winds and currents (hence "downwind"). H. William Menard was over-all expedition leader for the first half of the cruise, and Robert L. Fisher for the second half. The emphasis was on geology, geophysics, and chemistry.

The ships worked together and separately, carrying out seismic-refraction profiles, hydrographic casts, coring, dredging, continuous echo-sounding, current measurements with the "jog log," and collecting water samples for radio-isotope analysis, air samples for carbon dioxide content, plankton samples, bathythermograph readings, and ocean-floor temperature measurements. From their disadvantageous point far above the structure they sought, the

A Glimpse of the Expeditions

geologists explored the Clarion fracture zone, the East Pacific Rise, the Peru-Chile Trench, the discordant Nasca Ridge, and various guyots and seamounts.

The East Pacific Rise was of particular interest to Menard. That a shallow area existed between Tahiti and Chile had been known since the *Challenger* crossed it in 1873. Alexander Agassiz also found shallower soundings on a reconnaissance aboard the *Albatross* in 1900, from which the name Albatross Plateau was derived. After 1946, soundings from U.S. Navy ships traveling to and from Antarctica helped define the great bulge in the southeastern Pacific Ocean. Downwind Expedition crisscrossed the area and found that the rise was indeed a major distinctive topographic feature: a curving bulge from near New Zealand to the coast of North America, 2,000 to 4,000 kilometers wide, 13,000 kilometers long, and two to three kilometers high (with a few isolated volcanoes rising higher). Along the crest of the rise the earth's crust proved to be unusually thin, and the rate of heat flow from the earth's interior was unusually high; shallow earthquakes were already known from the region. Seismic studies showed that the mantle bulged upward beneath the rise. Crossing the rise transversely were a number of straight lines of mountainous topography — fracture zones — characterized by islands and submarine volcanoes. Menard theorized that a youthful convection current in the mantle could explain the upward bulging of the mantle; this would force the crust to arch upward and would create a system of tension cracks, susceptible to wrench faulting, parallel to the rise.

Where the land met the sea, the Downwind wanderers explored the coral reefs of Fakarava Atoll; the social life of Tahiti; the coral growths of the channel into the harbor at Rapa Island; Alexander Selkirk's cave on the Juan Fernández Islands; the cuisine and viniculture of Antofagasta, Viña del Mar, and Santiago, Chile, and the pisco of Callao and Lima,

Peru; the rocks of Sala y Gomez and its fortunately friendly sharks; and the trading potentialities of Easter Island.

That lonely rock was rumored to be a great place to acquire lava carvings simply in exchange for worn clothing. But an unusual rush of visitors to Easter Island shortly before the Downwind ships reached it had sadly depleted the supply of carvings. However, cooperative islanders sat up all night carving the soft rock into distinctive faces (rather like the giant carvings for which the island is famed), which they exchanged for the old – and new – clothing of the visitors. Meanwhile, "in best tradition Scripps moonlight tiki sneaking society covered other side [of] island like locusts. Travelled on foot, by borrowed jeep and, groan, on horseback. Examined pictographs, statues, craters, made ethnological studies."[27]

In Lima the oceanographers had discovered the whim of the public's interest in the ocean: reporters who met the two ships on 15 January expressed interest at first only in the hydrographic and biological aspects of the voyage, of keen interest to the country dependent upon fishing. The geologists took a back seat. But that afternoon an earthquake of magnitude 7 struck southern Peru and killed 28 persons in and near Arequipa. From that moment, said Fisher, "only geophysical and philosophical geology could get a line" from the representatives of the press.[28]

En route home the Downwinders on the *Baird* composed and recorded their odyssey in twenty-three verses and nine choruses, entitled "Downwind Calypso." (There was very little demand for copies – perhaps because of such outrageous rhymings as seven with haven, discover with rubber, and overjoyed with employed? Original poetry composed at sea by oceanographers has a limited market.)

The second Scripps IGY expedition was Dolphin, led by John A. Knauss, from March to June 1958; its purpose

A Glimpse of the Expeditions

was to study the deep eastward-flowing current beneath the westward-flowing South Equatorial Current. This work is cited in chapter 11.

Knauss led the third IGY expedition also, from 1 August to 30 September 1958. The *Horizon,* the *Spencer F. Baird,* and the *Stranger* all sailed on that trip to explore the currents in the region of light winds near the equator — which gave the name Doldrums Expedition. The oceanographers returned puzzled at the ocean's circulation, for this expedition had discovered an unexpected large flow of water deep beneath the equatorial countercurrent.

Biologists took advantage of the available transportation to do some studies for several weeks on Clipperton Island, where they supplemented their ship stores with "fish, wild spinach, tern eggs, coconut meat, coconut apples, and coconut palm salad." Conrad Limbaugh and the other divers studied shark behavior from what they called a shark cage, but which was really a diver cage, for the divers were inside taking notes on the sharks outside. They found the Navy's shark repellent to be at least 90 percent effective, they determined that the somewhat controversial fluorescent fabric for swimwear deterred sharks, and they observed that sharks find their prey chiefly by smell.

Entomologist Charles Harbison of the San Diego Museum of Natural History was in the Clipperton group and collected "so many insects that he would not even estimate the number." The biologists together collected invertebrates, vertebrates, and plants until they estimated that Clipperton had become "the best collected island in the eastern Pacific."[29]

Clipperton Island, "one of the loneliest, most isolated and smallest islands in the Pacific Ocean," is the only coral island in the eastern Pacific. It lies 670 miles southwest of Acapulco, Mexico, and is distinguished by one 65-foot-high guano-covered pinnacle in its five square miles of coral sand

Scripps Institution of Oceanography: Probing the Oceans

and rock. Mexico, France, Great Britain, and the United States have all claimed Clipperton at various times, but only Mexico ever located a garrison there, which it installed in 1908 and forgot in 1914; the handful of emaciated survivors were rescued by the U.S.S. *Yorktown* when it chanced by in 1917. King Victor Emmanuel of Italy, who had been asked by France and Mexico to determine the ownership in 1909, handed down a long-deferred decision — in 1931! — that Clipperton belonged to France.

Scripps biologists contributed slightly to international misunderstanding in 1958. They accepted France's claim to the isolated bit of rock, and they accepted botanist and Frenchwoman Marie-Hélène Sachet of the U.S. Geological Survey (also on the staff of the National Academy of Sciences) as a member of the land party, although Scripps was not really accustomed to having women on expeditions then. Differences of opinion arose between Limbaugh and Miss Sachet, essentially over who was in charge of what: it was, after all, *her* island; he was, after all, scientific leader. But Limbaugh had the last word, for he and three others stayed two weeks beyond the others, and during that time, partly because bad weather prevented them from accomplishing much scientific work around the island, the castaways occupied their time by transplanting young sprouting coconuts to the northeast side of the atoll, neatly set out to form the letters "U C."*

NAGA

Scripps did not wait long after the International Geophysical Year for another venture into international oceanography. The theoretical preliminaries were at the Ninth

---

*I haven't been able to find anyone who has been to Clipperton since, to tell me whether the planting survived.

A Glimpse of the Expeditions

Pacific Science Congress in Bangkok in 1957, on the premise that "a thorough investigation of the seas . . . is far beyond the resources of any one nation." Only through international cooperation and funding could the potential resources of the ocean become sufficiently studied to provide for the "starving millions."

In Bangkok, Harold J. Coolidge, Executive Director of the Pacific Science Board of the National Academy of Sciences, and Roger Revelle, as a member of the UNESCO International Advisory Committee on the Marine Sciences, conferred with Thai officials on the possibility of setting up a program to investigate the marine resources in the Gulf of Thailand and of the adjacent South China Sea. Revelle was very favorably impressed by the level of technical ability and the dedication of the Thai officers in the Royal Thai Navy Hydrographic Office.

Revelle went on to Vietnam, by invitation, to consult there with government officials. He was convinced that "real possibilities existed on the broad shelf southeast of Viet Nam for the development of extensive bottom fisheries," and he urged that a survey of the South China Sea be taken. UNESCO was interested in Revelle's recommendation, but felt that it probably would not be able to enlarge its marine program to the necessary financial stage for several years.

So Revelle proposed sending a Scripps ship to Vietnam and Thailand for a three-year survey, which would provide the double advantage of gathering the oceanographic information and simultaneously training scientists of the adjacent countries in research methods and equipment. Coolidge agreed with the idea and urged that a proposal be submitted to the U.S. State Department's International Cooperation Administration (ICA, now known as AID, Administration for International Development).

Revelle not infrequently put the cart before the horse:

Scripps Institution of Oceanography: Probing the Oceans

Early in January 1958 Dr. Revelle wrote Dr. Coolidge briefing him on the results of his trip to Viet Nam and giving him in some detail an outline of what he thought a U.S. sponsored program should attempt to accomplish. He cautioned Dr. Coolidge that he had not yet had an opportunity to discuss this proposed undertaking with either the Regents or the senior administrative officers of the University of California. Furthermore, he added that there was not yet wide agreement on the part of the Scripps staff that the Institution should attempt such a difficult enterprise in such a remote area of the world. However, he pointed out that the Scripps Institution had traditionally taken the position that scientific studies of the Pacific Ocean were its responsibility and that he felt they could not and would not dodge this responsibility when the need was urgent.[30]

Meetings and counter-proposals, correspondence and temporary setbacks occupied the next several months. The George Vanderbilt Foundation of Stanford University became a participant and for a while it appeared that Vanderbilt's yacht, the *Pioneer,* would become the survey ship. The United States Operations Missions in several countries in the proposed area were consulted. Coolidge journeyed out there to discuss the project and found "general agreement in principle," even much enthusiasm in places, but enough administrative setbacks to call forth the remark: "The battle of the South China Sea is still raging but I am afraid we are losing the engagement."

In June, however, the United States and Vietnam signed an agreement to undertake the oceanographic study. Thailand soon joined in the agreement. In due time ICA and Scripps worked out an acceptable proposal, and a grant from the U.S. Public Health Service and some funds at

John A. McGowan gathering squid — and ink — on Transpac Expedition, 1953; Robert Gilkey maneuvering dipnet in background.

Scripps Institution of Oceanography: Probing the Oceans

Scripps, chiefly from the Office of Naval Research, augmented the budget.

The objectives of what was to become a two-year project in southeast Asia were:

... to foster science in Southeast Asia in an acceptable form including:
To demonstrate the importance of oceanography and marine biology in relation to fisheries.

In cooperation with the governments concerned, to train oceanographic and fisheries scientists and technicians, to develop scientific understanding and appreciation, and to accelerate the progress of science in the Gulf of Thailand and adjacent portions of the South China Sea.

To lay scientific and administrative groundwork for early and continued development of marine resources in the Gulf of Thailand and adjacent portions of the South China Sea.

In mid-1958 Captain Faughn entered the picture, as project officer. James L. Faughn is a patient man, soft-spoken, and with a quick sense of humor. These attributes certainly helped as the preliminaries continued to drag over many more months. (The State Department is not the simplest organization for scientists to work with.) A major boost to the undertaking was Anton Bruun's acceptance of the post of scientific leader. Bruun was an internationally recognized Danish oceanographer who had participated in his country's *Dana* Expedition in 1928-30, had led the *Atlantide* Expedition in 1945-46, and had led the *Galathea* Expedition of 1951-52. He and Faughn both went to Vietnam and Thailand in January 1959 to discuss the forthcoming survey. They found much interest and cooperation, and they also found a name for the project: Naga Expedition,

A Glimpse of the Expeditions

named for the sea-serpent deity in Thai mythology, whose emblem was also used by the Thailand Ministry of Agriculture.

The Scripps ship *Stranger* was finally assigned to Naga Expedition (and was replaced at home by the *Hugh M. Smith* as a loan from the Bureau of Commercial Fisheries). Into the shipyard went the *Stranger* for repairs, alterations, and refitting, under the direction of Faughn, who added the duties of captain to those of project officer. Installed on the *Stranger* for its two years of duty, miles from home, were: a frame for handling equipment over the side of the ship instead of the stern; a side-mounted lift-net constructed by Marine Facilities personnel, who even cut the eucalyptus saplings necessary for booms; the trawl winch formerly on the *E. W. Scripps;* 7,000 meters of new three-eighths-inch wire for the trawl winch; two new reels of three-sixteenths-inch steel cable for the hydrographic winch; and two-meter rings for handling Stramin nets for plankton sampling (Scripps biologists, accustomed to one-meter nets, were skeptical of the unfamiliar Danish ones, but those who used them developed "a certain respect" and "a growing attachment" for them). Also, "at the urging of the Project Officer and with the approval of the Director, the existing dark colors of the vessel were changed to more suitable tropical white which had the added effect of giving her a much improved and yacht like appearance."

Personnel began to volunteer for Naga Expedition, beginning with the entire crew of the *Stranger,* as well as crew members from other Scripps ships, and the *Stranger*'s crew was assigned en masse to the outward journey. They and members of the scientific staff who expected to participate in parts of Naga Expedition attended a series of lectures on the cultural and political background of southeast Asia, and Thai language sessions as well.

In May 1959, Faughn set June 15 as the sailing date.

Scripps Institution of Oceanography: Probing the Oceans

Some said it couldn't be done. Thanks to Faughn, it was — although he gives credit to many others.

"The Stranger," said Faughn, "came out of the shipyard two days before the sailing date. . . . She looked very nice in her fresh coat of paint but the fact that it was still wet didn't help the loading and stowage problem much."

She sailed on schedule, ocean-hopped to Honolulu, Guam, Manila, Nhatrang (Vietnam), and reached Satahib Naval Base near Bangkok on 24 August 1959. Said her captain later:

> . . . By the time the vessel arrived in Bangkok all optimism about smooth sailing under tropical moons or sunlit skies was wholly confined to the uninitiated and to those too distant to be personally affected. The most charitable thing that can be said for full-blown monsoons and for unscheduled depressions travelling across the China Seas is that they serve effectively to temper both the enthusiasm of the overly romantic and the criticism of the overly dogmatic.

The vessel was very soon put into service for a ten-day orientation cruise from Bangkok for a group of local participants. That proved so successful that the decision was made to replace all but three of the *Stranger*'s crew with local men who would serve in the dual role of crewmen and scientific trainees.

In some ways [said Faughn] this was the most critical decision to be made in the following two years. The *Stranger* represented the most expensive, the most essential and the only irreplaceable piece of equipment of the program. On her continued and safe operation depended the entire endeavor and the lives of her crew.

A Glimpse of the Expeditions

The successful operation of the vessel on cruise after cruise repeatedly justified the confidence with which the above-mentioned decision was made and the resulting benefits were immediate and lasting. Overnight the *Stranger* ceased to be a stranger and the ship and the expedition became and remained a local country project. No better model for international cooperation at the "grass-roots" level could have been devised.

The seasonal northeast and southwest monsoons are the dominating factor in that part of the ocean surveyed by Naga Expedition. Determining the seasonal variations in the ocean itself and in the life within it was among the major objectives of the program. For the Gulf of Thailand area, cruises were planned as five cross-sections at varying intervals apart, each cruise to be about 2,000 miles long and to take 14 to 17 days. For the South China Sea the cruises were to include six lines about 100 miles apart, nearly perpendicular to the eastern coast of Vietnam and running out about 250 miles; these would take 30 to 40 days. The South China Sea cruises were set up to stop once at Saigon and twice at Nhatrang in Vietnam on each trip to accommodate two separate groups of trainees, as transportation for them to Thailand was not feasible.

The ten survey cruises that the *Stranger* carried out had generally the same programs, when the weather permitted. At each station the oceanographers and trainees took Nansen bottle casts, bathythermograph measurements, routine weather observations, plankton tows, surface-net hauls, midwater trawls, bottom-trawl samples, bottom-sediment samples, dipnet samples, surface-temperature measurements, and station-position determinations. Between stations, chemical analyses, biological classifying, and general observations of birds, fishes, current discontinuities, and various other phenomena kept everyone occupied.

Scripps Institution of Oceanography: Probing the Oceans

It was a successful and thoroughly international venture. "One of the most heartening aspects of the mutual labor and close association on board the *Stranger* was the remarkable ease with which friendships developed among the young Vietnamese, Philippine, Japanese, Indonesian, Korean, Thai and Hong Kong students and the rather effortless manner in which all accepted instruction from a staff consisting at times of nationals from the United States, France and Denmark."

Also, said Faughn, "Chief Petty Officer Wong Potibutra, of the Royal Thai Navy, managed somehow to solve the intricacies of an American research vessel's most complicated apparatus — her commissary department — to the satisfaction and deep gratitude of his multinational crew and shipmates."

By no means was all the work accomplished on the *Stranger*. Laboratory facilities were provided to the Naga participants at the Hydrographic Office of the Royal Thai Navy, at the Thai Department of Fisheries, and at Chulalongkorn University in Bangkok. In addition, local ships and small craft were provided by the Department of Fisheries, the Royal Thai Navy, and the Oceanographic Institute of Nhatrang.

Theodore Chamberlain of Scripps spent two months in Bangkok during Naga, teaching a course in marine geology. Edward Brinton spent a year there from March 1960, and he directed the analysis of biological field collections from the cruises at Chulalongkorn University. Margaret Robinson worked up the bathythermograph records and provided the cards for distribution; the next year she spent six months in Thailand under UNESCO sponsorship to teach the staff of the Thai hydrographic office how to analyze the temperature and salinity data gathered on Naga.

"From the data gathered," wrote participant Eugene C. LaFond shortly after the end of the expedition, "combined

A Glimpse of the Expeditions

with a knowledge of regional wind direction and speeds, it was possible to establish the most probable current or motion patterns in the Southeast Asian region. It was determined that during the summer, when southwest monsoon winds prevail, the current flows northward up the South Vietnam coast in the South China Sea, and a large clockwise eddy forms in the upper Gulf of Thailand. In winter, when northeast winds prevail, the circulation nearly reverses, with a southward flow off South Vietnam and a counterclockwise eddy in the upper Gulf. The other seasons, transition periods between the extremes of summer and winter, have variable circulation patterns."[31]

The area near Thailand surveyed by Naga Expedition proved to support a phenomenally large number of species of fishes and invertebrates, nourished by blooms of plankton brought on by the monsoons. Many of the species collected were previously undescribed forms. Even the fish markets and the fishermen's landing areas turned up unknown species, often gathered by fishermen quite close to shore. The South China Sea, however, was found to be much less productive, apparently because of fast-moving and shifting currents.

A great deal of the preliminary work on the oceanographic collections was done at the laboratories provided to Naga participants, especially the sorting of plankton samples, chemical analyses of water samples, and preparation of bottom-topography charts. Then much of the oceanographic, geologic, and topographic material was carried to Scripps for more detailed study, although some material was left in laboratories in Thailand; most of the fish collections were transferred to the George Vanderbilt Foundation at Stanford University. Preliminary reports were put out soon after the end of the expedition, and other Naga reports are still being published as the wealth of material continues to be analyzed.

Scripps Institution of Oceanography: Probing the Oceans

The training program was considered to have been a very successful part of Naga Expedition. As a fringe benefit, Bangkok, at least partly because of Naga, became a favorite Scripps stopping place, both as a port and as a welcome stopover on trips beyond.

On 24 June 1961 — two years and ten days after departure — the *Stranger* returned to home port, under the command of Frank Miller, who had taken over as captain the previous November. The homeward personnel included sixteen Thais, who had participated in training cruises in their own waters and were continuing their learning while sailing, as "postgraduates." Captain Faughn greeted the ship at the dock, having wound up Naga's details in Bangkok and arrived home just a few days ahead of the ship.

Expedition leader Anton Frederik Bruun died six months after Naga Expedition, on 13 December 1961, while delivering a lecture in Copenhagen. The death of this gentle man, defined by Carl L. Hubbs as "one of the world's leading ichthyologists, oceanographers, general biologists, and scientific statesmen,"[32] was a great loss to oceanography.

INTERNATIONAL INDIAN OCEAN EXPEDITION

Before the return of Naga Expedition, Scripps was participating in another international venture, for it had found another horizon to travel beyond.

The late 1950s were a time of international optimism, especially in the sciences. The International Geophysical Year contributed to the optimism. A desire to solve world problems through scientific programs led to the establishing of a number of international committees, councils, and congresses.

A Glimpse of the Expeditions

SCOR, the Special Committee on Oceanic Research, established in 1957, was one of these. It was "an international group, organized by the International Council of Scientific Union, and charged with furthering international scientific activity in all branches of oceanic research."[33] SCOR helped lead to the First International Oceanographic Congress and from there to the International Indian Ocean Expedition.

Roger Revelle was president of the International Oceanographic Congress, which was held at the United Nations headquarters in New York from 31 August to 11 September 1959. The sponsors of the well-attended affair were the American Association for the Advancement of Science, UNESCO, and SCOR. About 40 Scripps staff members attended parts of the congress, which drew 800 oceanographers from 38 countries. They spoke of many things: fluctuations in sea level, the composition of sea water, ocean currents, the habits of deep-sea creatures, the origins of life, and the sliding of continents. They discussed new techniques and new equipment, especially manned submersibles and improved deep-sea cameras. Many visited the Soviet research ship *Mikhail Lomonosov,* 330 feet long and conspicuously displaying a bow emblem of the earth encircled by a satellite. She had sailed into New York harbor bearing 40 Russian scientists to participate in the congress. Other research vessels were on display as well – five from American east-coast institutions – and Jacques-Yves Cousteau's *Calypso,* complete with her diving saucer.

At the congress two dramatic forthcoming projects were announced: a four-year international study of the scarcely studied Indian Ocean, and drilling a hole through the Mohorovičić discontinuity to the earth's mantle (see chapter 12). Both projects sent Scripps scientists to sea again.

That congress was credited with providing "the necessary impetus to arouse great public interest in oceanography and subsequently various national and international organizations have shown a greater willingness to sponsor and finance further research on an expanded scale."[34]

The scientists of the sea always have a faraway spot in mind when the "willingness to sponsor and finance" arises. There sat the Indian Ocean in 1959, one-seventh of the earth's surface: a potential source of food for the one-quarter of the world's population that lived along its shores; a region of monsoons, wind reversals that create upwelling and currents that impinge on all the oceanic circulation. There it sat, a mass of water scarcely probed by oceanic tools, a void – a holiday.

Arguments were marshaled so quickly for studying the Indian Ocean that just a few years later, when Warren Wooster "tried to find the genesis of the expedition," he "gave up after he had traced it to a conversation in the bar of the Commodore Hotel during the First International Oceanographic Congress in 1959"[35] – a not-at-all-unlikely setting for oceanographic genesis.

As with the IGY, nations joined in, with "big and elaborate" ideas. "The plans evoked indifference, in some cases hostility and in others open opposition, but they emerged in an environment that, historically, was peculiarly favourable to their development."[36] Early in 1960 SCOR appointed an International Indian Ocean Working Group of 28 members; the chairman was G. E. R. Deacon, Director of the National Institute of Oceanography in England, and vice-chairman was V. G. Kort of U.S.S.R. In the United States, five working groups were set up: marine geology, geophysics, and bathymetry; biological oceanography; physical and chemical oceanography; meteorology; and data handling and analysis. Thirteen other countries established programs and provided ships, while nine additional countries participated in projects ashore. "It was agreed from the outset that the international program for the Indian Ocean

A Glimpse of the Expeditions

should have a much stronger biological slant than had prevailed in the work at sea during the IGY."[37] Arrangements were made to found a biological collection center in Cochin, India, to serve as a repository for specimens and a training center for technicians from the Indian Ocean area.

Although called the International Indian Ocean Expedition, the project was actually a collection of separate expeditions by 40 ships from 1959 to 1965. As one participant noted during its interim, "It is by no means a tightly planned, thoroughly coordinated and directed project."[38]

For its part in IIOE, Scripps was soon ready. The institution had already planned to enter the Indian Ocean in 1960, as soon as its "new" ship was ready. From the Navy in 1959 it had acquired the *ARS-27,* quickly renamed *Argo,* which went into the shipyard to be converted to research use by rearranging to provide laboratory space, and by adding a six-ton crane and a winch that could handle 45,000 feet of heavy cable.

Meanwhile, the longest trip yet undertaken by Scripps was being planned: Monsoon Expedition. Robert L. Fisher, chairman of the U.S. Working Group on Geology and Geophysics, was also the overall coordinator for the institution's part in IIOE. Maps and charts brought forth a new list of ports to be savored: Darwin, Djakarta, Port Louis. Out came *National Geographics* and travel folders: "Isn't Bali where the dancers are?" "Have you heard about the Kandy dancers in Ceylon?" "Couldn't we visit the Seychelles?" "How about the temples at Mahabalipuram?" "At where?" Wives were soothed with promises of sapphires from Ceylon, silks from India — and everything in the world from Hong Kong. Not that the wives had any choice — for the descendants of Ulysses had that restless look again.

The *Argo* sailed on 23 August 1960, with Revelle wishing her "Godspeed, a safe return, and many discoveries." Her departure, he said, was "a great event in the history of

Scripps Institution of Oceanography: Probing the Oceans

our campus: a new ship venturing into a new ocean, and that the least known on earth." Because of delays during the ship's conversion, however, the *Argo* had to sail without a shakedown cruise to test everything new.* That lack became critical ten days later:

> ... things started breaking down. ... On one horrible morning the steering cables started to unravel at 3 AM, the big winch stripped all the teeth off its gears at 5, the smaller winches developed overheated bearings at 6, the compressor for the main meat-freezer blew its top at 7, and at about 9 one of the engineers reported that we were leaking salt water into a fuel tank at a high rate. By afternoon we had shifted all the meat into a refrigerator and reset its controls to make it a freezer, had patched the steering cables with lightweight wire, brought in the equipment that was on the end of the winch cables by hand, and had changed our plans and headed for Honolulu steering with the engines.[39]

Eleven days in Honolulu took care of "about 40 different repair jobs," and the ship went on to a rendezvous with an odd piece of lonely ocean:

> We of *Argo* send greetings from a point outside of space and time. At 0845 local time this day 25 or 26

---

*Besides shipyard delays, the *Argo* had an unusual last-minute sailing complication: of all things, counterfeiters. When the ship left the yard in Tacoma after conversion, some of her crew had been hired in that city. Two of those concluded that far-off ports would be ideal for passing homemade ten- and twenty-dollar bills. In San Diego before sailing, they grew restless, so they headed for the nearest foreign soil: Tijuana. Their first counterfeit bill was instantly spotted by an alert bartender, and they found themselves in the Baja California state penitentiary for six months until they waived extradition and were returned to the United States for trial. Their suitcases of bills had been removed from the *Argo* before sailing by Treasury Department officers.

A Glimpse of the Expeditions

September *Argo* stopped on a station at latitude 0, longitude 180* — neither north, south, east, west nor any particular day. Progress continues as we punish polliwogs and gather data on Scripps' longest shakedown cruise to date.

Neptunus Rex, [Captain Laurence] Davis, G. Shor

In Cairns, Australia (where Shor reported finding the natives friendly and the beer outstanding), negotiations were completed for chartering a local ship to make two-ship seismic-refraction lines possible. The tubby launch *Malita* (Bert Cummings, Master) thus temporarily joined the Scripps fleet for two months of work in the Australia-Indonesia area. The Scripps scientists who sailed on *Malita* considered it a memorable experience.

The *Argo* worked through Indonesian waters, zigzagged across the Indian Ocean to Mauritius, turned southward and then crossed back to Fremantle, on to New Zealand, and then dipped far south below the Antarctic circle for water samples and bottom cores before returning to warm up in Tahiti and head for home. In honor of her name, the ship sailed into home port, on 18 April 1961, triumphantly bearing a golden fleece draped across the bow. Harmon Craig, expedition leader on the final leg, ceremoniously presented the dyed sheepskin to Jeffery Frautschy at the dock.

The port lists for the three Scripps cruises for the International Indian Ocean Expedition read like a travel agent's dream — although they more often proved the agents' nightmare as plans changed constantly and equipment became lost or strayed. The two IIOE trips after Monsoon — Lusiad Expedition and Dodo Expedition — established a new pattern for Scripps trips: year-long cruises, with a periodic

*Sometimes called the west pole.

Scripps Institution of Oceanography: Probing the Oceans

rotation of crew members and a constant shift of scientific party and scientific program every month or two. Many from Scripps circled the globe as they joined, sailed with, and left the expedition halfway around the world.

Lusiad Expedition (May 1962 to August 1963) gained the record as the longest Scripps sea trip: 41,670 miles by the *Horizon** and 83,000 miles by the *Argo*. The veteran *Horizon,* the first Scripps ship to cross the equator (on Shellback Expedition in 1952), became on Lusiad the first Scripps ship to sail around the world. Her route to the Indian Ocean was through the Panama Canal, across the Atlantic Ocean and the Mediterranean Sea, and through the Suez Canal. Among the exotic loot aboard on her return on 19 February 1963 was a large female Aldabra tortoise, acquired somehow in the Seychelles Islands by Captain Marvin Hopkins for the San Diego Zoo. Although estimated by enthusiasts aboard ship to weigh close to a thousand pounds, she proved to be a mere 385. Madame Rupee endured the long sea trip very well and settled in easily at her new home.

The *Argo* on Lusiad sailed for the Indian Ocean in the opposite direction and spent ten months in that ocean before returning across the Atlantic and through the Panama Canal, to become the second Scripps ship to circumnavigate the globe. In addition to geological-geophysical reconnaissances in company with the *Horizon,* much of the time of the *Argo* in the Indian Ocean was spent on a concentrated study of the equatorial current system in a project with former Scrippsian John A. Knauss, who had become dean of the University of Rhode Island's Graduate School of Oceanography. The character of the Indian Ocean currents proved different from that of the Pącific and

---

*The first part of *Horizon's* voyage was called Zephyrus Expedition and became Lusiad when *Horizon* joined *Argo* in the Indian Ocean.

A Glimpse of the Expeditions

Atlantic oceans, most particularly in the greater variability of the flow.

The two ships worked both together and separately on Lusiad Expedition, and, as Fisher reported by radio, on one occasion reached very different ports simultaneously:

> Argo arrived Kerguelen 11 November holiday. First ship eleven months. [Norris W.] Rakestraw reports two cheek welcome by 60 inhabitants francaises. Collected rocks, mail. . . . Same day Argo gaulic frolic Horizon welcomed coolly by several hundred Saint Paul rockhopper penguins.

Dodo Expedition, again using the *Argo,* followed in 1964, placing its emphasis on geology and geophysics in the Indian Ocean area. It also worked with the British ship *Discovery* on a cooperative study of the Somali Current.

Scripps's part in the International Indian Ocean Expedition officially ended in July 1965, but that was only a beginning for the reported accomplishments. From the depth soundings gathered by the many IIOE expeditions, including the three Scripps cruises, coordinator Robert L. Fisher compiled a topographic chart of six million square miles of the western Indian Ocean. Geologists and geophysicists from a score of institutions defined the structure of the Indian Ocean region and fitted it into the emerging picture of sea-floor spreading, aided by drilling carried out there for the Deep Sea Drilling Project in 1972. Reports using data gathered by IIOE will continue to appear for many years.

A special program set up by the Food and Agriculture Organization of the United Nations was a review of oceanographic and meteorological information pertinent to fisheries development in the Gulf of Aden and adjacent parts of the Arabian Sea and the western Indian Ocean. This

Scripps Institution of Oceanography: Probing the Oceans

project, based on data gathered by IIOE ships, was contracted to the Institute of Marine Resources in 1966 and carried out by Warren S. Wooster, M. B. Schaefer, Margaret K. Robinson, and their assistants. The results were presented in 1967 as the "Atlas of the Arabian Sea for Fisheries Oceanography," in which it was concluded:

> Perhaps the most important finding of the Indian Ocean Expedition, so far as fishery oceanography is concerned, is the extremely high rate of primary productivity, and large standing crops of phytoplankton and zooplankton in the Arabian Sea, especially along the western side . . . one of the more productive parts of the World Ocean.[40]

The expeditions presented in these pages are only a sampling — early ones, elaborately organized ones, and a smattering of others — of the many carried out by the Scripps Institution. A few others have been cited in earlier pages within their disciplines.

Putting together one of the major expeditions requires patience, persistence, and long lists. H. William Menard presented effectively in *Anatomy of an Expedition*[41] how a typically complex trip such as Nova Expedition can originate and get under way. Diana Midlam and R. Nelson Fuller enumerated what was put aboard the *Argo* before the eight-month Zetes Expedition set out in 1966:

> Ten thousand pounds of fresh meat, 92,000 gallons of fuel oil, 100 gallons of chemicals, 1,000 pounds of granulated sugar, 55 mesh net tows for trapping ocean organisms, 47,000 gallons of fresh water, two electronic "fish" for sensing and recording the ocean's salinity, temperature, and depth; and. . . .
> Five buoys, 40 quarts of buttermilk, 400 pounds

When the work is all done: Fred Dixon, Chief Engineer Frank G. Fish, and John Sclater relaxing on the fantail of the *Thomas Washington*. Photo by Tom Walsh.

Scripps Institution of Oceanography: Probing the Oceans

of sand, eight cases of soap powder, 60 gallons of blue-gray paint, and 27 sets of Arctic foul-weather gear. . . .

. . . a case of cellulose sponges, 12 cans of brass polish, nine cases of paper towels, 20 gallons of pine oil disinfectant, 50 flashlight batteries, a gross of No. 3 lead pencils, 36 pads of yellow ruled paper, and 12 boxes of paper clips, not to mention tide and current tables and all the forms for handling navigational computations.

. . . Also stored are two scoop shovels, a bag of cement and the 400 pounds of sand, eight cases of ammonia, 16 cases of Purex, 24 brooms, two cases of Ajax, and three cases of Handi-cream. . . .

. . . Also on hand are 840 sheets, 840 bath towels, 420 pillow slips, 200 cooks' aprons, 500 dish towels, and 60 bedspreads.[42]

There were innumerable boxes of dry stores as well, and many other items — too much of some, not enough of others. Equipped with such paraphernalia, the Scripps ships go to sea year after year, racking up miles, sometimes gliding over glassy aquamarine water, sometimes wallowing in storm-spawned swells — and steadily wresting samples and specimens and answers from the enigmatic sea.

NOTES

1. *In* Helen Raitt, *Exploring the Deep Pacific* (New York: W. W. Norton, 1956), x.

2. *In* Robert C. Cowen, *Frontiers of the Sea* (Garden City, New York: Doubleday, 1960), 12. Copyright © 1960 by Robert C. Cowen. Reprinted by permission of Doubleday & Company, Inc.

3. Roger Revelle *in* Helen Raitt, *loc. cit.,* xi.

4. "The Research Ship *Horizon,*" SIO Reference 74-3 (1974), 4.

5. Letter of 18 January 1953.

6. *Hole in the Bottom of the Sea* (Garden City, New York: Doubleday, 1961), 138. Copyright © 1961 by Willard Bascom. Reprinted by permission of Doubleday & Company, Inc.

7. Roger Revelle, "Foreword," U.S. Geological Survey Professional Paper 260 (1954), iii.

8. "The Age of Innocence and War in Oceanography," *Oceans Magazine*, Vol. I, No. 3 (March 1969), 13-14.

9. Joint Task Force Memo, 11 May 1946, 15.

10. Proposal, 10 October 1949.

11. Manuscript, "We've Only Scratched the Pacific's Bottom."

12. Letter of 14 October 1950.

13. Letter of 15 October 1950.

14. Letter of 4 December 1950, to writer Milton Silverman.

15. Letter of 16 October 1950.

16. Radio message, 9 August 1951.

17. Radio message, 6 September 1951.
18. Radio message, 14 September 1951.
19. Radio message, 9 August 1951.
20. Radio message, 31 May 1952.
21. Radio message, 13 August 1952.
22. Letter to Francis P. Shepard, 18 January 1953.
23. *Hole in the Bottom of the Sea*, 39-40.
24. Milner B. Schaefer, "Introduction and summary," *Limnology and Oceanography*, Supplement to Vol. 7 (1962), iii.
25. *Life*, 15 July 1957, 19.
26. "The IGY and Mean Sea Level," *Naval Research Reviews* (May 1960), 22-23.
27. Radio message, 7 February 1958.
28. Radio message, 31 January 1958.
29. Conrad Limbaugh, "Introduction," IGY Clipperton Island Expedition, SIO Reference 59-13 (1959), 3.
30. Report on Naga Expedition by James L. Faughn; this section has been summarized and quoted from that report.
31. "Oceanography and Food," *Naval Research Reviews* (November 1961), 11-12.
32. *Copeia*, no. 2 (1962), 481.
33. Proceedings of SCOR, Vol. I, No. 1 (1965), iii.

34. Georg Wüst, "Proposed International Indian Ocean Oceanographic Expedition, 1962-1963," *Deep-Sea Research,* Vol. 6 (1960), 245.

35. Daniel Behrman, *The New World of the Oceans* (Boston, Toronto: Little, Brown & Co., 1969), 394.

36. J. V. Leyendekkers, "The International Indian Ocean Expedition," *Australian Journal of Science,* Vol. 27, No. 6 (December 1964), 153.

37. International Indian Ocean Expedition-United States Participation in the International Indian Ocean Expedition, 1961, 1.

38. Leyendekkers, *loc. cit.,* 153.

39. George Shor, letter of 26 November 1960.

40. Institute of Marine Resources, Annual Report for the Year Ending 30 June 1967, IMR Reference 67-17, 22.

41. New York, McGraw-Hill, 1969.

42. News release, January 1966.

The campus in 1938.

16. Back on the Beach

THE CAMPUS

In the first chapter is a glimpse of the Scripps Institution in 1936 – the place that would launch a thousand trips. There were then about thirty people at the laboratory by the sea, in three main buildings. Today there are slightly more than a thousand people at Scripps Institution, located – when in port – in a myriad of buildings on the campus and in an assortment of off-campus locations from Sorrento Valley to Point Loma.

The physical changes on the campus were most dramatic during the 1950s, when oceanography was expanding. Prior to that the only extensive building project was the repair of the pier in 1946, when buildings and grounds superintendent Carl Johnson supervised the jacketing of the pier pilings with steel and concrete and the redecking of the structure. In 1950 the Aquarium-Museum was built, as well as the north garage and the west garage in the service yard. The purchasing and storehouse building was added in 1953. The first addition to Ritter Hall followed in 1956, and it turned the face of the campus toward the sea. The

Scripps Institution of Oceanography: Probing the Oceans

experimental aquarium building was built in 1958, and in 1959 the "cafeteria" – called the general services building, now New Scripps Building.

By then, the School of Science and Engineering had been established, and the new general campus was being planned. The "cafeteria" was preempted for the office of the first chancellor, Herbert F. York, plus the Director's office, other administrative offices, and some laboratories. (The lunch stand was added on the northeast corner of New Scripps Building in 1961.)

The peak year for building was 1960, when the second addition to Ritter Hall, and Sverdrup Hall and Sumner Auditorium were all completed. The occasion was acknowledged in a historical ceremony on 18 May 1961, at which the name plaques of each building were appropriately unveiled: Sumner Auditorium by Mrs. Francis B. Sumner, widow of the building's honoree; Sverdrup Hall by Mrs. Harald U. Sverdrup, widow of the building's honoree; New Scripps Building by Mrs. J. G. Johanson, niece of George H. Scripps; and the new wing of Ritter Hall by Mrs. W. W. Hawkins, widow of Robert P. Scripps.

The flat area of the campus was suddenly brimful of buildings. The internal roads were rearranged (which eliminated a particularly attractive planting of succulents on the turn to the library), and most of the old cottages on the south end of the campus were removed. Revelle apologized in May 1960 that "there has been dust, mud, dirt, noise and a general mess caused by construction for a long time," even as he noted that the institution had to expand "in the national interest," and predicted that "all the people of La Jolla very shortly will be proud of Scripps Institution."[1]

Perhaps to avoid the building confusion, more Scripps people than ever before went out to sea in 1960; nine major expeditions that year logged more than 90,000 miles, and

Back on the Beach

for the first time Scripps ships entered the Indian Ocean and the Caribbean Sea.

In 1962 a small building was constructed near the landward end of the pier, and later a 60-foot steel tower was added; this facility was used until 1971 by researchers from Berkeley and UCLA on sea-water conversion methods. In the basement of that building were stored the Scripps seafloor rock samples. When the sea-water conversion project ended, Scripps researchers moved into the building, and in 1973 the Shore Processes Laboratory was built on its roof.

The only other addition on the flat area of the campus during the 1960s was the Physiological Research Laboratory west of the Aquarium-Museum in 1965. That group also built a facility to house dogs, horses, and sheep for research projects, just below radio station WWD in Seaweed Canyon* (so named because city trucks dumped seaweed there from La Jolla Shores and Scripps beaches). The building of the "farm" in 1965 put an end to the seaweed dumping, and to other extramural activities in Seaweed Canyon, such as the dumping of garden trimmings by nearby residents, the pistol-practice range of the campus police, and an archery range.

The obvious direction for further expansion was up the slope. Walter Munk chose a scenic site there for the Scripps laboratory of the Institute of Geophysics and Planetary Physics, built in 1963. The Hydraulics Laboratory, slightly uphill from IGPP, followed in 1964. It served first as an echoing setting for a lively farewell party for the Revelles; then the distinctive building with the wave-shaped roof was outfitted with a wave-and-tidal basin, a wind-wave channel, a wave-and-current channel, a granular fluid mechanics test facility, and a fluidizing channel.

During the 1950s and 1960s, various of the campus cottages were removed, and others were converted to offices

*Also called Snake Canyon, or Rattlesnake Canyon, for obvious reasons.

441

and laboratories as the tenants moved off campus.* In 1951 a group of Scripps staff members purchased the 40 acres of land comprising Scripps Estates Associates on the canyon rim above the institution and developed it into 42 homesites and a privately protected coastal-canyon preserve.

The graduate students set up their social center in the late 1960s, when they were given the use of T-8, a one-story house at the south edge of the campus that had been purchased when the land for the south parking lot was acquired. The center at first was under the auspices of the Dean of Student Activities of UCSD, until 1968 when a Scripps Student Committee was formed. Students renovated the building, which they call Surfside, into a recreation center, with a ping-pong room, pool table, change room, and storage rack for surfboards. Volunteer labor, some funds from the office of Student Affairs, and proceeds from vending machines on campus made the renovations and recreation facilities possible. TGIF — the weekly beerbust — began at Surfside in January 1968. As the student committee reported to staff luncheon in September 1969: "The purpose of this party was to provide a friendly atmosphere where the entire SIO community could meet and get better acquainted. This party has been very successful."

The uphill trend in construction has continued into the 1970s: the Deep Sea Drilling Building was completed in 1970, across La Jolla Shores Drive from the main campus; the Norpax building was completed in 1975, below the Fishery Oceanography Center (which was built by the federal government in 1964 and was renamed Southwest Fishery Center in 1970); the Carl Eckart Building to house the Scripps Library was completed in December 1976; and the Marine Biology Building was nearing completion at the close of 1976.

*The last tenants, George Bien and family, moved out of T-24 in 1969.

Back on the Beach

The library had long outgrown its space, even after extensive remodeling in the mid-1960s provided to it the entire building, which had also contained, at various times, a museum, an auditorium, some non-library offices, the mail room, and the telephone switchboard (a lively social center in its day). From 15,000 volumes in 1936, the library holdings had increased by the end of 1976 to include: 113,608 bound volumes, more than 26,000 maps, 3,843 microforms, 20,611 reprints, 27,312 reports, documents, and translations, 5,753 serial titles, and 120 linear feet of historical archives. Of necessity, during the 1970s some volumes had to be stored in other locations, some in the basement of IGPP and others in buildings at Camp Elliott.

The first full-time professionally trained librarian for Scripps, Roy W. Holleman, began in September 1950, soon after the retirement of longtime librarian Ruth Ragan. Besides extending the oceanographic collections, Holleman in the latter 1950s began assembling an all-subject general library for UCSD. Joseph Gantner succeeded Holleman from 1963 to 1966, when he transferred to the upper campus. William J. Goff, then assistant librarian at Scripps, and holding master's degrees in both geology and library science, succeeded Gantner in 1967.

For many years the Scripps library has been distinguished by its broad coverage of ocean-related literature — and equally distinguished by the cheerful helpfulness of its staff.

One long-discussed construction project that has not come to pass is the "Scripps Island." The concept of creating a unique replacement for the Scripps pier began in the early 1950s. The "Island" grew in conversation to incorporate a harbor, various underwater laboratories, aquarium facilities for research and holding purposes, data cables for relaying a number of continuous measurements, facilities for divers and their supplies, a sea-water intake, and more. Some hoped that the facility could provide for mooring the sea-going fleet. One of the earliest plans committed to paper was of a moderately small, crescent-shaped, rock island designed by Robert S. Arthur, Douglas L. Inman, and Admiral Charles D. Wheelock.

In 1964, shortly before leaving Scripps, Roger Revelle appointed a committee to consider the "Island," which became formalized as first, the Offshore Research Facility, and later, the Experimental Inshore Oceanographic Facility. Early in 1967, through the Foundation for Ocean Research and the city of San Diego, funds were provided for preliminary design studies of a research platform. As William A. Nierenberg pointed out:

> Whatever measurements one wants to make, whatever operations one would like to conduct, however one wishes to employ a man in the sea, the greatest fraction of the effort and the greatest source of danger is at the air-sea interface. . . . We visualize an Island connected by causeway to shore sufficiently far out past the surf zone and that much closer to the canyon area, so designed with particular installations and instrumentation, that the problem of inserting a man or his equipment into the sea and retrieving them become relatively trivial operations, thus reserving the maximum of the effort for engineering or scientific work.[2]

In November 1967, Robert H. Oversmith became the project engineer, and, with the engineering firm of Sverdrup and Parcel and Associates,* he prepared a preliminary design of a horseshoe-shaped laboratory, 300 feet long and 200 feet wide, to be located adjacent to Scripps submarine

*Headed by Leif J. Sverdrup, brother of Harald U. Sverdrup.

canyon, and connected to shore by a 2,400-foot curved bridge. The regents of the university in 1969 approved the proposal for building the facility, and in 1972, the city of San Diego leased to Scripps a square mile of sea floor for the island. The construction was estimated at $18,600,000.

Some Scripps researchers were beginning to question the cost involved, especially in relation to the research benefits. A poll in 1972 showed that "only a very small fraction [of the Scripps staff] would make intensive use of the facility, two-thirds would use it only occasionally, if at all." The "Island" was shelved. Douglas L. Inman, one of the early proponents, commented that *"not* building the island made us learn how to develop the technology of working in the open ocean, at which we have been very successful."

THE ADMINISTRATION

Throughout the years of greatest expansion of the Scripps Institution, the dominant figure on campus was Roger Revelle, a man who has been described as physically and temperamentally designed for the study of the deep oceans. Physically he *is* big – six feet, four inches tall, and with oversized hands and feet. Temperamentally he is broad: he has a wide grasp of knowledge and assimilates material quickly. There is a quiet self-confidence about this big grave man, yet also humility, and he listens with interest and sympathy.

Revelle's history at Scripps, as noted throughout these pages, was a long one: from 1931 to 1964, from graduate student to director and dean.

He was born on 7 March 1909, in Seattle, where his father practiced law. When he was seven, the family moved to Pasadena, California. Revelle attended Pomona College in Claremont, where he intended to study journalism, but

Scripps Institution of Oceanography: Probing the Oceans

he was turned toward geology by Professor Alfred Woodford and went on to graduate work at Berkeley. Seafloor cores that had been saved from the research ship *Carnegie,* which exploded in Samoa in November 1929, had been sent to Scripps Institution, which queried Berkeley for a geologist to help analyze them. Revelle was sent from Berkeley in 1931. La Jolla was, fortuitously, the birthplace of his bride, Ellen Clark — a great-niece of Ellen Browning Scripps — whom he had met while she was attending Scripps College, not far from his own alma mater. They were married in 1931.

After earning his Ph.D. at Scripps in 1936, with a dissertation on the *Carnegie* cores, Revelle spent a year in Europe, chiefly at the Geophysical Institute in Norway, and returned to Scripps as an instructor. In 1941 he became an assistant professor and joined the U.S. Naval Reserve, through which he was assigned to duty at UCDWR (see chapter 2). As he told it long afterward:

> . . . a year and a half later [Commander Rawson] Bennett [head of the sonar design section of the Bureau of Ships] arranged for my transfer to Washington, ostensibly to the Hydrographic Office. Unfortunately, the Hydrographer, a charming and gentle admiral, was somewhat out of touch. He could not think of anything for an oceanographer to do, except to examine some old echo soundings that showed shoals and banks off the west coast of Central America where later ship crossings had found only deep water. I appealed to Bennett to give me a job that might be more relevant to the real world, and he set me up as officer in charge of what amounted to an oceanographic subsection in the Bureau of Ships. . . . From this vantage point I was able to help guide and foster the Navy's growing involvement in oceanography

throughout the latter part of the war and the early postwar period.³

Immediately after the war, in the U.S. Naval Reserve, by then with the rank of Commander, Revelle served as head of the geophysics branch of the Office of Naval Research, and was in charge of the oceanographic investigations of Operation Crossroads in 1946 (see chapter 15). Much later he recounted his return to Scripps in 1948:

> Harald Sverdrup, the little great man who had already begun to transform the institution, felt he must return to Norway, and that is when a struggle began as to who should be his successor. Harald Sverdrup and Carl Eckart were determined that I should have the job, even though I was still in Washington, working for the Office of Naval Research. But there was a great deal of equally determined opposition.
> It was resolved by Carl [Eckart] agreeing to become the Director; he asked me to be his Associate Director. Although I didn't realize it at the time (he never let on), this was actually a ploy on Carl's part, with Harald's encouragement, to create a situation in which the opposition to my appointment might be diminished. As soon as Carl judged that this had happened, he resigned, and I took his place, first as Acting Director and then as Director.
> After I assumed the job, I rapidly gained a reputation as a poor administrator. But in some ways, compared to Carl, I was an administrative genius. The difficulty was that he took the job too seriously. No detail was too small, no problem too unimportant, for him to attack it with meticulous and elaborate care, giving attention to every detail, and examining every

alternative. The rigor in definition and precision of thought, and the inability to leave any loose strings untied, which were his great strengths as a scientist, were just what was *not* needed as an administrator. I remember he spent a good deal of time trying to tidy the Scripps Institution up; it was quite a messy place in those days and this was a completely frustrating job for him.[4]

Revelle, as the director, had his defenders and his detractors. He also had big ideas, which he was eager to carry through, sometimes too quickly. "I think I'm undiplomatic," he once said. "I just bull things through."[5] Eckart, however, felt that Revelle had "the ability to put knowledge and enthusiasm together in words that excite people."[6] Not always a good speaker, he was nevertheless at times almost an orator on the possibilities in oceanography during his Scripps days. His enthusiasm came through in a mellow resonant tone in measured, sometimes hesitant, phrases.

Big projects appealed to Revelle, and he was intent upon keeping Scripps in the forefront while other institutions were moving into oceanography. The most intense booster for the institution, Revelle was also the keenest recruiter — of people whose imaginative approach to their discipline appealed to him and who could contribute to oceanography. Some of these arrived, fired with his enthusiasm, to find that they did not even have an office, certainly not a laboratory. Assistant Director Jeffery Frautschy or other campus officials would scurry to find them space and funds.

Revelle had — and has — a great respect for the capabilities of science in helping to solve world problems. While at Scripps, he was drawn into a great many international committees: various UNESCO appointments; Atoms for Peace; the International Geophysical Year; the International

Back on the Beach

Indian Ocean Expedition; the Special Committee on Ocean Research, which, with the American Association for the Advancement of Science, organized the First International Oceanographic Congress in 1959, of which Revelle was president; the U.S. National Committee for the International Biological Program; the United States-Japan Committee on Scientific Cooperation; the International Association of Physical Sciences of the Ocean; and others. There were national committee appointments as well. No wonder that some Scripps staff members grumbled that Revelle was too often elsewhere and was not tending to affairs at home base.

Through his committee obligations, Revelle knew a great many oceanographers throughout the world. He had a way of persuading them to take on projects that required a great deal of time and energy. An example is the summary treatise, *The Sea: Ideas and Observations on Progress in the Study of the Seas.*[7] The classic text of 1942, *The Oceans,* by Sverdrup, Johnson, and Fleming, was indeed somewhat out of date twenty years later. Maurice N. Hill of the Department of Geodesy and Geophysics in Cambridge, England, said: "Revelle suggested that we produce another such volume containing ideas and observations concerning the work accomplished during the twenty years since this masterpiece. It was suggested that this new work should not attempt to be a textbook but a balanced account of how oceanography, and the thoughts of oceanographers, were moving."[8] Hill served as general editor of the new work until his death in 1966, when Arthur E. Maxwell of Woods Hole Oceanographic Institution took on the task. The first volume of the treatise came out in 1962, "from the pens of many authors," noted Hill. By 1974 the project had reached five volumes (in six books) in an invaluable summary of all the vast field of oceanography except marine biology. Naturally, many of the contributors were Scrippsians.

Scripps Institution of Oceanography: Probing the Oceans

Revelle has always had an impressive impact:

> ... At a sleepy meeting in Paris, a big slouching man rose to his feet and began to speak of geology at sea [wrote Daniel Behrman]. His voice and presence filled the committee room.
>
> He told of oily uncomfortable ships, of the great grinding mills that destroy the sea floor in the deep trenches, of the maps of this realm that were no better than the land maps of the seventeenth century. More than any other single figure, Revelle is responsible for the introduction of oceanography into public affairs. He began as a marine geologist and geophysicist; he has evolved into a statesman of science.[9]

At home base during the 1950s Revelle began envisioning a new kind of university, one that started at the top. He carried on a long campaign, and he gained some political foes along the way; but the result was the establishment, first, of the Institute of Technology and Engineering, then the School of Science and Engineering, housed on the Scripps campus from 1958 to 1963, and from that the entire campus of the University of California at San Diego.* Revelle has credited Carl Eckart as being his co-worker and goad throughout the early planning, and he has pointed out that many others also devoted a great deal of time to founding the new campus, but the drive was certainly Revelle's. The innovative features were the multi-college concept within the one big campus, and opening the college at the graduate-student level first, taking off essentially from Scripps Institution into other fields of science and later,

---

*The campus was first known as the University of California at La Jolla, and one Scripps dissertation came out under that name, in 1961. But pressure from the rest of the city led to changing the name. La Jolla is, after all, within the city of San Diego.

humanities. To Revelle can go a great deal of credit for drawing to UCSD some of the top names in their fields. As Walter Munk said: "It was an interesting experiment to watch a bunch of sailors start a university."[10] In addition to a great many hours of planning, those sailors contributed the dormitory names for the first college — for noted exploration ships*; they donated the trident symbol to UCSD; and they awarded the name, Revelle College, for the sailor who started it all.

Ten years after Revelle's idea had become a full campus of the university, Chancellor William J. McGill looked back at its founding:

> This is the end of an era at the University of California, San Diego. The era began with Sputnik and with the national panic which that little beeping Soviet satellite created. Sputnik's effects on American education were on the whole remarkably positive, and they were in full ascendancy when UCSD was born a decade ago. Only in such an era could the extraordinary beginnings of UCSD have been conceived, much less attempted, and only in such an era could the attempt have been brought off successfully. It was an era of unparalleled national investment in education, especially in science. The Russian Sputnik fathered many new American university campuses and caused the sudden dramatic expansion of many old ones. It was a time of bold educational planning, of sudden affluence for professors, and a pervading sense of limitless vistas of academic excellence, both here and throughout the country — but especially here.[11]

Revelle's last few years at Scripps Institution were interrupted times at his home base. From October 1961 to

---

*Helen Raitt, a participant on Capricorn Expedition, compiled the list of dormitory names.

Scripps Institution of Oceanography: Probing the Oceans

February 1963 he served as science advisor to Secretary of the Interior Stewart Udall. In that post he headed a panel of experts in both natural and social sciences in a valuable study and analysis of land and water development in the Indus River basin of West Pakistan. Revelle then returned to Scripps and simultaneously became University Dean of Research for all campuses. In 1964 he accepted the appointment as director of the Center for Population Studies at Harvard, to which he was enticed by some of the academics with whom he had worked on the Pakistan project.

Revelle is remembered fondly by those who have been long at Scripps, one of whom said just after his departure: "It is a great tribute to Roger, I think, that he has retained the cordial regard and friendship even of those who have felt that his shortcomings as an administrator were serious." Part of his popularity was no doubt due to his enthusiasm for oceanography just as that field was undergoing its greatest expansion. Revelle knew and liked oily ships, from the 64-foot *Scripps* and 94-foot *E. W. Scripps* to the 213-foot *Argo,* and he understood oceanographers, himself included. "The chief motivation of most oceanographers I know," he once said, "is the sheer excitement of finding out what has never been known before."[12] He went on to speak of what oceanography could accomplish for the world, but, as always when Revelle spoke, his theme was "Oceanography is fun." That phrase, which he used oft-times, has been joked about — almost every time the seas start sloshing over the fantail or the winch jams — but the joking is done by men who knew Revelle at Scripps and who go back to sea themselves again and again. As one of those said later, "When Roger was around, things were always exciting."

Revelle's fiftieth birthday is remembered fondly too. The idea of a surprise party for the occasion of 7 March 1959 was mainly John A. Knauss's, who provided his house for the event and established the theme of *Cannery Row.* (A

Back on the Beach

few days after the invitations were sent out, every bookstore in town had run out of Steinbeck's book.) With appropriate quotations from the book, Lynne and John Knauss urged the guests to bring presents — "homemade, or at least something you yourself found or caught"; to bring liquid refreshment, which, "in best Monterey tradition," would be dumped into one large punch bowl; and to keep the secret from Roger.

"The conspiracy grew and there were visits back and forth" — as people thought of ingenious gifts and ideas. The Shors' house across the street from Knausses' was decorated with "models" to represent the Bear Flag Restaurant in *Cannery Row*. All of Revelle's family, except himself, knew of the event, so it was arranged that Roger and Ellen would have dinner at their daughter and son-in-law's (George and Anne Shumway's) house, about a block away from Knausses', while the party assembled. At 8:30 the crowd strolled to the Shumways', where Scripps police officer Howard ("Mac") McKelvey led off with his siren, and Roger was summoned forth — to his complete surprise. Serenaded by the wheeze of a calliope (played by chemist-musician Charles D. Keeling), he was seated upon a litter and borne up the street on the shoulders of an honor guard, amid confetti and banners, to be deposited in the midst of the festivities.

It was a great party. The gifts represented Scripps ingenuity at its best. The punch was remarkable. And it seemed as if everyone at Scripps was there. The finale was the arrival of a giant box, which opened to reveal "Texas Bobbie" Roberts (a striptease artist, about six feet tall). Revelle vanished into the box and the two were carried off triumphantly. Texas Bobbie's breathless comment later was: "I never knew there was so much to learn about oceanography!"

One who deserves a great deal of credit for having kept Scripps running smoothly during Revelle's absences is

Nan Limbaugh (left) and Thea Schultze, at the farewell party for the Revelles in the Hydraulics Laboratory, 1964.

Back on the Beach

Jeffery D. Frautschy, who became assistant director on 1 August 1958. As mentioned in chapter 2, Frautschy began his oceanographic career with UCDWR, in 1942. After World War II he was a student at Scripps, then spent two years with the U.S. Geological Survey and in graduate work at the University of Southern California before joining the staff of the institution in 1949. He headed the research support shop, directed the Scripps portion of the International Geophysical Year, and as assistant director served as troubleshooter everywhere on campus. During the construction boom of the 1950s and 1960s, Frautschy always knew where the utility connections ran through the campus, and he has long kept track of the history of the institution's structures and ships and people. By training a geologist, in practice Frautschy is an all-round engineer who has contributed a great many ingenious improvements to shipboard equipment and techniques. One of his most widely used early contributions was a coreless three-strand wire rope that endured much longer than the seven-strand wire with a central core ("aircraft cord") that had been previously used. Since 1972 he has served very actively on the California Coastal Conservation Commission, and in 1975 he became Associate Director of the Institute of Marine Resources, with responsibility for the University of California Sea Grant program, headquartered at Scripps.

Fred N. Spiess, director of the Marine Physical Laboratory, stepped in as acting director of the institution in 1961. This native Californian, born in Oakland in 1919, had graduated from Berkeley in physics in 1941 and had immediately entered Navy submarine school. During World War II he made thirteen submarine patrols. Spiess received his M.S. at Harvard in communication engineering and his Ph.D. at Berkeley in physics. In 1952 he joined the staff of MPL and in 1958 he became director of that laboratory.

Scripps Institution of Oceanography: Probing the Oceans

As acting director at Scripps from 1961 to 1963, while Revelle was science advisor to Udall, Spiess worked closely with UCSD's first chancellor, Herbert F. York, to retain the autonomy of the institution, in its long-established role of both teaching and research, as the new undergraduate campus was being established. A long-time Scrippsian said of Spiess's term as acting director: "I have never known a period when affairs of the Institution were run so smoothly. Matters were settled promptly and with the use of remarkably fine judgement."

When Revelle left in 1964, Spiess became director of Scripps, partly as an interim measure while a second chancellor for UCSD was being sought. Spiess helped complete the negotiations for establishing the Physiological Research Laboratory (see chapter 8), and he expanded the programs in biology and chemistry at the institution. His valuable contributions to the development of *Flip* and the Deep-Tow instrument package are cited in chapter 4.

In July 1965, William A. Nierenberg became director of Scripps Institution. Physicist Nierenberg was born in New York city in 1919. He received his B.S. from City College of New York (with one year at the University of Paris), and his Ph.D. from Columbia University. He was a participant in the Manhattan Project during World War II. After the war he taught at the University of Michigan, and in 1950 became professor of physics at Berkeley. In 1953-54, on leave from Berkeley, he was Project Director of Columbia University's Hudson Laboratories, then returned to Berkeley until his appointment at Scripps. From 1960 to 1962 he served in Paris as the Assistant Secretary General for Scientific Affairs of NATO, and he was simultaneously Professor Associé at the University of Paris.

At Berkeley Nierenberg established the Atomic Beam Laboratory on the main campus and the Atomic Beam

Back on the Beach

Research Group at Lawrence Radiation Laboratory. His researches there included atomic-beam measurements of electronic and nuclear properties of radioactive atoms, gaseous-diffusion theory and experiments, cascade theory, atomic and molecular beams, the measurement of nuclear spins, magnetic and electric quadrupole moments, hyperfine anomalies with particular application to radioactive nuclei, and similar applications to atomic electronic ground states.

This hoarse-voiced, staccato-talking man is disconcertingly able to talk and listen simultaneously. His interest in ocean research and his pride in the Scripps Institution of Oceanography are unending. He is a strong advocate of using the right — and best — equipment for the job at hand, which he often ensures by negotiating for the necessary funding. The task has not been easy, as the money available for increasingly expensive marine research has become much more difficult to obtain during the 1960s and 1970s.

One of the earliest programs that Nierenberg advocated when he became director was computers for the larger Scripps ships. The Deep-Tow group of the Marine Physical Laboratory actually carried the first Scripps computer system — a PDP-8 computer — to sea on the *Thomas Washington* in the fall of 1966, for use in calculating navigation for the acoustic transponder and to provide a digital-data logging capability for the magnetometer and the precision echo-sounder of the Deep-Tow system. In the following year an IBM 1800 Data Acquisition and Control System — the "Red Baron" — was installed on the *Thomas Washington*, through joint sponsorship of Scripps and International Business Machines. At first IBM provided personnel to service the system, but Scripps soon decided to hire its own technicians, engineers, and programmers, to service the shipboard equipment and programs and also to participate in other underway projects. The Shipboard Computer

Scripps Institution of Oceanography: Probing the Oceans

Group, headed by J. Lynn Abbott since 1966, routinely provides personnel for the computers on the expeditions of the *Thomas Washington* and the *Melville* (the *Argo* carried a computer on Circe Expedition in 1968 and on Scan Expedition in 1969). The group also operates and maintains the Scripps computer facility, located in Ritter Hall. On expeditions the computer continuously handles programs for ship navigation and underway measurements of bathymetry, salinity-temperature-depth data, magnetometer readings, surface-water measurements, and more, as well as certain special programs for individual researchers on board. The navigation program is set up to handle input from a satellite receiver; the first such installation for Scripps was on the *Argo* in 1968.

As Nierenberg said in 1969: "In many ways I feel that I am reliving my life, watching the development of the computerized ship and its effect on oceanographic research," for he had been deeply involved in the development of the applications of computers to nuclear physics and particle-physics research while at the Lawrence Radiation Laboratory.

Nierenberg also established the Applied Ocean Engineering Laboratory at Scripps in 1969, under the financial support of the Advanced Research Projects Agency. For this unit a steering committee of distinguished scientists was established, to select significant projects in advanced engineering marine research. Marion W. Johnson was the project manager until 1971, when Gerard H. Fisher succeeded him; both of them had previously been at Columbia University's Hudson Laboratories. The several programs supported by ARPA at Scripps for several years were: instrumentation and installation of mid-ocean buoys, under the direction of John D. Isaacs and in a cooperative project with the Applied Physics Laboratory of Johns Hopkins University; research on stable floating platforms, under the direction of Fred N.

William A. Nierenberg (right) and Ed Coughran (then with IBM) inspecting the new shipboard computer system, 1966.

Spiess, which led to constructing and testing scale models of multi-leg platforms for oceanographic research; adapting a quartz vertical accelerometer for deep-ocean measurements, under Walter H. Munk, Robert D. Moore, and William A. Prothero; devising equipment for recording subsurface pressures in the ocean caused by earthquakes, under Hugh Bradner and John Isaacs; investigating the water-sediment interface near the breaker zone and the velocity field of breaking waves, under Douglas L. Inman and William G. Van Dorn; and constructing equipment to measure radio signals scattered from the sea surface in order to determine the directional spectrum of ocean waves, under Nierenberg and Munk, in a cooperative project with the Center for Radio Astronomy at Stanford University. For this project Nierenberg set up the computer programs for handling the complex data. Although AOEL has ended, several of the engineering projects, some of them bearing significantly on future ocean technology, have been incorporated into other units at Scripps.

Also established at Nierenberg's instigation was the Center for Marine Affairs, which began in 1970 under a grant from the Ford Foundation. Warren S. Wooster first headed the group, which brought together specialists from the social sciences, law, government, and oceanography to explore the conditions for freedom of oceanic research and for determining international pollution policy. In 1973 the center, then headed by Gerald L. Wick, was placed in the Institute of Marine Resources, and, in cooperation with Mexican agencies, also undertook a study of long-range marine resource management issues related to desert coasts, with emphasis on the Baja California coastline.

In service to the nation and its oceanography, Nierenberg presided as chairman of the National Advisory Committee on Oceans and Atmosphere (NACOA) from October 1971 until February 1975, and he has continued as a member

Back on the Beach

since then. NACOA's goal, as defined in its first annual report (1972) has been "to help clarify what is good husbandry of the resources of the sea and air and what this can mean to the United States of America." The committee has advised the President and the Congress on law of the sea, United States fisheries, weather modification, coastal zone management, resource management, and energy.

Nierenberg was also chairman of the Department of Defense Advisory Board for Project AGILE from 1969 until 1972. He has served as an Advisor-at-Large to the Department of State, as a member of the United States Commission for UNESCO, the JASON Divison of the Institute for Defense Analyses, the National Academy of Sciences' Space Applications Board, the California Advisory Commission on Ocean Resources, the President's Science and Technology Advisory Group, the Advisory Committee on Law of the Sea, and a number of other national and international panels and boards. He sometimes saves committee-commuting time by piloting an airplane himself to and from meetings, having become a proficient pilot during the 1970s.

Like most of his predecessors as director, Nierenberg has tried to organize the diversified marine institution into a logical administrative unit, and he has admitted that Scripps doesn't lend itself to being organized.

For years the administrative organization of Scripps has been joked about ("the only state institution that is run by the inmates"), and occasionally deplored. "The Scripps Institution has never been planned," wrote former director Eckart in 1965, "and many of the organizational features have just happened." Perhaps the most honest appraisal of the organization appeared in the thin volume, *Manual of Rules and Procedures* for the institution, in 1951:

<blockquote>
Organization Chart of the SIO
To be supplied later
</blockquote>

Scripps Institution of Oceanograpny: Probing the Oceans

An outsider who was speaking kindly of the institution noted, in 1973, that it has "an organizational and management structure that has remained relaxed and flexible. The very informality of its management has been an important factor in its past accomplishments."

The flexibility has resulted from the institution's dual role as a graduate teaching school and a research organization. Appointments overlap within the research divisions and the organized laboratories, with inevitable confusion to the uninitiated (and sometimes to the accounting office). The academic staff includes both a professorial series and a parallel research series — some of whom also teach classes.

Over the years Scrippsians have grumbled at rules and procedures imposed from "above" — Berkeley, or UCLA (to which Scripps was attached for administrative academic affairs until UCSD was established), or UCSD. Spiess deplored in 1962: "We are already tending in the large scale administration of the University to rely too readily on use of ratios, formulae, IBM machines and broadly applied rules as substitutes for direct knowledge and sensitivity on the part of those who must make decisions." The freedom created originally by the awareness that Scripps was an independent research laboratory from 1903 until 1912, and furthered by the distance from Berkeley to Scripps during the early years, has never been readily yielded. When aggravated, Scripps officials are inclined to suggest that a certain recommended action might be contrary to the original deed of transfer in 1912 and to remind reformers that certain university policies are stated by the Regents of the University to be "not applicable to . . . the Scripps Institution of Oceanography."* And memories at Scripps are long.

*Similarly exempted from certain policies, as defined in 1961, were the Agricultural Experiment Station, the Lick Observatory at Mt. Hamilton, the Lawrence Radiation Laboratories at Berkeley and Livermore, and Los Alamos Scientific Laboratory.

Back on the Beach

Example: Sverdrup was sharply criticized in the spring of 1947 by President Sproul for not having notified him of the early negotiations to establish the Marine Life Research program. The reprimand was undeserved, for Sverdrup had indeed provided the information, but others in the university administration had failed to get it to Sproul. Sverdrup received an apology.

Example: Revelle drew a reprimand from Sproul in 1954 over the negotiations to transfer the U.S. Fish and Wildlife Service offices from Stanford to Scripps, which Revelle had discussed fully with Sproul in person earlier. "I believe what we have done so far is kosher," replied Revelle, "and a necessary preliminary to formal negotiations." He was allowed to proceed.

Example: In the mid-1950s one of the university regents suggested that the Scripps Institution should be moved to Santa Barbara, possibly over a twenty-year period. The university campus there was in the process of moving from its downtown Santa Barbara location to its present site nine miles west of the city on the shore. Scripps administrators were appalled. Revelle's masterful reply (mostly derived by Frautschy) logically explained the impossibility – even absurdity – of such a disruption to Scripps. The lack of ship berthing and of sources for marine supplies and equipment, the absence of cooperating Navy facilities, the shortage of available personnel in the smaller city, and even the "somewhat poorer weather conditions" of Santa Barbara were cited as "serious disadvantages," as well as the expense of moving equipment and personnel and the tragedy of breaking off the half-century record of scientific observations from the Scripps pier. It was even noted that the Santa Barbara faculty might not indeed "welcome the influx of a rather specialized laboratory with a budget and staff twice as big as their own." The cost of transfer was boldly estimated as

$2,338,000. To the relief of Scripps administrators, the subject was dropped.

SOCIAL ACTIVITIES

The formality that was typical during T. Wayland Vaughan's era at Scripps began to change in the 1930s, and the changes continued after the war. Shirt sleeves became more common than suit coats; Hawaiian and Tahitian shirts were special favorites. Military-style crew haircuts were in vogue after World War II and, until the late 1960s, only an occasional beard was seen, usually an indication that the "bearded wonder" had recently returned from a long sea trip. The lunch area became known as "Bikini plaza."

Almost everyone addressed one another by first name, from director to graduate student. As a crew member said, "It's kind of hard to call a guy 'Mister' when you've passed the salt to him at mess aboard ship."[13] The students, many of them veterans, were part of the extended family, and for a long time, as it was not uncommon for a student to take six to ten years to complete his studies and his dissertation. Students were often expedition leaders, on long as well as short cruises.

For many years, the women's group, Oceanids, was the tone-setter for campus social activities, and was also the service group for special events and visitors. In the 1940s Mrs. Francis B. ("Mom") Sumner had been the leader of what was first known as "Scripps Wives," a sewing and social group that devoted much time to projects for servicemen. In 1946, according to Helen (Mrs. Russell W.) Raitt, "a group of enthusiastic young student wives took the initiative and organized a wives' club," which was also endorsed by the faculty wives. In April 1952, Mrs. John S. (Sally)

Back on the Beach

Bradshaw and other student wives, "with renewed vim and vigor," again organized a Scripps women's group, "to foster social affairs for all persons on the Scripps Campus and to provide a means by which new students and new personnel may become acquainted."[14] Denis Fox proposed the name Oceanids, for the Greek ocean nymphs, the daughters of Oceanus and Tethys. After voting on a constitution, the group's first organized activity was "an informal dance for all hands in the library of Scripps Institution," in honor of the return of Shellback Expedition.

The organization was open to all women connected with the institution: staff wives and student wives, women employees, and women students. Not all eligibles attended the activities, of course, but there was a broad scattering of regulars throughout all disciplines and all ranks. The members provided and served refreshments at all special occasions on campus, served as ushers for campus-sponsored lectures, and hosted a Christmas party (originally for the children, in a carry-over of the custom established by Gudrun Sverdrup), an Easter-egg hunt, dances, potluck dinners, and some lectures. Funds for these events were usually raised through ticket sales to a night at the Old Globe Theatre. The organization set up special interest groups — book review, bridge, folk dancing, sewing, and others. Helen Raitt and Carol Schultz were the prime movers for starting Oceanids' information-filled monthly newsletter, "Bear Facts," in 1962. Newcomers were welcomed cordially, as potential cookie-bakers and ushers as well as new members of the oceanographic family. Lonesome wives whose husbands were at sea could find news items, a respite from loneliness, and a word of sympathetic understanding at Oceanids' gatherings. The organization has become the women's group of the entire UCSD campus.

Members of Oceanids plus many staff members contributed to the performance "Flip" in 1960; this original

Scripps Institution of Oceanography: Probing the Oceans

musical review was composed by draftsman Madeleine Miller (now Mahnken) and directed by secretary and musician Lorayne Buck and her husband Frederick. The theme was to prove that "oceanography is fun," and, put to music, it certainly was fun. There were mermaids and grunion hunters, sons of the beaches, girls from the "Friendly Islands," and a Rube Goldberg sort of machine that erupted bubbles; there were also some pointed digs at certain campus figures. The highlight of the choreography was a spectacular dance, titled "Flip, Flop, Flip," by nimble terpsichoreans in swim-fins.*

A much-appreciated welcoming custom was established by Edith Nierenberg in 1965, when her husband became director: a coffee party and introduction to the campus for all new wives and women employees of Scripps. The event usually has included a walking tour of several of the institution's laboratories, to provide a glimpse of the variety of researches. The hostesses — who are "oldtimers" — find that the gathering helps them understand their campus better.

A well-attended summer event throughout the 1950s was the family beach picnic, with games and prizes and beer. This was enthusiastically sponsored by the local chapter of the California State Employees Association, of which Ben Cox was a longtime official. (The annual beach picnic that was resumed in 1969 is hosted by graduate students and funded anonymously.)

Town and gown turned out to celebrate the University's Charter Day every March, often at Scripps in the form of an open house and public lecture. The 1951 Charter Day presented the dedication of Vaughan Aquarium-Museum (see chapter 9). In 1952 the *Horizon* and the *Crest* carried a hundred visitors out to sea for a demonstration of oceanographic equipment at work; to the delight of Carl L.

---

*"Flip II" and "Flip III" were presented in 1961 and 1962, but they were not so thoroughly oceanic in theme.

Hubbs, on that one-day jaunt the new Isaacs-Kidd midwater trawl added to the fish collection the first adult *Dolichopteryx* (a spookfish) recorded from the North Pacific and the third threadfin slickhead *(Talismania bifurcata)* ever found — duly reported by a seasick newsman. In 1954 the five ships of the research fleet were open to public display at the dock. The 1956 Charter Day became a university-wide celebration of Robert Gordon Sproul's twenty-fifth year as president of the university. At Scripps he dedicated the first addition to Ritter Hall, attended a two-day symposium on biology, and presided at an evening lecture. Sproul observed that during his quarter-century as president, Scripps had increased its budget 25-fold, its staff tenfold, and its buildings only threefold. The construction spree was just beginning.

In 1957 a display of oceanographic equipment was set out on the Scripps pier for visitors to view, and a continuous film program was offered in the library. The institution's first robed academic procession distinguished the evening ceremony that year, when Robert Maynard Hutchins addressed an overflow crowd on the subject of "Science and People." And in 1960, Eleanor Roosevelt, 75 years old, erect and firm-voiced, exhorted an even larger audience in San Diego's Russ Auditorium: "We have to wake up, change our views, and realize this is a different world today."

It was the world of Sputnik and space — and "inner space," as oceanography was beginning to be called. In that era, while envying the money being expended in the space program, Scripps scientists were wont to observe: "We know less about the ocean's bottom than the moon's behind."*

*This remark, which Bob Fisher believes may have been first used by Athelstan F. Spilhaus, appeared many times in speech and print by oceanographers, often oddly bowdlerized. In 1940, long before Sputnik, Harald Sverdrup wrote: ". . . we live on the shores bordering the largest ocean on earth, an ocean less charted than the surface of the moon." ("Research in Oceanography," *California Monthly,* December 1940)

The campus in 1949.

The campus in 1963.

Scripps and its offspring, UCSD, in November 1975. Photo by Phil Stotts.

Back on the Beach

Other special events on campus brought out the institution's families and many townspeople. On 22 April 1952, a sizable crowd greeted the *Galathea,* called "one of the world's outstanding pieces of oceanographic scientific equipment." The 80-meter-long frigate was homeward bound on the Danish Deep Sea Round the World Expedition when she paused for several hours off the Scripps pier. Visitors went out in smallboats to consume Danish pastries (and Carlsberg beer, the Carlsberg Foundation being the chief financial support of the two-year expedition), and to admire the oceanographic accoutrements, especially the giant winch that could handle seven miles of cable. Expedition leader Anton F. Bruun told of gathering great numbers of sea anemones, sea cucumbers, worms, clams, and crustaceans from ocean depths greater than six miles in the Philippine Trench.

In October 1958 the campus greeted newly inaugurated University President Clark Kerr, who reviewed the Scripps fleet, which was emblazoned in the university's colors, blue and gold. With some tricky maneuvering, *Horizon, Spencer F. Baird, Stranger, Paolina-T, T-441, Buoy Boat,* and even minute *Macrocystis* lined up, bow toward the beach, just beyond the Scripps pier, to honor the visitor.*

The end of the International Geophysical Year in 1958 was climaxed at Scripps by the visit of the Soviet research ship *Vityaz* to San Francisco. Americans were still chagrined at the success of Sputnik I in 1957. What might Soviet scientists be capable of doing in oceanography? For the *Vityaz* visit a flurry of letters began in October 1958, from Revelle to IGY and state department officials, and finally mutual invitations by radio flitted between the institution and the ship (which was not allowed to visit San Diego). When the 363-foot vessel sailed into San Francisco on 18

---

*Only *Orca* was unable to participate, as she was in the shipyard.

Scripps Institution of Oceanography: Probing the Oceans

December, she was met by a large crowd that included 20 Scripps scientists who had flown up to ask questions and to take the grand tour. The following day the *Horizon* sailed into San Francisco, bearing some of the Scripps seismic group to join the fun. The American oceanographers were much relieved to find that Soviet methods of exploring the ocean were similar to, and no better than, their own. John Tyler summed it up: "Walk into any /*Vityaz*/ lab and you will find differences in technique and in detail, with some ideas you like and some you don't like."[15] The great size of the Soviet research ship, its twelve separate laboratories, a scientific party of 65 people, and the many concurrent programs were considered by Scripps people to be generally not advantages (especially as all participants were expected to stay for the entire expedition, many months long). From San Francisco ten of the Soviet oceanographers were flown to Miramar Naval Air Station in San Diego. The guests were hosted for dinner, in twos and threes, at Scripps homes, with other staff families as additional guests — a form of entertainment that especially appealed to the Soviets, who asked many questions about American homes and family life. Their own interpreters, a few Russian-speaking Scrippsians, and considerable arm-waving kept the conversations flowing.

On 27 September 1966, Vice President Hubert H. Humphrey visited Scripps Institution, in his capacity as chairman of the National Council on Marine Resources and Engineering Development. He was provided with what he called "a very exciting tour" of the institution and some of its novel equipment, after which he declared: "We're on the threshold of a new age of exploration. . . . Our dreams for the oceans are not those of the poets and the prophets. They are practical dreams. . . . We intend to develop the bountiful resources of the sea to serve man's pressing needs."

Back on the Beach

In the same capacity three years later, on 23 October 1969, Vice President Spiro T. Agnew arrived at Scripps for what newsman Bryant Evans called "probably the shortest and most concentrated survey course in oceanography in history."[16] En route to a Palm Springs vacation, Agnew was on the campus for only an hour, and was shown sediment cores taken by the Deep Sea Drilling Project and deep-sea tide capsules of IGPP. Campus officials had only scanty warning of the brief visit, but Director Nierenberg was able to fly home from England (via Washington, D.C.) to welcome the visitor. (On the airplane he spilled soup on his only necktie and had to borrow one from Carl L. Hubbs — patterned, typically, in whales.)

After very elaborate preparations, Emperor Hirohito of Japan visited Scripps Institution on 9 October 1975, during his first trip to the United States. Biologist Hirohito was very interested in the marine specimens set out for his inspection, and was presented with a fine specimen of the "living fossil" *Neopilina* collected on Southtow Expedition in 1972.

The brunt of arranging details of the visits of such distinguished visitors fell to the Scripps branch of the Public Affairs Office, headed by R. Nelson Fuller from 1965. For some years that office has been responsible for providing information on Scripps researches to news media and for handling official visits to the institution in cooperation with the director's office, as well as answering requests for oceanographic information that arrive at the institution by letter and telephone by the dozens each week. Since the mid-1960s the Scripps Public Affairs Office has also coordinated, edited, and published the annual reports of the institution.

Until the late 1940s, the publicity duties of the institution were handled chiefly by the director's office, aided by various staff members. For example, during the 1930s a weekly news feature, called Institution Notes, which reported the comings and goings of the campus community, was provided to local newspapers by W. E. Allen (Scripps' first "publicity secretary," who started in 1919), Denis L. Fox, or Claude E. ZoBell. As public interest in inner space increased, the task of handling news and visitors grew larger, so that in 1949 a Public Relations Committee was established, and in May 1951 an Office of Public Relations was formed (under the committee). Carl L. Hubbs was appointed chairman of the committee, then director of the office. From the time he joined the staff of the Aquarium-Museum in 1946, Sam Hinton devoted a great deal of time to answering inquiries and conducting visitors about. The staff of the Scripps library also served — and still do serve — as an information center for the public.

In 1950 Thomas A. ("Lon") Manar joined the staff to prepare the reports of the Marine Life Research program. In the latter 1950s he became the public information officer for Scripps, and handled coordination with the news media as well as institution-sponsored events, such as Charter Day programs. For some years Scripps offered a lecture series annually in cooperation with the La Jolla Theatre and Arts Foundation, which required considerable coordination on details by Manar. When UCSD was established, the Scripps public information office came under the auspices of the upper campus.

And Now . . .

In 1940 Harald Sverdrup wrote:

> Oceanography may sometime in the distant future give us a many-colored picture of the oceans, but at the present time we have only started working with a large and intriguing puzzle game, each one of us trying

Back on the Beach

to put together pieces of similar color, hoping that all the fragments can be joined into one complete picture. It is as yet far too early to guess what this picture will look like, but it is not difficult to visualize oceanography as a unified field of research.[17]

The picture is not yet complete, but a great many pieces of the puzzle have indeed been put into place. The unified field of research has grown into a complexity of institutions and disciplines. It utilizes equipment that Sverdrup could not have envisioned 40 years ago: sea-going computer systems, satellites for navigation, research platforms that stand on end, electron microscopes, devices that glide from the depths of the sea when called.

The Scripps Institution of Oceanography no longer stands alone, but shares its field with dozens of other oceanographic research organizations. While there is a certain rivalry amongst them, there is also a great deal of cordial cooperation. In fact, the trend in oceanography since the latter 1960s has been toward multi-institutional research projects. As Nierenberg said in 1971: "The developing attitude on the part of government managers is the growing dependence on the 'larger' programs. These are important programs that require major concentrations of manpower and money because of either synoptic or engineering considerations."[18]

Such projects — which simultaneously involve a number of researchers at several institutions — have occupied a great deal of the director's time and that of the principal investigators: the Deep Sea Drilling Program, discussed in chapter 12; developing Norpax, which expanded into a major research effort from the North Pacific Buoy Project; establishing Geosecs, which has drawn in a wide range of oceanographic and meteorologic researchers; the *Alpha Helix* program, which has attracted hundreds of researchers

Scripps Institution of Oceanography: Probing the Oceans

to spending short periods of time aboard that vessel; the Sea Grant Program, which has become a statewide, multi-campus effort; and other units which are in the embryonic or adolescent stage. Such broad programs, too close to the present to be yet put into perspective, will become a history to be recounted at some future date.

"These projects," continued Nierenberg, "make extraordinary demands on the limited manpower of the oceanographic community, but they are very rewarding in their results and applications."

That seems to be it: Oceanography is not only fun, as Revelle said; it is also very rewarding.

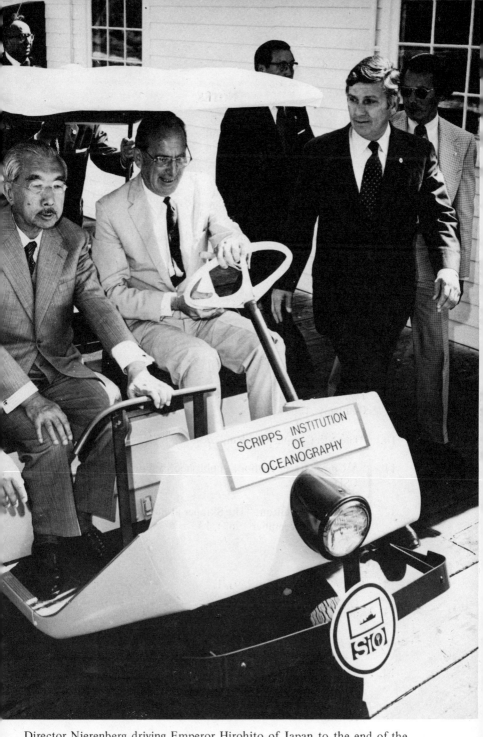

Director Nierenberg driving Emperor Hirohito of Japan to the end of the Scripps pier, 9 October 1975.

NOTES

1. *San Diego Union,* 10 May 1960.

2. Letter of 11 May 1967.

3. "The Age of Innocence and War in Oceanography," *Oceans Magazine,* Vol. I, No. 3 (March 1969), 10.

4. Talk at memorial service for Carl Eckart, 3 November 1973.

5. Mary Harrington Hall, "Revelle," *San Diego & Point Magazine,* Vol. 13, No. 7 (May 1961), 43.

6. *Ibid.*

7. New York and London, Interscience Publishers.

8. *The Sea,* vii.

9. *The New World of the Oceans* (Boston: Little, Brown and Company, 1969), 398.

10. "The Nth Campus Problem," *Bear Facts* (May 1966), 2.

11. "A University in Motion," published by Friends of the Library, UCSD, 1969, 6.

12. *In* Andrew Hamilton, "The Skipper at Scripps," *Think Magazine* (November-December 1963), 13.

13. *Ibid.*, 12.

14. *Bear Facts* (January 1969), 1.

15. Summary by T. A. Manar, 15 January 1959.

16. *San Diego Union,* 24 October 1969.

17. "The Unity of the Sciences of the Sea," *Sigma Xi Quarterly,* Vol. 28, No. 3 (Autumn 1940), 105.

18. SIO Annual Report, 1971, 7.

APPENDIX 1

Dates of Significance to the Scripps Institution of Oceanography

Summer 1892 — William E. Ritter conducted his first summer field program, at Pacific Grove, California.

26 September 1903 — The Marine Biological Association of San Diego was formed, to found and endow a biological institution "to carry on a biological and hydrographic survey of the waters of the Pacific Ocean adjacent to the coast of Southern California."

10 August 1907 — Pueblo Lot 1298, approximately 170 acres on the north edge of La Jolla, was acquired at city auction for $1,000 by the Marine Biological Association for the institution.

12 July 1912 — The land and buildings of the institution were transferred to the University of California, which conferred the name Scripps Institution for Biological Research.

14 October 1925 — The name of the institution was changed to Scripps Institution of Oceanography, and its program into ocean research began to change accordingly.

24 August 1956 — The Regents of the University authorized an expanded graduate program in La Jolla, which led to the establishment of the Institute of Technology and Engineering, which became the School of Science and Engineering, followed by the University of California at La Jolla, which was rechristened the University of California, San Diego.

Summer 1963 — Buildings on the present UCSD campus became available for the personnel of the general campus, who moved from their temporary locations at Scripps Institution.

APPENDIX 2

Scripps Institution Personnel in the National Academy of Sciences (and year of election)*

George E. Backus (1969)
Andrew A. Benson (1973)
Milton N. Bramlette (1954)
Theodore H. Bullock (1963)
Carl Henry Eckart (1953); died 1973
Walter M. Elsasser (1957); left Scripps 1962
Albert E. J. Engel (1970)
Robert M. Garrels (1962); left Scripps 1971
J. Freeman Gilbert (1973)
Carl L. Hubbs (1952)
John D. Isaacs (1974) — who is also the only Scripps member of the
                                          National Academy of Engineering
Henry William Menard (1968)
Walter H. Munk (1956)
William A. Nierenberg (1971)
Roger R. Revelle (1957)
Per F. Scholander (1961)
Francis B. Sumner (1937); died 1945
Harald U. Sverdrup (1945); died 1957**
T. Wayland Vaughan (1921); died 1952

FOREIGN MEMBERS
Edward C. Bullard (1959)
Devendra Lal (1975)

---

    *A. Baird Hastings was elected to the Academy in 1939; he is a research associate in the UCSD Medical School and has had an office at Scripps for many years.
    **Sverdrup resigned from N.A.S. in 1951, when he had lost American citizenship by maintaining residence in Norway for three years; he was elected a foreign member in 1952.

APPENDIX 3

Buildings on the Scripps Campus

	Year completed
Aquarium (original structure; removed 1951)	1915
Deep Sea Drilling Building	1970
Deep Sea Core Building	1973
Director's House (T-16)	1913
Eckart Building (for Scripps Library)	1976
Experimental Aquarium	1958
Addition	1965
Geodesic Domes	1959
Hydraulics Laboratory	1964
Institute of Geophysics and Planetary Physics	1963
Library (old)	1916
Lunch stand	1961
Marine Biology Building (under construction in 1976)	
Norpax Building	1975
Physiological Research Laboratory	1965
Pier and pier-end buildings	1916
Radio Station WWD	1952
Research Support Shop	1960
Ritter Hall	1931
First addition	1956
Second addition	1960
Scripps Building	1910
Addition ("New Scripps")	1959
Seawater tower (removed December 1932)	1910
Service Yard	
West garage	1950
North garage	1950
Purchasing-Storehouse	1953
Shore Processes Laboratory	
Built as Seawater Test Facility and Core Storage	1962

Remodeling	1973
Sumner Auditorium	1960
Surfside, converted from former house	1966
Sverdrup Hall	1960
Vaughan Aquarium-Museum	1950

Twelve cottages for staff members and visitors were built in 1913, and nine more, plus a commons building, were built in 1915-16. Some of these have been removed, and some have been moved from their original location. Other "temporary" buildings have been added by construction or by moving barracks buildings onto the campus. All these buildings have been numbered in the T series, which also includes the former director's house.

The "Fisheries Building" was built in 1964 by the federal government on land leased to it by Scripps Institution. First named the Fishery Oceanography Center, the building was renamed Southwest Fishery Center in 1970. It is under the jurisdiction of the National Marine Fisheries Service.

Facilities located off the campus include: Nimitz Marine Facilities, located on six acres of land at Point Loma, leased from the Navy from 1965 to 1975, when the land was given to Scripps; most of the buildings of the Marine Physical Laboratory, which are in Navy buildings on Point Loma; the buildings of the Visibility Laboratory, rented from the Navy on Point Loma; the Mount Soledad facility in La Jolla for radioisotope researches; several offices rented in Sorrento Valley for the Geosecs project; the transmitters of radio station WWD and other facilities on land formerly part of the Navy's Camp Elliott, east of Miramar Naval Air Station; and a calibration barge for the use of the Marine Physical Laboratory at San Vicente reservoir.

APPENDIX 4

Graduate Students
and the List of Ph.D. and M.S. Recipients

The first graduate students "at Scripps" were enrolled at Berkeley, a state of affairs that continued until the late 1930s. From 1938 the graduate administration of the institution was handled through UCLA, as part of the Southern Section of the university. The procedures and requirements of the dissertation-producing system seemed mystical — and remote — to the Scripps students in those days. They drove to UCLA as rarely as possible and dwelt in a maze of half-information until the moment of truth, when the dissertation was finally delivered, by hand, to the archivist of the UCLA library. For a number of years in the 1950s Norris W. Rakestraw served as the dean of students at Scripps, and he helped many a student avoid the administrative pitfalls.

"With the establishment of the School of Science and Engineering and UCSD, three [teaching] departments were formed [at Scripps] in the early 1960's. Some of the geologists and geochemists of the Institution spearheaded the establishment of a Department of Earth Sciences, which then became a component of the School of Science and Engineering. At the same time, a group of biologists at Scripps, having previously obtained approval for a curriculum in marine biology, succeeded in establishing a Department of Marine Biology within Scripps in 1962. Other scientists of the Institution, including a full spectrum from physical oceanographers to earth scientists, biologists and chemists, still felt the need for continued graduate training which would emphasize the interactions between the different aspects of marine science. They took the initiative in obtaining approval of the formal establishment of the Department of Oceanography within SIO."*

These three teaching departments functioned separately until 1967, when they were reorganized into a single Graduate Department of the Scripps Institution of Oceanography. Within that department are seven curricular groups: biological oceanography, marine biology, marine chemistry, geological sciences, geophysics, physical oceanography, and applied ocean sciences. Warren S. Wooster became the first chairman of the combined department, succeeded in 1968 by Edward L. Winterer, who served effectively in the post for five years. In 1973 Joseph R. Curray became department chairman and served until January 1976, when he left on sabbatical, and Fred N. Spiess took the position.

In 1936 five students were registered for graduate work at Scripps Institution. At the opening of the 1976 academic year, 185 students were registered for graduate work at the institution. Throughout that interval the students have been a very significant raison d'être of the laboratory by the sea. In the pages of this book, "colleagues" and "co-workers" have often included graduate students. They have led major and minor expeditions, have published extensively, have served actively on committees, have worked alongside their professors, and have pleased and saddened them. Most of them have gotten their feet wet, and a lot of them have been seasick. One distinguished himself by failing the UCLA-imposed language requirement examination three times before going on to receive his Ph.D.; another, by allowing 14 years to elapse between enrolling and receiving his Ph.D.; a third, by receiving the only Ph.D. awarded by "the University of California at La Jolla."

The records of the Scripps alumni have been noteworthy. Some have joined the staff of their alma mater to train the next generation, and many others have become professors and researchers at other oceanographic labora-

*Memorandum from Joseph R. Curray to Paul Saltman, 22 November 1974.

tories. Several have been directors or administrative officers at such institutions. Some have advised presidents — and at least one has been in jail.

Raitt and Moulton, in *Scripps Institution of Oceanography: First Fifty Years,* listed 137 Ph.D. recipients who "did the bulk of their work or all at Scripps Institution of Oceanography" from 1912 through 1965.* Since then, an additional 215 Ph.D. degrees have been awarded by Scripps Institution through the end of 1976, as follows:

1966-67

Tim P. Barnett
Abner Blackman
Anthony J. Bowen
William Fisher Busby
George Frederick Crozier
Jean Henri Filloux
Donald V. Helmberger
Thomas Sterling Hopkins

Kern Kenyon
Stanley A. Kling
Albert M. Kudo
Jimmy C. Larsen
Keith Brian Macdonald
Erk Reimnitz
Helmuth Sandstrom

1967-68

Norman G. Banks
John R. Booker
Yan Bottinga
Thomas A. Clarke
Brian G. D'Aoust
Everett L. Douglas
Frank J. Hester

Karl P. Kuchnow
David L. Leighton
Rabindra Prasad
Vann Elliott Smith
Herbert Windom
Robert L. Zalkan

1968-69

Wolfgang H. Berger
Francis A. Dahlen
Walter M. Darley
Bruce W. Frost
Frederick P. Healey
G. Ross Heath
James H. Jones
Bruce P. Luyendyk

Theodore C. Moore, Jr.
Elizabeth V. Murray
Thomas R. Osborn
David S. Piper
Robert H. Stewart
A. H. Mark H. Wimbush
Chi Shing Wong

*The year 1966 as given in Raitt and Moulton is omitted from the above.

485

1969-70

Marvin H. Beeson
Jonathan Berger
Lo-Chai Chen
William E. Farrell
Nick Fotheringham
Richard W. Grigg
Willis B. Hayes
Alice C. Jokela
Daniel E. Karig
Paul D. Komar
Ching-Ming Kuo

Roger L. Larson
James W. McBeth
David F. McGeary
Charles Meyer, Jr.
Charles B. Miller
Gerald B. Morris
William R. Normark
Michael L. Richards
Eli A. Silver
John M. Wells

1970-71

William F. Blankley
Gerald W. Bowes
Clement G. Chase
James L. Congleton
John B. Corliss
Mary R. Eaton
John M. Edmond
Jean M. Francheteau
Jeffrey B. Graham
Michael C. Gregg
Alan R. Hargens

Douglass F. Hoese
Peter M. Kroopnick
Richard F. Lee
Alfred R. Loeblich III
James J. McCarthy
Robert W. Piddington
Mary W. Silver
Robert J. Tait
Ray F. Weiss
David R. Young

1971-72

Katherine Y. Bowen
Yu-Chia Chung
John P. Greenhouse
Barbara B. Hemmingsen
James D. Irish
David A. Johnson
Leonard E. Johnson
Robert Karl Johnson
Dan H. Kerem
Lawrence A. Klapow
Ronald K. Lam

Robert D. Nason
Robert W. Owen, Jr.
Michael R. Petersen
Charles F. Phleger
Thomas B. Scanland
Joseph N. Suhayda
Cornelius W. Sullivan
Leighton R. Taylor, Jr.
Wayne H. Wilson, Jr.
John H. Wormuth

1972-73

Roger N. Anderson
Ross O. Barnes
Charles K. Barry
David W. Behringer
K. Gopalakrishnan
John A. Grow
Peter W. Hacker
Ann C. Hartline
Jed Hirota
David C. Judkins
Daniel L. Kamykowski

Dale A. Kiefer
Bert N. Kobayashi
John Douglas Macdougall
Robert C. May
John E. McCosker
Joane S. Molenock
Christopher J. Platt
Genelle W. Renz
Edward R. Sholkovitz
Rosemary A. Thompson
Paul M. Yoshioka

1973-74

Arthur M. Barnett
James L. Cairns
Gregory F. Dreyer
Edgardo D. Gomez
Eric O. Hartwig

Loren R. Haury
Robherd E. Lange
Richard J. Seymour
Robert R. Warner
Arthur A. Wolfson

1974-75

Raymond J. Andersen
Terrance G. Barker
Edward B. Brothers
Kenneth W. Bruland
David A. Clague
John E. Cromwell
Thomas F. Dana
Bonnie J. Davis
John R. Dingler
Allan F. Divis
Gary H. Dobbs III
John G. Duman
Thomas H. Ermak
Jimmie L. Greenslate
Robert T. Guza
Robert F. Hartwick
William D. Ivers
Richard D. Jarrard

Thomas C. Johnson
Peter A. Jumars
Kim D. Klitgord
Peter F. Lonsdale
Philip S. Low
Jon S. Mynderse
Douglas W. Oldenburg
Mary J. Perry
Robert Pinkel
Harald S. Poelchau
Francis J. Rokop
Roy A. Schroeder
Gary B. Smith
Martha O. Stallard
Mia J. Tegner
Brian E. Tucker
Gordon O. Williams

1975-76

Michael A. Barnett
Raymond T. Bauer
Raymond Peter Buland
Douglas R. Diener
Manuel E. Fiedeiro
Ronald A. Fritzsche
James Douglass Hauxhurst
Barbara M. Hickey
Stephen P. Huestis
Lawrence A. Lawver
Cynthia L. Lee
George Shalett Lewbel
James T. Natland
Richard Kiichi Nishimori
Ron S. Nolan

John A. Orcutt
John S. Patton
John Sheldon Paul
Stephen E. Pazan
Barbara B. Prezelin
James A. Raymond
Lloyd A. Regier
Michael S. Reichle
Harry Francis Ridgway, Jr.
Bruce Ray Rosendahl
Richard L. Salmon
Vernon P. Simmons
John Thomas Turk
Theodore Carl Tutschulte
Clark R. Wilson

July-December 1976

Ralph J. Archuleta
John Edward Burris
David Michael Gardiner
Greg Holloway
John Bodley Keene
Harold William Lyons

Arne Croizat Joseph Mortensen
Ian Reid
George Frederick Sharman
Eric Shulenberger
Gérard G. Stock
Nancy Wallis Withers

Masters' Degrees (1965 through 1976)

1965

Norman Guy Banks
John Ratcliffe Booker
Elaine Ruth Brooks
Donald Vincent Helmberger
Jon Richard Kirkpatrick

Carl Hubert Malone
Michael Lee Richards
Charles Elmer Roberson
Herbert Lynn Windom
Robert Libman Zalkan

1966

Peter Wolfgang Hacker
Kurt Hecht
Jacquelin Neva Miller
Jose Hipolito Monteiro

Daniel Agusto Rodrigues
Jack Irving Simmons
Ray Franklin Weiss

1967

Jonathan Berger
Jean E. Brenchley
Francis Anthony Dahlen, Jr.
Manuel E. Fiadeiro
Rene Eugenio Gonzalez, Jr.

Janet Evelyn Preslan
Vernon Pitkin Simmons
James Fielding Smith
Bernard John Zahuranec

1968

Christian Eric Abranson
Hillel Gordin
Greg Holloway
Robert James Huggett
Leonard Evans Johnson
Itamar Perath
Charles Frederick Phleger

Philip Andrew Rasmussen
William Albert Richkus
Martin Gregory Sattler
Andrew Soutar
Barbara Ann Symroski
Robert Bruce Williams

1969

Karen Leialoha Achor
Scott Knight Anderson, Jr.
Michael Alvin Barnett
Robert David Bowlus
Yi-Maw Chang
Charles Hovey Clifford
Peter Jay Delmonte
Douglas Charles Eaton
Calvin C. Fong
Barbara Mary Hickey

James D. Irish
Gerald Fulton Johnson
Barry Stephen Kues
Heryl Gordon Kroopnick
George Shalett Lewbel
Robert Pinkel
Francois Albert Revel
Peter Dragutin Sertic
Anne Spacie

1970

Joellen Sarner Barnett
Michael James Lees
Harold William Lyons
Arnold W. Mantyla
Dwight Dufresne Pollard
Alberto Ramirez-Flores
James Edward Smith

Albert Mills Soldate, Jr.
Forrest Eugene Steber
David Alan Summerville
Alina Margarita Szmant
Timothy Perkins Whorf
Samuel Thomas Wilson

1971

Neil J. Adler
Gerard Yves Conan

Robert Thomas Guza
John Murrah Harding, Jr.

Timothy Clay Curtis
Heide Sing Davis
John Joseph Dickinson
Ralph Vincent Dykes
Eric Rodholm Ernst
Kakkala Gopalakrishnan

Malladi Venkata Lakshmana Rao
Wayne Lewis Olsen.
Hugh Paul Slawson
Otto Frederick Steffin
Stephen Eberly Thompson, Jr.
Dennis Dale Todd

1972

Alan Floyd Chatfield
David Milton Checkley, Jr.
Lawrence E. Deysher, Jr.
Isabel Foster Downs

Gordon John Lusk
Gary Parker Owen
Sheldon Morton Sanders
George D. Wilson

1973

Tarsicio Jerez Antezana
John Wayne Hill
James Melven Schweigert

Kin Hing Tsang
Clark Roland Wilson
David A. Yuen

1974

James Alan Bailard
Robert Morey Cutler
Robert Sherman Detrick, Jr.
Robert Edward Fricks
Fei Sophia Hu
Stephen Paul Klein

Dudley Wade Leath
Ronald Lee Oda
William Leslie Preslan
Richard Wendel Robinson
Jeffrey Donald Rude

1975

Michael Duane Applequist
Bryan Reeder Burnett
Jeffrey Thomas Dillingham
Craig Louis Etka
Diana Jean-Marie Gabaldon
Lloyd Leslie Green

Yolanda Delores Montejano
Steven Elden Moran
Paul Andrew Spudich
Eduardo A. Valenzuela Ayala
Gregory Wallace Withee

1976

Victor Chow
Timothy Gerrodette

Roger P. Hewitt
Jan Leslie Hillson

Index

Abbott, J. Lynn, 458
Acapulco Trench Expedition, 405
Agassiz, Alexander, 354, 411; see also Alexander Agassiz
Agnew, Spiro T., 473
air-sea interaction, 258-65
Alexander Agassiz (first), 7, 354, 356
Alexander Agassiz (second), 66, 345, 356, 366
Alexis, Carl, 301
algae, 32, 127, 230, 232-33; see also kelp studies
Allanson, Audley A., 134, 407
Allen, Winfred E., 8, 11, 23, 213-14, 228-30, 232, 474
Allison, Edwin C., 221
Alpha Helix, 170, 172-76, 178-79, 232, 345-46, 356-57, 475
Alvariño, Angeles, 72
American Miscellaneous Society (AMSOC), 301-303, 306, 308
American Petroleum Institute, 225, 275, 278; see also API Project 51
Amundsen, Roald, 9
Anderson, Ernest C., 331
Anderson, Victor C., 87, 89, 105-106
anisotropy, 94-95
Antipode Expedition, 211, 298, 300, 338
antisubmarine warfare, 25, 89
API Project 51, 144, 223, 275-78, 324
Applied Ocean Engineering Laboratory, 458, 460
Applied Oceanography Group, 258-62
Aqua-lung, 127-28; see also Scuba
Aquarium-Museum, 12-14, 22, 33, 124, 169, 185-98, 212, 235, 439, 441, 466, 474; docents, 195
archaeological studies, 108, 134, 208, 335-37
Argo, 138, 255, 289, 297-98, 353, 357, 370, 427, 428 & n, 429-32, 452, 458
Aries Expedition, 74
Arnold, James R., 331 & n, 339
Arrhenius, Gustaf O. S., 285-86, 290, 339
Arthur, Robert S., 244, 444
Atlantis (myth), 393
Atwater, Tanya, 299-300
Ault, James P., 8
Austin, Thomas S., 230

Backus, George E., 153, 162
bacteria studies. See microbiology
Bada, Jeffrey L., 340-41
Bainbridge, Arnold E., 333, 338
Baird, Spencer Fullerton, 367; see also Spencer F. Baird
Baker, Charlotte, 1
Baker, Fred, 1, 6, 13
Ball, Eric G., 168
Ball, Henry, 367
Barbados Oceanographic and Meteorological Experiment (BOMEX), 105, 262
Barnhart, Percy S., 12, 186, 189, 192, 197, 206, 235
Barr, Edward S., 246, 351
Bartholomew, George A., 208
Bascom, Willard, 31, 57, 262, 301, 303-304, 306, 378, 400-402
Bates, Charles, 32
bathythermograph (BT), 27, 245-51, 422
Beal, M. Allen, 410
Beers, John R., 141
Behrman, Daniel, 158, 450
Bennett, Rawson, 446
Benson, Andrew A., 182, 221-22
Benthic Laboratory, 106-107, 135
Berberich, Frank, 52, 386
Berger, Jonathan, 160
Berger, Wolfgang, 285
Bering Sea Expedition, 176, 178, 232
Berkner, Lloyd V., 406
Berner, Leo D., 72
Bernstein, Elmer, 362
Bien, George, 324, 332-33, 335, 442 n
Billabong Expedition, 175-76, 232
biological studies, 11-13, 26, 32, 106, 122-27, 136-43, 191-92, 201-37, 276, 389, 405, 421, 427; see also individual subjects
Biomass Laboratory, 75
Bjerknes, Vilhelm, 9
Blackburn, Maurice, 136-37
Black Douglas, 46, 50
Blandy, W. H. P., 380
Blinks, Lawrence R., 168
Block, Barry, 162
Bluefin, 11-12, 251
Boden, Brian P., 219
Boden, Elizabeth. See Kampa, Elizabeth

491

Boegeman, Dwight E., Jr., 107
Bolster, C. M., 401
bongo net, 73
Bostwick, La Place, 13
Bowen, Anthony J., 279
Bradner, Hugh, 133, 160-61, 460
Bradshaw, Sally, 464-65
Brain Research Institute, 169
Bramlette, Milton N., 223, 283-84
Brandel, Gus, 348
Branson, Peter S., 348
Briley, Charles H., 369
Brinton, Edward M., 72, 74, 219, 422
Bronk, Detlev, 189
Bronson, Earl D., 104
Brown, Daniel M., 71, 73
Brune, James N., 161
Brunot, Kenneth E., 311
Brush, Birchard M., 280
Bruun, Anton Frederick, 418, 424, 471
BT. *See* bathythermograph
Buck, Frederick and Lorayne, 466
Buffington, Edwin C., 286
Bullard, Edward C., 58, 97-98, 100, 153, 163-64, 302, 305, 388, 390, 392
Bullard probe, 96-97, 163, 390, 392; *see also* heat flow studies
Bullock, Theodore H., 127, 168-69, 234
bumblebee buoy, 263
buoy boats, 358, 471
Burnette, Julian G., 44
Burt, Wayne V., 246, 252
Bush, Irving T., 360
Buzzati-Traverso, Adriano A., 204, 212

California Cooperative Oceanic Fisheries Investigation (CalCOFI), 47, 49-50, 56-58, 62, 65-68, 72, 75, 117-18, 213, 232, 252, 383
California Current, 19, 45, 49, 65, 67, 75, 143, 251-52, 255, 262-63, 382
California Fish and Game, 11-12, 23, 44-46, 56, 60, 123-24, 235, 251
Campbell, William Wallace, 235
Capricorn Expedition, 92, 98-99, 134, 283-85, 290, 292, 294, 352-53, 377, 400-405
Carlucci, Angelo F., 141
Carmarsel Expedition, 278
Carnegie, 7-8, 10, 13, 446
Carson, Rachel, 352

Carter, Arthur B., 54, 384, 386
Center for Marine Affairs, 144, 254, 460
Chamberlain, Theodore H., 280, 422
Chambers, Stanley W., 11
Chapman, Wilbert M., 44-45, 118-19
Chase, Thomas E., 288
chemical studies, 12, 19, 321-41
Cheng, Lanna. *See* Lewin
Chinook Expedition, 368, 405
Chow, Tsaihwa J., 325, 339-40
Christensen, R. J., 26, 404
Chubasco Expedition, 404
Chung, Yu-chia, 337
Churchward, James, 393
Circe Expedition, 298, 300, 458
Clark, Frances N., 44, 67
Clarke, Thomas A., 135
Clendenning, Kenneth A., 123, 126
Clipperton Island, 413-14
Cochrane, Edward L., 79
Cochrane, John, 246
Coe, Wesley R., 213-14
Colbeth, Clifford W., 348
Committee on Oceanography of National Academy of Sciences, 7-9
Compton, Karl T., 113-14
Controlled Ecosystem Pollution Experiment (CEPEX), 143
Coolidge, Harold J., 415-16
coring and core studies, 20, 23, 97, 225, 227-28, 277, 283-86, 290, 314-15, 325, 388, 396, 398
Corry, Charles, 97
cottages, 14, 22, 154, 440; *see also* director's house
Couper, B. King, 247, 250
Cox, Ben, 198, 466
Cox, Charles S., 264
Craig, Harmon, 144, 297, 324, 331-32, 335-39, 429
Craig, Valerie, 337
Crandall, Wesley C., 123, 354, 359
Crest, 37, 49, 66, 346, 358, 363, 466
Crocker, Templeton, 230
Croker, Richard, 44
Cromwell, Townsend, 254, 287 n
Cromwell Current, 253-54
Crossroads. *See* Operation Crossroads
Crouch, Donal C., 54
Crowell, John, 32
Cummings, Bert, 429
Cupp, Easter Ellen, 12, 229
Curran, Frank, 197
Curray, Joseph R., 277-78, 298, 484

currents, studies of, 12, 19-20, 35, 71, 74, 243-44, 251-57, 275, 389, 400, 413, 423, 430-31; *see also* California Current
Curtis, C. C., 366
Cusp Expedition, 404
CUSS I, 304-306, 308

Danish Deep-Sea Round-the-World Expedition, 226, 471
Darwin, Charles, 221, 326, 377, 388
Data Collection and Processing Group (DCPG), 51, 68, 75, 139, 338
David Starr Jordan, 54, 349
Davis, Laurence, 429
Davis, Russ E., 262
Dayton, Paul K., 217, 219, 237
DCPG. *See* Data Collection and Processing Group
Deacon, George E. R., 426
deep scattering layer, 27, 60, 72, 87, 89, 219-20, 404
Deep Sea Drilling Project, 54, 197, 286, 301, 304, 308-16, 431, 473, 475; building, 311, 314, 442
Deep Tow, 102, 107-108, 162, 456-57
DeVries, Arthur L., 181
Dexter, Steve, 408
diatoms, 11, 141, 219, 228-30, 233-34, 323, 325
Dietz, Robert S., 20, 23, 273, 286, 385-87
Dill, Robert F., 286, 295
director's house, 14, 35, 47
diving. *See* Scuba
Dixon, Fred S., 297, 338
Dodo Expedition, 298, 429, 431
Doldrums Expedition, 413
Dolphin, 359
Dolphin Expedition, 254, 412
Downwind Expedition, 290, 292, 410-12
Drake, Charles, 310 n
dredging, 20, 277, 290, 293-95, 385
Dubois, Carl, 97
Duntley, Seibert Quimby, 111-115
Durant, Russell Clifford, 360
Durant, William Crapo, 360

earthquake studies, 160-62, 337, 460
Easter Island, 292, 412
East Pacific Rise, 98, 161, 285-87, 411
Eastropac Expedition, 138-39, 255

Eastropic Expedition, 254, 405
Eckart, Carl, 30, 45, 80-81, 83-84, 87, 156, 366, 447-48, 450, 461
Eckart Building, 442
ECR layer, 27
Edgar, N. Terence, 311
Ellen Browning, 359
Ellen B. Scripps, 158, 345, 359-60
Elsasser, Walter M., 324 & n
Elsner, Robert, 169, 181
Emery, Kenneth O., 20, 23-24, 279
Emiliani, Cesare, 309
Engel, Albert E. J., 293-94
Engel, Celeste G., 293, 295, 300
Enns, Theodore, 169, 234
Enright, James T., 219, 230
Epel, David, 234
Eppley, Richard W., 141
Equapac Expedition, 405
Eurydice Expedition, 295
Evans, Bryant, 275, 473
Ewing, Gifford C., 208, 261, 405
Ewing, Maurice, 90, 303, 309, 312
E. W. Scripps, 16, 19-20, 23, 25, 33, 37, 49, 52, 92, 186-87, 230, 251, 273, 324, 345-46, 348-49, 351, 360-61, 380, 419, 452
expeditions, 373-433; SIO first, 20, 378, 380; *see also* individual names
Experimental Aquarium, 193, 440
Eyring, Carl F., 26, 404

Fager, Edward W., 203, 216-17, 219
Farrell, W. E., 163
Farwell, Charles, 197
Faughn, James L., 37, 174-76, 346, 350, 383, 386, 394, 418-20, 422, 424
Faulkner, D. John, 340
Fenn, Wallace O., 168
Ferris, Noel, 350
Fiadeiro, Manuel, 337
Field, A. J., 311
First International Oceanographic Congress, 301, 425-26, 449
Fisher, Frederick H., 104-105
Fisher, Gerard H., 458
Fisher, Robert L., 291-95, 298, 300, 378, 404, 410, 412, 427, 431, 467 n
fisheries studies, 117-24, 126, 136-40, 206, 213, 252-54, 262, 431-32; *see also* sardine project
Fishery Oceanography Center, 47, 137, 139, 442

493

fishes, studies of, 11-12, 124, 180-81, 186, 190-94, 204-206, 209, 211-13, 215, 234, 330, 398, 400, 413, 423, 467; *see also* fisheries; sardine project
Flechsig, Arthur O., 132, 134-35, 206, 216-17
Fleming, Guy and Margaret, 35
Fleming, Richard H., 12, 24-25, 251, 449
Fleminger, Abraham, 72
"Flip" (musical), 465-66
Flip (platform), 102, 104-105, 157, 262, 349, 361-62, 365, 456
Flynn, Errol, 206
Flynn, T. Thomson, 206
Focke, Alfred B., 84, 405
Folsom, Theodore R., 144, 324, 327, 329-30, 336
Food Chain Research Group, 73, 140-43
foraminifera, 6, 11, 284-86, 335
fouling studies, 32, 192-93, 225
Fox, Denis L., 12, 14, 23, 32, 189, 192, 214-16, 225, 465, 474
Franklin, Dean, 169
Frautschy, Jeffery D., 25, 90, 99, 250, 349, 353, 385, 394, 407, 429, 448, 455, 463
free-vehicle instruments, 69-72, 157, 264-65
Fuller, R. Nelson, 432, 473

Galathea, 226, 277, 418, 471
Gantner, Joseph, 443
Garey, Walter F., 179
Genter, Tillie, 22
Geochemical Ocean Sections Study (Geosecs), 337-38, 475
Geological Data Center, 101, 136, 288
Geological Society of America, 20, 271
geological studies, 14, 20, 108, 128, 269-316, 394-96, 398, 410-11, 431
geophysical studies, 86, 90-102, 149-64, 410-11
George, Julian, 198
Geosecs. *See* Geochemical Ocean Sections Study
Gianna, 362
Gibson, Carl H., 105, 262
Gieskes, Joris M., 340
Gilbert, Freeman, 162

Glomar Challenger, 54, 286, 311-12 314-15, 354
Glosten, Lawrence M., 104, 170, 172
Goff, William J., 443
Goldberg, Edward D., 279, 290, 324-27, 339
Golden One, 362
Gorczynski, Wladyslaw, 242
Goro, Fritz, 305
Graham, Jery B., 71
Grant, U. S., IV, 29
Green, Cecil H. and Ida, 154, 156
Greenberg, D. M., 323
Greenberg, Daniel S., 306, 308
greenhouse effect, 330-32
Grigg, Richard W., 135
grunion, 187, 209, 211
Gulf of California, 161, 192, 194, 276-77; expeditions of 1939-40, 20, 22-23, 225, 230, 351, 361, 380

Hagiwara, Susumu, 169
Haines, Robert B., 398
Hamilton, Andrew, 387, 394
Hamilton, Edwin L., 286
Hammel, Harold T., 181
Hancock, G. Allen, 230
Hanley, Franklin B., 25
Hanor, J. S., 286
Harbison, Charles, 413
Harvey, George, 132
Hastings, A. Baird, 168, 170, 180
Haubrich, Richard A., 153, 160, 162
Hawkins, James W., 300, 339
Hawkins, Mrs. W. W., 440
Haworth, Leland, 156
Haxo, Francis T., 232
Haymaker, Frank, 127, 273-74
heat flow studies, 96-99, 385, 390 & n, 392
Hedberg, Hollis, 306, 308
Hedgpeth, Joel W., 204, 220, 366
Heiligenberg, Walter F., 212
Hemmingsen, Edvard A., 169, 181
Hendershott, Myrl C., 266
Hersey, J. Brackett, 307, 309, 310 n
Hess, Harry Hammond, 94, 101, 301, 303, 307
Hessler, Robert R., 222
Hill, Maurice N., 449
Hilo Expedition, 307
Hinton, Sam, 58, 186-87, 189-91, 195, 400, 474
Hirohito, Emperor, 404, 473
Hoctor, Fred, 351
Hohnhaus, George W., 297

494

Holland, Nicholas D., 234
Holleman, Roy W., 443
Holm-Hansen, Osmund, 141
Hopkins, Marvin, 364, 430
Hord, Donal, 154
Horizon, 37, 49, 55, 61, 66, 277, 289-90, 298, 305, 345-46, 358, 362-63, 367, 374, 376, 383-88, 394-95, 399-401, 404, 410, 413, 430-31, 466, 471-72
Horning, Otto, 407
Horrer, Paul L., 56, 62, 405
Howard, Gerald V., 287
Huang, Joseph C. K., 264
Hubbs, Carl L., 45, 60-61, 69, 144, 202, 205-206, 208-209, 219, 237, 261, 335, 366, 424, 466-67, 473-74
Hubbs, Laura C., 206, 261
Hugh M. Smith, 254, 363-64, 399, 419
Hughes, L. H., 368
Humphrey, Hubert H., 472
Hunt, Frederick V., 88
Hutchins, Robert Maynard, 467
Hydraulics Laboratory, 160, 193, 198, 266, 279, 441

IATTC. *See* Inter-American Tropical Tuna Commission
ichthyology. *See* fishes; fisheries
IGPP. *See* Institute of Geophysics and Planetary Physics
IMR. *See* Institute of Marine Resources
Ingram, Jonas H., 29
Inman, Douglas L., 144, 279-80, 444-45, 460
Institute of Geophysics, 99, 149-52, 277, 383, 393
Institute of Geophysics and Planetary Physics, 149-64, 258, 441, 443, 473
Institute of Marine Resources, 57, 117-45, 201, 275, 287, 290, 364, 432, 455, 460
Institute for Pure and Applied Physical Sciences, 286
Institute for the Study of Matter, 285
Inter-American Tropical Tuna Commission (IATTC), 137-39, 252-53, 255, 287 n, 399
internal waves, 105, 158, 251, 266
International Committee on the Oceanography of the Pacific, 6-7

International Decade of Ocean Exploration, 136, 338, 340
International Geophysical Year, 150, 204, 253-54, 265, 290, 332, 334-35, 405-14, 424, 426-27, 448, 455, 471
International Indian Ocean Expedition, 291, 357, 424-32, 448-49
International Phase of Ocean Drilling (IPOD), 315
invertebrates, 213-23, 234, 340, 413; *see also* deep scattering layer; Marine Life Research Program; plankton studies
Irving, Laurence, 176
Isaacs, Irwin, 44
Isaacs, John D., 31, 45, 56-57, 60-63, 69-70, 72, 121, 145, 262, 327, 392, 394-95, 400, 458, 460
Isaacs-Kidd midwater trawl, 60, 66, 71, 212, 398, 467

Jasper, 26, 369
Jensen, David, 234
Johanson, Mrs. J. G., 440
Johnson, Carl, 226, 439
Johnson, Marion W., 458
Johnson, Martin W., 12, 14, 16, 24-25, 27, 33, 45, 72-73, 189, 219, 449
Johnson, Walter S., Jr., 361
Joint Oceanographic Institutions Deep Earth Sampling (JOIDES), 309-10, 315
Jones, Alan C., 368
Jordan, David Starr, 205
Junior Oceanographers Corps, 195, 197

Kampa, Elizabeth, 219-20
Kampmann, Patricia A., 195
Kaplan, Joseph, 406
Karig, Dan, 300
Keeling, Charles D., 325, 333-34, 453
Kelco Company, 123-24
Kellogg, William C., Jr., 81
kelp studies, 32, 122-27
Kennedy, John F., 305
Kerr, Clark, 471
Kessler, N. A., 369
Kidd, Lewis W., 60, 69
King, Bernice K., 195
Kirby, Harold L., 258-59
Kirk, Paul, 323
Kiwala, Robert S., 194

Klein, Hans, 68, 75
Knauss, John A., 252-54, 302, 333, 412-13, 430, 452-53
Knauss, Lynne, 453
Koczy, Fritz, 310 n
Kofoid, Charles Atwood, 1
Koide, Minoru, 326
Kooyman, Gerald, 181
Kort, V. G., 426
Kuenen, Ph. H., 283

Ladd, Harry, 303
LaFond, Eugene C., 24-25, 79, 245, 247, 250, 279, 422
La Jolla Radiocarbon Laboratory. *See* radiocarbon dating
Lal, Devendra, 326, 338-39
Lamont (-Doherty) Geological Observatory, 90, 97, 303, 307, 309, 310 & n, 312, 314-15
Lawrence, Ernest, 25
Lawson, Howard B., 362
Lee, Richard F., 222
Leighton, David, 126
Leipper, Dale, 245, 265
Lemche, Henning, 277
Lewin, Lanna Cheng, 73, 222
Lewin, Ralph A., 204, 233
Lewis, Fred, 368
Libby, Willard F., 149-50, 163, 334
library, 14, 22, 33, 443, 474; building, 13, 185-86, 442
Liebermann, Leonard N., 81
Lill, Gordon, 301, 303, 305
Limbaugh, Conrad, 123-24, 126-28, 130-32, 236, 413-14
Loma, 378
Longhurst, Alan R., 138
Lǫvberg, Ralph, 160
lunch stand, 440
Lusiad Expedition, 298, 429-31
Lyman, John, 323, 360

MacDonald, Gordon, 153
Mack, Phillip R., 250
Macrocystis (boat), 126, 364, 471
Macrocystis pyrifera. *See* kelp studies
magnetic studies, 99-102, 108, 162, 300
Mahnken, Madeleine. *See* Miller, Madeleine
Malinowski (IGY observer), 408
Malita, 429
Mammerickx, Jacqueline, 288
Manar, Thomas A., 474

manganese nodules, 108, 197, 286, 290, 325, 388, 396, 398
marine algae. *See* algae
Marine Biological Association of San Diego, 1, 6, 13, 123, 185, 354
marine biology. *See* biological studies
Marine Biology Building, 442
marine chemistry. *See* chemical studies
Marine Facilities, 85, 175, 345-50, 419
Marine Foraminiferal Laboratory, 284
marine geology. *See* geological studies
marine geophysics. *See* geophysical studies
marine invertebrates. *See* invertebrates
Marine Life Research Program, 43-76, 79, 117, 120, 139, 189, 201, 212, 222, 252, 255, 258, 262, 338, 346, 356, 363, 382, 463, 474
marine mammals, 168, 170, 178, 180-81, 206, 208, 212, 261
Marine Neurobiology Facility, 169, 234
Marine Physical Laboratory, 79-108, 135, 162, 197, 260, 276-77, 346, 349, 362, 383, 405, 455, 457
marine plants, 228-33; *see also* algae, kelp studies, plankton studies
Marine Research Committee (MRC), 44-45, 47, 49, 56, 62, 65, 117
Marine Technician's Handbook, 297
Marr, John C., 44, 47, 56, 67
Martinac, Joseph M., 174
Mason, Ronald G., 99-100
Massey, John, 386 & n
Matiota, Neemia, 408
Matthews, Drummond H., 101
Maxwell, Arthur E., 97-98, 163, 303, 385, 390, 392, 449
McAlister, Edward D., 259-60
McCarthy, Joseph, 86
McCosker, John E., 197, 211
McCoy, Floyd, 405, 410
McEwen, George Francis, 11, 32, 189, 241-42, 251, 265, 359
McGehee, Maurice S., 107
McGill, William J., 451
McGowan, John A., 72-73
McHugh, J. Laurence, 45, 51
McKelvey, Howard, 453
McKenzie, Dan P., 163, 299
McMillan, D. J., 383

Melville, George Wallace, 364
Melville, 338, 346, 353, 364, 458
Menard, H(enry) William, 4, 121, 136, 269, 286-87, 289-90, 294, 297-98, 363, 370, 374, 387, 410-11, 432
Mero, John L., 290
microbiology, 12, 24, 223-28, 389 & n
microseisms, 160-61
Midlam, Diana, 432
Midpac Expedition, 92, 97, 149, 227 n, 246, 251, 283, 286, 350-51, 363, 380, 382-94, 401 n
Mid-Pacific Mountains, 221, 385, 387, 393, 395
Mielche, Hakon, 226-27
Miles, John W., 160
Miller, Frank, 424
Miller, Gaylord, 153
Miller, Madeleine, 466
Miller, Robert C., 44
MLR. *See* Marine Life Research Program
Moberg, Erik Gustaf, 7-8, 11, 143, 251, 321, 323-24
Mohanrao, G. K. J., 329
Mohole Project, 300-309, 425
Mohorovičić discontinuity, 92, 98, 302 & n, 303, 306
Monsoon Expedition, 361, 427-29
Montgomery, R. B., 254
Moore, David G., 278, 287, 298, 311
Moore, Robert D., 162, 460
Moorehead, William C., 44
Moriarty, James R., 134
Morita, Richard, 227 n, 389 & n
Morris, Gerald, 95
Morse, Robert W., 85
Mount Soledad laboratory, 336
MPL. *See* Marine Physical Laboratory
Mrak, Emil, 118
MRC. *See* Marine Research Committee
Mu, 393
Mudie, John D., 102
Mukluk Expedition, 405
Mullin, Michael M., 141
Munk, Judith, 154, 157
Munk, Walter H., 23, 25, 30, 35, 57, 111, 115, 149-50, 152-54, 156-58, 243, 266, 301-303, 305, 402, 441, 451, 460
Murphy, Garth I., 56
Murray, Earl A., 134
museum. *See* Aquarium-Museum

Nafe, John, 307
Naga Expedition, 346, 363, 369, 414-24
Namias, Jerome, 63, 263
National Aeronautics and Space Administration, 113
National Defense Research Committee, 24
National Marine Fisheries Service, 47, 54, 137, 213, 349
National Oceanic and Atmospheric Administration (NOAA), 47, 113, 132, 144
National Oceanographic Data Center, 68, 139, 250
National Science Foundation, 113, 124, 153, 156, 158, 167-68, 178, 301, 303, 306, 310-11, 338, 349, 357, 406
Navy. *See* U.S. Navy
Newman, Dave, 408
Newman, William A., 221, 278, 327
Nierenberg, Edith, 311, 466
Nierenberg, William Aaron, 3, 258, 298, 310, 444, 456-58, 460-61, 473, 475-76
Nimitz, Chester W., 350
Nimitz Marine Facilities. *See* Marine Facilities
Nixon, Alfreda Jo, 358
NOAA. *See* National Oceanic and Atmospheric Administration
Norpac Expedition, 55-56, 255, 405
Norpax, 264, 475; building, 442
Norris, Kenneth S., 206
North, Wheeler, 124, 131
Northern Holiday Expedition, 61, 251-52, 285, 290, 394-99
North Pacific Buoy Program, 262, 264, 475
North Pacific Experiment. *See* Norpax
Nova Expedition, 297-98, 337, 432
nuclear test studies, 25, 84, 263, 327, 329, 333, 380-82, 400-401, 405

Oceanids, 464-65
Ocean Research Buoy. *See* ORB
Oceans, The (book), 24, 35, 449
Oconostota, 345, 364-65
Oglebay, Florence, 153
Olcott, Harold S., 121, 144
Oliver, David, 44
O'Neil, Stanley J., 404
Opening and Closing Paired Zooplankton Net. *See* bongo net

497

Operation Crossroads, 57, 230, 261, 380-82, 386, 388, 405, 447
Operation Ivy, 400-401
Operation Midpac. *See* Midpac Expedition
Operation Wigwam, 84, 327, 405
optical studies. *See* Visibility Laboratory
ORB, 105-106
Orca, 66, 305, 365, 471 n
Orton, Grace L., 212
Outler, Finn W., 81, 83, 346, 383
Oversmith, Robert H., 444

Palmer, Claude, 187
Paolina-T, 37, 49, 66, 89, 92, 115, 346, 365-67, 404, 471
Parker, Frances L., 284
Parker, Robert H., 223
Parker, Robert L., 162-63, 299
Parsons, William S., 380
Patullo, June G., 407-408
PCE(R)-855, 83, 346
PCE(R)-857, 83, 92, 346, 383-87, 399
Peterson, Melvin N. A., 311
Peterson, Robert O., 359
Phister, Montgomery, 44
Phleger, Fred B, 223, 261, 276, 284
physical oceanography, 201, 241-66
Physiological Research Laboratory (PRL), 167-82, 193, 201, 234, 357, 441, 456
Piccard, Jacques, 291
pier. *See* Scripps pier
pigment studies, 13, 23, 192, 205, 215-16
Pioneer, 99-100
Piquero Expedition, 74, 255, 300
plankton studies, 11-13, 19-20, 50, 55-56, 58, 69, 72-75, 137, 139, 141, 143, 177, 228-30, 232, 323, 398, 432
plate tectonics, 270, 295, 299-300
Plaza Lasso, Galo, 400
Point Loma facilities, 24, 47, 80, 85, 112, 137, 260, 349
Potibutra, Wong, 422
Powell, John D., 198
Pritchard, Donald W., 252
PRL. *See* Physiological Research Laboratory
Prothero, William A., 161, 460
Public Affairs (and information) Office, 187, 473-74

radiocarbon dating, 134, 208, 275, 329, 332-35
radiolaria, 285-86, 325
radio station (WWD), 52, 54, 101, 170, 386, 441
Radovich, John, 56
Raff, Arthur D., 99-100
Ragan, Ruth, 14, 443
Raitt, Helen, 352, 401 & n, 464-65
Raitt, Russell W., 26, 81, 87, 90, 92, 94-96, 283, 298, 304, 307, 385-86, 388, 404
Rakestraw, Norris W., 324, 332, 431, 483
Rand, William W., 311
Rechnitzer, Andreas B., 212
Red Lion, 366
Reid, Joseph L., Jr., 55, 57, 253, 255-56
Remote Underwater Manipulator. *See* RUM
Revelle, Ellen Clark, 351, 360, 446, 453
Revelle, Roger R. D., 4, 13, 23-25, 35-37, 45, 47, 56-57, 68, 79, 83, 98-99, 101, 117-20, 145, 150, 152, 156, 168, 189, 195, 204, 225, 247, 291, 302-303, 305, 309-10, 323-24, 326-27, 331-34, 336, 346, 351-52, 373, 378, 380-84, 386-88, 389 n, 392-94, 401 & n, 406, 415-16, 425, 427, 440-41, 444-53, 456, 463, 471, 476
Revelle College, 451
Rex, Robert W., 326
Richardson, Park, 410
Richter, William H., 366
Riedel, William R., 223, 285-86, 305-306, 310 & n, 311, 314, 396
Riley, Harold, 236
rip currents, 23-24, 244, 279
Ritter, William Emerson, 1, 4, 6, 67, 185, 201, 236, 241
Ritter Hall, 13, 22, 190, 193, 206, 215, 439-40, 458, 467
Roberts, "Texas Bobbie," 453
Robinson, Margaret, 246, 250, 422, 432
Rockefeller Foundation, 202-204
Roosevelt, Eleanor, 467
Rosenblatt, Richard H., 191, 206, 211, 221
Rosendahl, Bruce, 295
Ross, Murdock G., 367
Rubey, William, 303
Rudnick, Philip, 88, 104

498

RUM, 105-106

Sachet, Marie-Hélène, 414
Saluda, 92
Sammuli, Harold, 93
sardine studies, 19-20, 37; *see also* Marine Life Research Program
Sargent, Marston C., 32, 230, 232, 250
Saur, J. F. T., 363
Saxby, Don T., 45
Scan Expedition, 95, 312, 351, 458
Schaefer, Milner B., 121-22, 137, 139-40, 145, 405, 432
Schick, George B., 69-70, 263
Schmidt-Nielsen, Knut, 168
Scholander, Per F., 167-70, 172-75, 179-82, 203, 357
Scholander, Susan, 172-73, 357
School of Science and Engineering, 153, 168, 440, 450, 483
Schultz, Carol, 465
Schwartzlose, Richard A., 62, 70
Scientific Committee on Oceanic Research (SCOR), 253-54, 425-26, 449
Sclater, John G., 97-98, 300
Scripps, Edward Willis, 1, 6, 8, 153, 189, 359, 378; *see also E. W. Scripps*
Scripps, Ellen Browning, 1, 6, 354, 360, 446; *see also Ellen B. Scripps*
Scripps, George Henry, 440
Scripps, Robert Payne, 8, 15, 345, 360, 440
Scripps (boat), 7-8, 11-12, 14, 143, 186, 230, 324, 366-67, 452
Scripps Clinic and Research Foundation, 168, 170 & n, 180
Scripps Estates Associates, 442
Scripps, E. W. (ship). *See E. W. Scripps*
Scripps Field Annex. *See* Point Loma facilities
Scripps Institution, founding, 1, 4, 123
Scripps Institution for Biological Research, 6, 11, 201
Scripps "Island," 443-45
Scripps Laboratory, 13-14, 22, 52, 185, 193; "New" Scripps, 440
Scripps library. *See* library
Scripps pier, 14, 33, 107, 130, 135, 157, 186, 191 & n, 193, 195, 197, 214, 229, 236, 243-44, 329-30, 334, 340, 439, 443, 463, 467, 471

Scripps Seamount, 395
Scripps Shoreline-Underwater Reserve, 195, 235
Scripps submarine canyon, 13, 127, 197, 216, 273, 444-45
Scripps Tuna Oceanography Research (STOR), 136-40
Scuba, 123-24, 127-36, 194, 216, 273-74, 277, 279-80, 287, 297, 402
Sea, The (book), 449
sea-floor spreading, 94, 98, 101, 161, 277, 283, 285-86, 295, 298-99, 316, 337, 431
Sea Grant program, 122, 144-45, 197, 455, 476
Sealab II, 106-107, 135
seasick pills, 58
Seaweed Canyon, 170, 441 & n
seaweeds. *See* algae, kelp studies
Sefton, Joseph W., 365
seismic studies, 90-96, 105, 108, 277-78, 298, 381, 383, 385-86, 388
self-contained underwater breathing apparatus. *See* Scuba
Serena, 15, 360
service yard, 439
Sessions, Meredith, 70, 263
Sette, Oscar E., 44, 63
Seven-Tow Expedition, 300
sewage studies, 124
Seymour, Richard J., 145
Shellback Expedition, 252, 254, 399-400, 465
Shepard, Elizabeth, 274, 351
Shepard, Francis P., 13, 20, 23-25, 29, 79, 127, 212, 271, 273-79, 293-94
Shimada, Bell, 287 & n
Shipboard Computer Group, 457-58
Shipek, Carl J., 287, 290, 389
ships, 345-70; *see also* individual names
Shor, Elizabeth N., 356
Shor, George G., Jr., 86, 92, 94-96, 304, 307, 310 n, 359-60, 429, 453
Shore Processes Study Group, 279-80, 282, 441
Show Expedition, 94-95, 308
Shumway, George and Anne, 453
signal processing, 81, 86-89
Silverman, Maxwell, 99, 352-53, 359
Skewis, Ed, 174
Slichter, Louis B., 149-50, 152, 393

499

Smith, Hugh McCormick, 363; *see also Hugh M. Smith*
Smith, Stuart M., 288
Snatch, 357
Snodgrass, Frank E., 157-58
Snodgrass, James M., 57, 97, 128, 247, 249, 385, 390, 392
Snyder, John Otterbein, 209
Somero, George N., 234
Soutar, Andrew, 68
South-Tow Expedition, 222
Southwest Fishery Center, 47, 193, 442
Special Developments Division, 57, 128, 332
Spencer F. Baird, 55, 254, 277, 287 n, 305, 352, 365-67, 400-402, 404, 410, 412-13, 471
Spiess, Fred Noel, 84, 102, 104, 107, 310, 455, 458, 460, 462, 484
Spiess, Sarah W., 47 n, 362
Spilhaus, Athelstan F., 245, 467 n
Sproul, Robert Gordon, 8, 29, 79, 117, 119, 189, 205, 463, 467
ST-908, 366
staff luncheon, 34
Steinbach, H. Burr, 168
Steinbeck, John, 304-305
Step-1 Expedition, 252-53
Stetson, Henry, 271
Stevenson, Merritt, 255
Stewart, Harris B., Jr., 293, 295, 376-77, 395, 399, 401
Stewart, James R., 131, 134-35
Stommel, Henry, 337
Stone, Lewis, 360
STOR. *See* Scripps Tuna Oceanography Research
Stose, Clemens W., 346, 348
Stover, Allan J., 208
Stranger, 55, 66, 363, 367-69, 413, 419-22, 424, 471
Strickland, John D. H., 140-41
Stroup, E. D., 254
Styx Expedition, 93, 221, 256
submarine canyons, 274, 280; *see also* Scripps submarine canyon
Suess, Hans E., 208, 275, 324-25, 331-32, 334-36
Sumner, Francis Bertody, 11-12, 204, 215
Sumner, Mrs. Francis B., 440, 464
Sumner Auditorium, 440
Sumnernoon, 233
Surfside, 442
Sverdrup, Gudrun, 14, 22, 440, 465

Sverdrup, Harald Ulrik, 9-10, 13-15, 19, 22-25, 29-32, 34-36, 44-46, 57, 83, 149, 201, 204, 242-44, 246, 251, 271, 346, 367, 378, 393, 444 n, 447, 449, 463, 467 n, 474-75
Sverdrup Hall, 112, 440
Swan Song Expedition, 254
Swedish Deep Sea Expedition, 220, 284-85, 390 n
Sweeney, Beatrice, 232

T-16. *See* director's house
T-441, 143, 369, 405, 471
Taft, Bruce A., 139, 255
Tasaday Expedition, 96
Taylor, Leighton R., Jr., 197
Thaddeus, 7, 367
thermocline, 26-27, 260, 280
Thomas, Baron, 93, 377
Thomas Washington, 96, 138, 175, 255, 346, 353, 369-70, 457-58
tide studies, 11, 14, 157-58, 242, 266
Todd, Michael, 361
Tonga Trench, 71, 99, 290-91, 404
Torrente, Michele and Guiseppe, 365
Tracey, Joshua, 303
Transpac Expedition, 220, 252, 404
Trapani, Peter G., 348
trenches, studies of, 99, 227-28, 290-91, 295, 300; *see also* Tonga Trench
Tschudy, Robert H., 32
Tseng, Cheng Kwai, 32, 127
tsunami researches, 153, 160, 265, 274-75, 407
tuna studies. *See* fisheries; Scripps Tuna Oceanography Research
Tuthill, Carr, 134, 191
Tyler, John E., 112, 472

Udall, Stewart L., 121, 452, 456
underwater camera, 70, 72, 108, 123-24, 274, 389
underwater sound, 25-27, 33, 72, 79-81, 86, 88-89, 102, 104, 245, 273, 386
U.S. Air Force, 29, 32, 112-13, 154
U.S. Bureau of Commercial Fisheries, 19, 136, 138, 230, 253, 287-88, 419
U.S. Coast and Geodetic Survey, 12, 99, 230, 288
U.S. Coast Guard, 7, 23, 50, 57, 113, 176, 229, 245, 348, 353

U.S. Fish and Wildlife Service, 23, 44-47, 50, 52, 56, 254, 363, 365, 367, 380, 399
U.S. Geological Survey, 6, 278, 284, 293, 303, 324, 334, 380, 382, 414, 455
U.S. Naval Undersea Center, 278, 327
U.S. Navy, 7, 23-25, 27, 33-34, 36-37, 47, 49, 52, 79-80, 85-86, 88, 90, 93, 102, 112, 128, 160, 230, 236, 243-45, 249-52, 255, '258-59, 265, 269, 288, 291, 348-50, 353-54, 357-59, 361, 364-67, 369-70, 374, 382, 384, 386, 411, 413, 427, 446-47, 455, 463
U.S. Navy Bureau of Ships, 27, 36, 79-80, 84, 121, 225, 247, 446
U.S. Navy Electronics Laboratory, 80-81, 83-84, 232, 245, 278, 286, 346, 349, 352, 383, 387, 389; see also U.S. Naval Undersea Center
U.S. Navy Fleet Sonar School, 27
U.S. Navy Hydrographic Office, 243, 446
U.S. Navy Oceanographic Office, 250
U.S. Navy Office of Naval Research, 36, 84, 111, 115, 135, 204, 232, 265, 301-303, 349, 361-62, 418, 447
U.S. Navy Office of Research and Inventions, 36, 111
U.S. Navy Ordnance Laboratory, 99, 105
U.S. Navy Pacific Submarine Force, 27
U.S. Navy Radio and Sound Laboratory, 24
University of California, Berkeley, 1, 6, 12, 23, 57, 118, 120-22, 128, 133, 144, 214, 223, 290, 321, 350, 400, 441, 446, 455-56, 462
University of California, charter day, 22, 187, 189, 466-67, 474
University of California Division of War Research (UCDWR), 24-27, 29-31, 37, 52, 79-80, 83, 90, 187, 245, 247, 271, 273, 361, 369, 446, 455
University of California, Los Angeles (UCLA), 23, 52, 81, 87, 119-21, 127, 149, 151, 153, 160, 168-69, 208, 211, 219, 221, 273, 277, 283-84, 286, 383-84, 393, 406, 441, 462, 483

University of California Natural Land and Water Reserve System, 235-37
University of California, San Diego, 3, 112, 122, 141, 151 & n, 153, 156, 168-69, 182, 191, 234, 237, 283, 286, 324-25, 327, 336, 339, 360, 442-43, 456, 462, 465, 474, 483; founding, 450-51
Urey, Harold C., 336 & n
Utility Boat, 370

Vacquier, Victor, 97-98, 100-101, 302
Van Allen, James A., 405-406
van Andel, Tjeerd H., 276, 309, 310 & n, 311
Van Atta, Charles W., 105, 262
Van Dorn, William G., 265, 407, 460
Van Orden, M. D., 250
Vaughan, T(homas) Wayland, 1, 6-8, 10-11, 14, 33, 35, 189, 214-15, 221, 235, 237, 258, 321, 323, 378, 464
Vaughan Aquarium-Museum. See Aquarium-Museum
Venrick, Elizabeth, 73-74
Vermilion Sea Expedition, 277
Vine, Allyn, 102
Vine, Fred J., 101
Visibility Laboratory, 111-115, 258
Vitousek, Martin, 408
Vityaz, 471-72
Volcani, Benjamin E., 204, 233-34
Von Herzen, Richard P., 97-98

Walker, Boyd, 127-28, 209, 211
Walker, Theodore J., 212
Walsh, Don, 291
Washington, Thomas, 370; see also *Thomas Washington*
Watson, Charles, 362
wave-forecasting project, 29-32, 34, 157, 243
weather studies and forecasting, 11, 76, 242, 258, 261-64
Weiss, Herbert V., 326
Weiss, Ray F., 337-38
Wells, Morgan, 135
Wheelock, Charles D., 121, 444
Whitaker, Thomas W., 32
Wick, Gerald L., 460
Wickline, LaRoy B., 364
Wigwam. *See* Operation Wigwam
Wilkie, Donald W., 191-95, 197
Wilkins, Hubert, 10
Williams, Peter M., 141

501

Winterer, Edward L., 286, 300, 484
Winterer, Jacqueline. *See*
 Mammerickx, Jacqueline
Wisner, Robert L., 212
Wohnus, J. Frederick, 32, 123
Woodford, Alfred, 446
Woods, Baldwin M., 119-20
Woods Hole Oceanographic Institution, 27, 31, 81, 98, 102, 180, 222, 245, 261, 271, 284, 307, 309, 315, 324, 353, 381-82, 449
Wooster, Warren S., 138, 144, 252-54, 377, 395, 398-99, 404, 426, 432, 460, 484
Work Projects Administration, 13-14, 242, 244
World War II, researches, 24-33, 113-15, 225, 243, 245
Worzel, J. Lamar, 312
Wright, Charles S., 83-84, 99
WWD. *See* radio station

XBT. *See* bathythermograph

Yayanos, A. Aristides, 181
York, Herbert F., 149, 440
Young, Clement C., 236
Young, Robert T., 13
Young, Robert W., 81

Zenkevich, L., 204
Zetes Expedition, 255-56, 432
ZoBell, Claude E., 12, 24, 32, 189, 223, 225-28, 389, 474